Problems in
Classical Mechanics

Dedicated to
my brother B. N. Katkar
whose foresight enlightened the path of my career

Preface

It is hoped that students must learn mathematics not only to become a competent mathematicians but also skilled users of mathematics in the solution of the realistic problems. They must learn how to use their mathematical knowledge in solving the problems of the real world. I believe that through the study of classical mechanics, students will learn something about an art of applying mathematical knowledge to solve such problems. Comprehensive account of the mathematical artifact and numerous examples in this book will help the aspirants to develop an ability to understand and apply mathematics.

I have great pleasure in presenting a thoroughly new book on classical mechanics. The material is the standard post-graduate syllabi of most of the universities. *Problems in Classical Mechanics* has been written for the use of students preparing for post-graduate examinations of universities and competitive examinations. An enormous number of examples (around 200) have been worked out with complete mathematical theory in the form of fifty seven theorems spread over six chapters to aid in the understanding of the underlying theory and grasping its intricacies. Another set of more than seventy examples with answers is given as an exercise at the end of each chapter to test and develop students' understanding of the subject. Efforts have been made to put the subject matter in as lucid and comprehensive manner as it is essential. Various reference books by the eminent authors have been utilized in the preparation of the text and the author is gratefully indebted to them. I have streamlined the examples and exposition, making the book easier to teach and learn from. It is hoped that the teachers, the students and large number of entrants to the competitive examinations from both physics and mathematics departments will be benefited with the subject matter of this new book.

It is indeed a great pleasure for me to thank my wife Geeta and daughters Kirti and Trupti for their patience and assistance in several ways throughout the preparation of the manuscript. Finally I take this opportunity to thank Mr. N. K. Mehra, the Managing Director of Narosa Publishing House, for his personal interest and excellent co-operation in the publication of this book.

Any constructive suggestions for the improvement of the subject matter will be highly appreciated.

L. N. Katkar

Contents

Canonical Equations in Poisson Brackets; Poisson Theorem;
Exercise

1 Lagrange's Formulation

1. INTRODUCTION

One of the great mathematicians of the 20th century Hilbert said that any truth can be expressed in the language of mathematics. In particular, truth of the physical world is expressed in the frame work of mathematics through applied mathematics while truth which is independent of physical world is expressed through pure mathematics. Once the truth is expressed in the language of mathematics then it will be absolute. It means that it is independent of observer. Truth we mean it here as the laws of nature- the physical realities. Physicists express the laws of nature in the principles of physics which latter express in the language of mathematics so as to be invariant. Thus to express any truth whether depends upon the physical world or not in the language of mathematics, mathematicians use some symbols such as coordinates. But the co-ordinates however, just play a role of markers or codes and will no way influence the truth. These are just mathematical tools or the instruments in the hands of a mathematician through which the physical realities are expressed. For example, suppose a particle is fired from a fixed point with initial velocity u, not exactly vertically upward but making an angle α with the horizontal. Then what instruments do mathematicians need to discuss its motion? *i.e.,* to find the position, the velocity and the distance of the particle at some time latter, and also the path followed by the particle at the end of its journey.

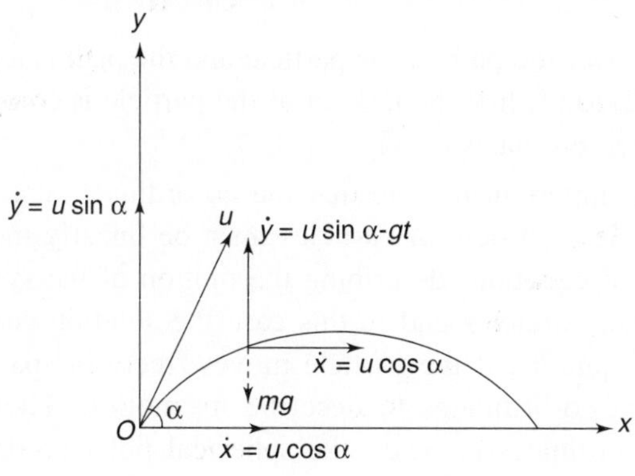

Well, to answer such questions, mathematicians do not need any meter stick to measure the distance covered by the particle at the instant, they don't need any speedometer to find its speed at any instant t, nor they need any clock to record the time taken by the particle to cover a particular distance. In fact, they need not have to do any such experiment. What they need to describe the motion of the particle is simply the co-ordinates. Hence the single most important notion in mechanics is the concept of co-ordinates. With the help of these co-ordinates, the motion of a particle or system of particles can completely be described. For instance, we need two coordinates to discuss the motion of the particle in a plane. To illustrate this let P be any point on the path of the particle and (x, y) be its coordinates. The only force acting on the particle is the gravitational force in the downward direction. Resolving this force horizontally and vertically, we write from Newton's second law of motion the equations of motion as

$$\ddot{x} = 0, \ddot{y} = -g. \qquad \qquad \ldots(1)$$

Integrating the above two equations and using the initial conditions we readily obtain

$$\dot{x} = u \cos \alpha,$$

$$\dot{y} = u \sin \alpha - gt, \qquad \qquad \ldots(2)$$

where u is the initial velocity of the particle when $t = 0$. Integrating equations (2) once again and using the initial conditions we obtain

$$x = u \cos \alpha \cdot t, \qquad \qquad \ldots(3)$$

$$y = u \sin \alpha\, t - \frac{1}{2} gt^2. \qquad \qquad \ldots(4)$$

Equations (2) determine the velocity of the particle at any time t, while equations (3) and (4) determine the position of the particle at that instant. Further, eliminating t from equations (3) and (4), we get

$$y = x \tan \alpha - \frac{1}{2} \frac{gx^2}{u^2 \cos^2 \alpha}. \qquad \qquad \ldots(5)$$

This equation gives the path of the particle and the path is a parabola. We see from equations (2) to (5) that the motion of the particle is completely discussed only through the co-ordinates (x, y).

However, it is important to note that the co-ordinates used to describe the motion of a particle or system of particles must be linearly independent. If not then the number of equations describing the motion of the system will be less than the number of variables and in this case the solution cannot be uniquely determined. For example, if the particle moves freely in space, then we need three independent co-ordinates to describe its motion. These can be either the Cartesian co-ordinates (x, y, z), the spherical polar co-ordinates (r, θ, ϕ)

or any other coordinates. However, if a particle is moving along a curve in the 3-dimentional space, then all the three co-ordinates are not independent. In fact these coordinates are related by the equation of the curve; hence these three co-ordinates cannot be used for its description. Of these coordinates one coordinate is dependent on other two independent coordinates. Only the two independent coordinates can be used for the description of the motion of the particle.

Similarly only one coordinate is used to discuss the motion of a particle moving along a co-ordinate axis. Because along this coordinate axis only one co-ordinate varies and other two are constants and only the varying co-ordinate can be used for the description of the motion of the particle. These independent coordinates have special status in mechanics and can be discussed in detail in the sequel.

Equation of Motion and Conservation Theorems

Equation of motion of a particle

Consider a particle of mass m whose position vector with respect to some fixed point is $\bar{r}$. If $\bar{F}$ is a force applied on the particle then the equation of motion of the particle is given by Newton's second law of motion

$$\bar{F} = \dot{\bar{p}}, \qquad \qquad ...(6)$$

where $\bar{p} = m\dot{\bar{r}}$ is the linear momentum of the particle. The force is defined to be

$$\bar{F} = \text{mass} \cdot \text{accel.}$$

$$\bar{F} = m\bar{a}.$$

Hence equation (6) becomes

$$\ddot{\bar{r}} = \bar{a}. \qquad \qquad ...(7)$$

Integrating this equation, we get

$$\dot{\bar{r}} = \bar{a}t + c, \qquad \qquad ...(8)$$

where c is the constant of integration and is to be determined. Now applying the initial conditions, we have when $t = 0$, the initial velocity $\dot{\bar{r}} = \bar{u}$.

$$\Rightarrow \qquad \qquad c = \bar{u}.$$

Hence the velocity $\bar{v} = \dot{\bar{r}}$ at any time t is given by the equation (8) as

$$\bar{v} = \bar{u} + \bar{a}t. \qquad \qquad ...(9)$$

Integrating the equation (9) we get

$$r = ut + \frac{1}{2} at^2 + c_1,$$

where c_1 is the constant of integration. At $t = 0$, $r = 0 \Rightarrow c_1 = 0$. Hence we have

$$r = ut + \frac{1}{2} at^2. \qquad \qquad \text{...(10)}$$

This equation gives the distance covered by the particle at any time t. One can combine equations (9) and (10) and write

$$v^2 = u^2 + 2ar. \qquad \qquad \text{...(11)}$$

This equation determines the velocity of the particle at a given distance. Equations (9), (10) and (11) are the algebraic equations of motion and are derived from the equation (6) namely

$$\overline{F} = \dot{\overline{p}}. \qquad \qquad \text{...(12)}$$

This is the differential equation of motion. It follows from equation (12) that if the applied force is zero then the linear momentum of the particle is conserved.

Equation of Motion of a System of Particles

Consider a system of n particles of masses $m_1, m_2,..., m_n$ having position vectors $\overline{r}_i$, $i = 1, 2,..., n$ relative to an arbitrary fixed origin. Any particle of this system will experience two types of forces viz.,

(i) external forces on the system $\overline{F}_i^{(e)}$, $i = 1, 2,..., n$ and

(ii) internal forces $\overline{F}_{ji}^{(\text{int})}$.

Thus the total force acting on the ith particle of the system becomes

$$\overline{F}_i = \overline{F}_i^{(e)} + \sum_j \overline{F}_{ji}^{(\text{int})},$$

where $\sum_j \overline{F}_{ji}^{(\text{int})}$ is the total internal force acting on the ith particle due to the interaction of all other $(n - 1)$ particles of the system. Thus the equation of motion of the ith particle is given by

$$\overline{F}_i^{(e)} + \sum_j \overline{F}_{ji}^{(\text{int})} = \dot{\overline{p}}_i. \qquad \qquad \text{...(13)}$$

The equation of motion of the whole system is obtained by summing over i the equation (13). Thus we have

$$\sum_i \overline{F}_i^{(e)} + \sum_i \left(\sum_j \overline{F}_{ji}^{(\text{int})} \right) = \sum_i \dot{\overline{p}}_i.$$

We write this equation as

$$\sum_i \overline{F}_i^{(e)} + \sum_{i,j,i \neq j} \overline{F}_{ji}^{(\text{int})} = \sum_i \dot{\overline{p}}_i. \qquad \qquad \text{...(14)}$$

The term $\sum\limits_{i,j,i \neq j} \overline{F}_{ji}^{(\text{int})}$ represents the vector sum of all the interacting forces due to the presence of remaining $(n-1)$ particles. However, there is no self interacting force, hence $\overline{F}_{ji} = 0$, $\forall\ i = j$. Also the internal forces obey the Newton's third law of motion. That is the action of one particle on the other is equal but opposite to the action of second on the first. This implies that the mutual interactions between the ith and the jth particles are equal and opposite. *i.e.*,

$$\overline{F}_{ji}^{(\text{int})} = -\overline{F}_{ij}^{(\text{int})}.$$

This gives $\qquad \sum\limits_{i,j,i \neq j} \overline{F}_{ji}^{(\text{int})} = 0.$

Thus the equation of motion (14) of the system becomes

$$\sum_i \overline{F}_i^{(e)} = \sum_i \dot{\overline{p}}_i,$$

$$\overline{F}^e = \dot{\overline{p}}, \qquad \qquad \text{...(15)}$$

where $\overline{P}$ is the total momentum of the system and $\overline{F}^e$ is the total external force acting on the system.

Conservation Theorem of Linear Momentum of the System of Particles

Theorem 1 If the sum of external forces acting on the particles is zero, the total linear momentum of the system is conserved.

Proof: Proof follows immediately from equation (15). *i.e.*, if

$$\overline{F}^e = 0 \Rightarrow \overline{P} = \text{const.}$$

Angular Momentum of the System of Particles

Consider a system of n particles of masses $m_1, m_2, ..., m_n$ having position vectors $\overline{r}_i$, $i = 1, 2, ..., n$ relative to an arbitrary but fixed origin. The angular momentum of the ith particle of the system about the origin is given by

$$\overline{L}_i = \overline{r}_i \times \overline{p}_i.$$

Thus the total angular momentum of the system about a point is equal to the vector sum of the angular momentum of individual particles about the point. Hence we have

$$\sum_i \overline{L}_i = \sum_i \overline{r}_i \times \overline{p}_i. \qquad \qquad ...(16)$$

If $\overline{N}$ is the total torque acting on the system, then equation of motion of the system is given by

$$\overline{N} = \frac{d\overline{L}}{dt} = \frac{d}{dt}\left(\sum_i \overline{r}_i \times \overline{p}_i \right).$$

$$\overline{N} = \frac{d\overline{L}}{dt} = \sum_i \dot{\overline{r}}_i \times \overline{p}_i + \sum_i \overline{r}_i \times \dot{\overline{p}}_i. \qquad \qquad ...(17)$$

But we have

$$\sum_i \dot{\overline{r}}_i \times \overline{p}_i = \sum_i \dot{\overline{r}}_i \times m_i \dot{\overline{r}}_i = 0. \qquad \qquad ...(18)$$

Now consider

$$\sum_i \overline{r}_i \times \dot{\overline{p}}_i = \sum_i \overline{r}_i \times \left(\overline{F}_i^{(e)} + \sum_j \overline{F}_{ji}^{(int)} \right),$$

$$\sum_i \overline{r}_i \times \dot{\overline{p}}_i = \sum_i \overline{r}_i \times \overline{F}_i^{(e)} + \sum_i \overline{r}_i \times \sum_j \overline{F}_{ji}^{(int)},$$

$$\sum_i \overline{r}_i \times \dot{\overline{p}}_i = \sum_i \overline{r}_i \times \overline{F}_i^{(e)} + \sum_{i,j} \overline{r}_i \times \overline{F}_{ji}^{(int)}, \qquad \qquad ...(19)$$

However, the term $\sum_{i,j} \overline{r}_i \times \overline{F}_{ji}^{(int)}$ can be expanded for $i \neq j$ as

$$\sum_{i,j} \overline{r}_i \times \overline{F}_{ji}^{(int)} = (r_2 - r_1) \times F_{12}^{(int)} + (r_3 - r_1) \times F_{13}^{(int)} + (r_3 - r_2) \times F_{23}^{(int)} + ...$$

$$\sum_{i,j} \overline{r}_i \times \overline{F}_{ji}^{(int)} = |r_i - r_j| \times F_{ji}^{(int)},$$

$$\sum_{i,j} \overline{r}_i \times \overline{F}_{ji}^{(int)} = r_{ij} \times F_{ji}^{(int)}, \text{ for } r_{ij} = |r_i - r_j|. \qquad \qquad ...(20)$$

Interchanging i and j on the right hand side of equation (20), we get

$$\sum_{i,j} \overline{r}_i \times \overline{F}_{ji}^{(int)} = r_{ji} \times F_{ij}^{(int)},$$

$$\sum_{i,j} \overline{r}_i \times \overline{F}_{ji}^{(int)} = -r_{ij} \times F_{ji}^{(int)}. \qquad \qquad ...(21)$$

Adding equations (20) and (21), we get

$$\sum_{i,j} \overline{r}_i \times \overline{F}_{ji}^{(int)} = 0. \qquad \qquad ...(22)$$

Consequently on using the equations (18), (19) and (22) in the equation (17) we readily obtain

$$\overline{N} = \frac{d\overline{L}}{dt} = \sum_i \overline{r}_i \times \overline{F}_i^{(e)}. \qquad \text{...(23)}$$

This equation shows that the total torque on the system is equal to the vector sum of torques acting on the individual particles of the system.

Conservation Theorem of Angular Momentum of the System of Particles

Theorem 2   If the total external torque acting on the system of particles is zero, then the total angular momentum of the system is conserved.

Proof: Proof follows immediately from equation (23). *i.e.,* if

$$\overline{N} = 0 \Rightarrow \overline{L} = \text{const.}$$

Some Definitions

Constraint Motion

Many times we see that the motion of a particle or a system of particles is not free motion, but it is limited by putting some restrictions on the position co-ordinates of the particle or system of particles. The motion under such restrictions is called constraint motion or restricted motion. The mathematical relations establishing the limitations on the position co-ordinates are called as the equations of constraint. Mathematically the constraints are thus the relations between the position co-ordinates and the time t. Consequently, all the co-ordinates are not linearly independent; constraint relations relate some of them. Thus in general the constraints on the motion of a particle or a system of particles are always possible to express in the form $f_r(x_i, y_i, z_i, t) \leq$ or $\geq$ or $= 0$, where $r = 1, 2,...,k$ represents the number of constraints and (x_i, y_i, z_i) are the position co-ordinates of the ith particle of the system, and $i = 1, 2,...n$.

Examples of motion under constraints

(1) The motion of a rigid body, (2) The motion of a simple pendulum, (3) The motion of a particle on the surface of a sphere, (4) The motion of a particle along the parabola $x^2 = 4ay$, (5) The motion of a particle on an inclined plane etc.

Holonomic and non-holonomic Constraints

If the constraints on a particle or system of particles are expressible as equations in the form

$$f_r(x_i, y_i, z_i, t) = 0, \ r = 1, ,..., k \qquad \text{...(24)}$$

then constraints are said to be holonomic otherwise non-holonomic constraints. A system of particles is called respectively holonomic or non-holonomic system if it involves holonomic or non-holonomic constraints.

For example: Constraints involved on the motion of a rigid body, and simple pendulum are examples of holonomic constraints, while constraints involved in the motion of a particle on the surface of a sphere, the motion of gas molecules inside the container are the examples of non-holonomic constraints. However, this is not the only way to describe the non-holonomic system. A system is also said to be non-holonomic, if it corresponds to non-integrable differential equations of constraints. Such constraints cannot be expressed in the form of equation of the type $f_r(x_i, y_i, z_i, t) = 0$. Hence such constraints are also called non-holonomic constraints. Obviously, holonomic system has integrable differential equations of constraints expressible in the form of equations. The holonomic constraints are further classified into two parts viz., Scleronomic and Rheonomic Constraints.

Scleronomic and Rheonomic Constraints

When the constraints relation do not explicitly depend on time are called scleronomic constraints. While the constraints, which involve time explicitly are called rheonomic constraints. The examples cited above are all scleronomic constraints. A bead moving along a circular wire of radius r with angular velocity ω is an example of rheonomic constraint and the constraint relation is given by $y = x \tan \omega t$.

WORKED EXAMPLES

Example 1: Consider a system of two particles joined by a massless rod of fixed length l. Suppose for simplicity, the system is confined to the horizontal plane xy. Suppose further that the system is so constrained that the centre of the rod cannot have a velocity component perpendicular to the rod. Show that the constraint involved in the system is non-holonomic.

Solution: Let (x_1, y_1) and (x_2, y_2) be the positions of the two particles connected by the mass-less rod of length l. The system is shown in the figure.

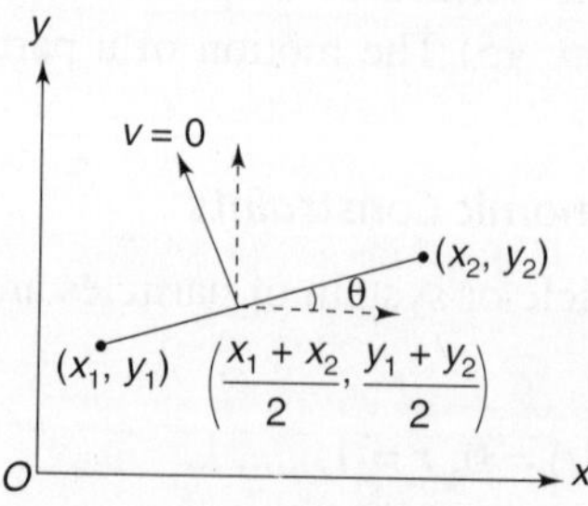

Since the length between the two particles is constant, clearly one of the constraint relations is

$$(x_1 - x_2)^2 + (y_1 - y_2)^2 = l^2, \qquad \ldots(25)$$

where
$$x_2 - x_1 = l \cos \theta, \; y_2 - y_1 = l \sin \theta. \qquad \ldots(26)$$

The constraint (25) is clearly holonomic. The other constraint is such that the centre of the rod cannot have velocity component perpendicular to the rod. Mathematically this is expressed as

$$(\dot{x}_1 + \dot{x}_2) \cos (90 + \theta) + (\dot{y}_1 + \dot{y}_2) \cos \theta = 0,$$

$$(\dot{x}_1 + \dot{x}_2) \sin \theta = (\dot{y}_1 + \dot{y}_2) \cos \theta. \qquad \ldots(27)$$

This constraint cannot be integrated and hence the constraint is non-holonomic and consequently, the system is non-holonomic.

Example 2: Consider a circular disc which always remains vertical while rolling on a horizontal plane without sliding. Find the equations of constraints and show that they are non-holonomic.

Solution: Let 'a' be the radius of the disc. The constraint is such that the plane of the disc is always vertical. This implies that the axis of the disc is perpendicular to the vertical z-direction. Thus the velocity of the disc has magnitude given by $v = a\dot{\phi}$. Since the direction is perpendicular to the axis of the disc, then the components of velocity along x-axis and y-axis are given by

$$v_x = \frac{dx}{dt} = v \sin \theta \text{ and } v_y = \frac{dy}{dt} = -v \cos \theta$$

$$\Rightarrow \qquad \frac{dx}{dt} = a \frac{d\phi}{dt} \sin \theta \text{ and } \frac{dy}{dt} = -a \frac{d\phi}{dt} \cos \theta$$

$$\Rightarrow \qquad dx - a \sin \theta d\phi = 0 \text{ and } dy + a \cos \theta d\phi = 0 \qquad \ldots(28)$$

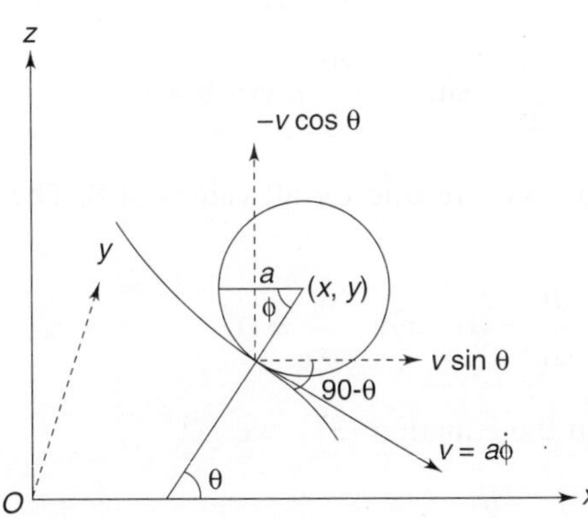

These equations fix up the constraints of the motion of a circular disc of radius 'a'.

Now we claim that these constraints are non-holonomic. That is we prove that the equations (28) are not integrated. It means that we cannot find an integrating factor $f(x, y, \theta, \phi)$ that will turn either of the equations in to a perfect differential and hence the constraints cannot be reduced to the form $f(x, y, \theta, \phi) = 0$. In order to show this, let us assume that there exists a function $f(x, y, \theta, \phi)$ which makes equations (28) integrable.

$$\Rightarrow \qquad df = \frac{\partial f}{\partial x}\, dx + \frac{\partial f}{\partial y}\, dy + \frac{\partial f}{\partial \theta}\, d\theta + \frac{\partial f}{\partial \phi}\, d\phi = 0. \qquad \ldots(29)$$

On using the equation (28) in the equation (29), we get

$$df = \frac{\partial f}{\partial x}\, a \sin\theta\, d\phi - \frac{\partial f}{\partial y}\, a \cos\theta\, d\phi + \frac{\partial f}{\partial \theta}\, d\theta + \frac{\partial f}{\partial \phi}\, d\phi = 0,$$

$$\Rightarrow \qquad \left(\frac{\partial f}{\partial x}\, a \sin\theta - \frac{\partial f}{\partial y}\, a \cos\theta + \frac{\partial f}{\partial \phi} \right) d\phi + \frac{\partial f}{\partial \theta}\, d\theta = 0, \qquad \ldots(30)$$

$$\Rightarrow \qquad \frac{\partial f}{\partial x}\, a \sin\theta - \frac{\partial f}{\partial y}\, a \cos\theta + \frac{\partial f}{\partial \phi} = 0 \text{ and} \qquad \ldots(31)$$

$$\frac{\partial f}{\partial \theta} = 0. \qquad \ldots(32)$$

Differentiating the equation (31) partially with respect to θ and using the equation (32), we get

$$\frac{\partial f}{\partial x}\, a \cos\theta + \frac{\partial f}{\partial y}\, a \sin\theta = 0. \qquad \ldots(33)$$

Differentiating the equation (33) partially with respect to θ again we get

$$-\frac{\partial f}{\partial x}\, a \sin\theta + \frac{\partial f}{\partial y}\, a \cos\theta = 0. \qquad \ldots(34)$$

Equations (33) and (34) are true for all values of θ. The equation (34) is true if

$$\frac{\partial f}{\partial x} = 0 \quad \text{and} \quad \frac{\partial f}{\partial y} = 0. \qquad \ldots(35)$$

Substituting this in the equation (31), we get

$$\frac{\partial f}{\partial \phi} = 0. \qquad \ldots(36)$$

Hence we have $\dfrac{\partial f}{\partial x} = 0, \dfrac{\partial f}{\partial y} = 0, \dfrac{\partial f}{\partial \theta} = 0$ and $\dfrac{\partial f}{\partial \phi} = 0$. This implies that f does not involve x, y, θ, ϕ. This implies that the relation of the form $f(x, y, \theta, \phi) = 0$ does not exist which is consistent with the constraint equations. Hence the equations of constraints are not integrable and hence are non-holonomic.

Degrees of Freedom and Generalized Co-ordinates

Consider the motion of a free particle in a space. To describe its motion we need three independent co-ordinates, such as the Cartesian co-ordinates x, y, z or the spherical polar co-ordinates r, θ, ϕ etc. The particle is free to execute motion along any one of the axes independently with change in only one co-ordinate. In this case we say that the particle has three degrees of freedom. Thus we define

$\mathcal{Definition:}$ The least possible number of independent co-ordinates required specifying the motion of a particle or a system of particles completely by taking into account the constraints is called degrees of freedom.

Example: For a system of N particles free from constraints moving independent of each other has $3N$ degree of freedom as each particle of the system requires three coordinates to specify its position.

Generalized co-ordinates

A system of N particles free from constraints has $3N$ degrees of freedom. If however, there exists k holonomic constraints expressed in k equations of the form

$$f_i(r_1, r_2, \dots, r_n, t) = 0, \ i = 1, 2, \dots, k, \qquad\qquad \dots(37)$$

then $3N$ co-ordinates are not all independent but related by k equations of constraints given in the equation (37). We may use these k equations of constraint to eliminate k of the $3N$ co-ordinates, and we are left with $3N - k = n$ (say) independent co-ordinates. These are generally denoted by $q_j, j = 1, 2\dots, n$ and are called the generalized co-ordinates and the system has $3N - k$ degrees of freedom.

$\mathcal{Definition:}$ A set of linearly independent variables $q_1, q_2, q_3, \dots q_n$ that are used to describe the configuration of the system completely by taking into account the constraints forces acting on it is called generalized co-ordinates.

Thus in general we have

No. of degrees of freedom – No. of constraint's = No. of generalized co-ordinates.

Note: The generalized co-ordinates need not be the position co-ordinates, which have the dimensions of length, breath and height, but they can be angles, charges or momentum of a particle.

Transformation Relations

It is always possible to express the position co-ordinates of a particle or a system of particles in terms of generalized co-ordinates and vice versa. This expression is called the transformation relation. If r_i, $i = 1, 2, 3,...n$ are the position vectors of the n particles of the system and q_j, $j = 1, 2,...n$ are the generalized co-ordinates, then there exists a relation

$$r_i = r_i(q_1, q_2, q_3,... q_n, t), \qquad\qquad ...(38)$$

is called the transformation relation. One should note that if there is explicit time dependence in some or all of these functions defined in the equation (38), the system is generally called rheonomic, otherwise the system is called scleronomic.

Work: Let a force $\overline{F}$ be acted on a particle at $\overline{r}$. Suppose the particle is displaced through an infinitesimal distance dr due to the application of force $\overline{F}$. Then the work done by the force $\overline{F}$ is given by

$$dW = \overline{F}\, d\overline{r}$$

If the particle is finitely displaced from point $P(r_1)$ to $P(r_2)$ along any path, then the work done by $\overline{F}$ is given by

$$W = \int_{r_1}^{r_2} \overline{F}\, d\overline{r}. \qquad\qquad ...(39)$$

Conservative Force: The work done by the force calculated in the expression (39) is in general depends on the extreme positions of the particle and also the path along which it travels. If a force is such that the work done depends only upon the positions P_1, P_2 and not on the path followed by the particle, then the force $\overline{F}$ is called conservative force, otherwise non-conservative.

We show that the gravitational force is conservative. Let therefore a particle of mass m move along a curve PQ under gravity. The only force acting on the particle is its own weight w in the downward direction. Therefore, work done by the force is given by

$$W = \int_{P}^{Q} \overline{F}\, d\overline{r}.$$

If $\overline{F} = Xi + Yj$ and $\overline{r} = xi + yj$, where X and Y are the components of the force along the co-ordinate axes. We see that $X = 0$, and $Y = -w$. Hence we have

$$W = \int_{a}^{b} - w\,dy,$$

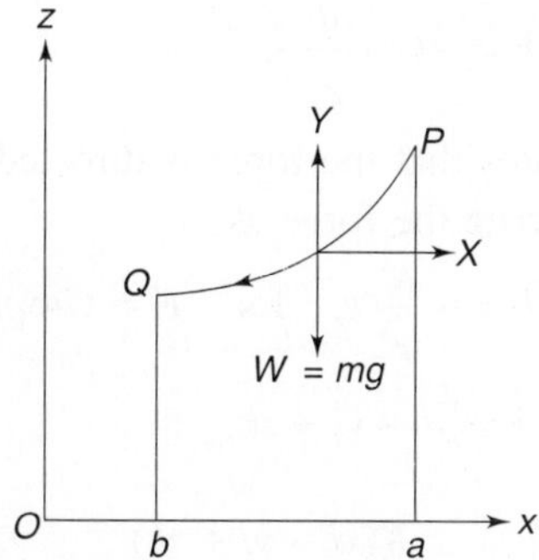

where a and b are the ordinates at points P and Q respectively. Solving the integral, we obtain

$$W = w(a - b).$$

The equation shows that the work does not depend upon the path but depends on the extreme points. Hence the gravitational force is conservative. Alternately, we say that, the force $\overline{F}$ is conservative if the work done by it around the closed path is zero.

i.e. $\qquad\qquad\qquad \overline{F}$ is conservative iff $\oint \overline{F}\, d\overline{r} = 0.$ $\qquad\qquad$ ...(40)

However, by Stokes theorem, we have

$$\oint \overline{F}\, \overline{dr} = \int_s \nabla \times \overline{F} \cdot ds, \qquad\qquad ...(41)$$

where ds is an arbitrary surface element. Thus from the equations (40) and (41) we have

F is conservative iff $\nabla \times \overline{F} = 0.$ $\qquad\qquad$ ...(42)

However, from the vector identity $\nabla \times \nabla V = 0$ and the equation (42) we have

$$\overline{F} = -\nabla V \text{ or } \overline{F} = -\frac{\partial V}{\partial r},$$

where V is a potential called potential energy of the particle and is a function of position only. Thus the force $\overline{F}$ is conservative if

$$\overline{F} = -\nabla V \qquad\qquad ...(43)$$

and conversely. The negative sign indicates that $\overline{F}$ is in the direction of decreasing V. We use the definition (42) to show that the inverse square law of attractive force (central force) is conservative. Consider the inverse square law of attractive force between two particles given by

$$\overline{F} = -G\,\frac{m_1 m_2}{r^2}, \qquad \qquad \ldots(44)$$

where negative sign indicates that the force is directed towards the fixed point, hence it is attractive. We write the force as

$$\overline{F} = -\frac{K}{r^3}\,\overline{r}, \quad \text{for} \quad K = Gm_1 m_2. \qquad \qquad \ldots(45)$$

For
$$\overline{r} = xi + yj + zk,$$

we have
$$\overline{F} = -\frac{K(xi + yj + zk)}{\left(x^2 + y^2 + z^2\right)^{\frac{3}{2}}}. \qquad \qquad \ldots(46)$$

For conservative force we have $\nabla \times \overline{F} = 0$.

Consider therefore

$$\nabla \times \overline{F} = \begin{vmatrix} i & j & k \\ \dfrac{\partial}{\partial x} & \dfrac{\partial}{\partial y} & \dfrac{\partial}{\partial z} \\ -\dfrac{Kx}{r^3} & -\dfrac{Ky}{r^3} & -\dfrac{Kz}{r^3} \end{vmatrix},$$

$$\nabla \times \overline{F} = -K\left[i\left(\frac{\partial}{\partial y}\left(\frac{z}{r^3}\right) - \frac{\partial}{\partial z}\left(\frac{y}{r^3}\right)\right) + j\left(\frac{\partial}{\partial z}\left(\frac{x}{r^3}\right) - \frac{\partial}{\partial x}\left(\frac{z}{r^3}\right)\right) + k\left(\frac{\partial}{\partial x}\left(\frac{y}{r^3}\right) - \frac{\partial}{\partial y}\left(\frac{x}{r^3}\right)\right) \right],$$

$$\nabla \times \overline{F} = -K\left[i\left(\frac{-3yz}{r^5} + \frac{3yz}{r^5}\right) + j\left(\frac{-3xz}{r^5} + \frac{3xz}{r^5}\right) + k\left(\frac{-3yx}{r^5} + \frac{3yx}{r^5}\right) \right],$$

$$\Rightarrow \qquad \qquad \nabla \times \overline{F} = 0.$$

This shows that the inverse square law of attractive force $\overline{F}$ is conservative.

Some other examples of conservative forces are the electrostatic force between two charge particles, the mutual interaction between two particles, the reaction and the tension in the string, while the forces which try to destroy the motion of the particle are the non-conservative forces. The friction force and the air resistance are some of the examples of non-conservative forces because the work done by such forces around a close path is not zero. For instance suppose a particle is moved against a constant frictional force $\overline{F}$ through a distance s on a horizontal plane then the work done by the frictional force $\overline{F}$ is $-\overline{F}s$, where negative sign indicates the direction of the force opposite to the direction of motion. If the particle is brought back to its original position on the same path, the amount of work done is $-\overline{F}s$. Therefore, the total work done by the force in bringing the particle back to its original position is $-2\overline{F}s$. Thus the frictional force is not conservative.

Virtual Work

If the system of forces acting on a particle be in equilibrium then their resultant is zero and hence the work done is zero. Thus in the case of the particle be in equilibrium there is no motion, hence there arises no question of displacement. In this case we assume the particle receives a small virtual displacement (the displacement of the system which causes no real motion is called as virtual or imaginary displacement) and it is denoted by δr_i. Virtual displacement δr_i is assumed to take place only in the co-ordinates and at fixed instant t, hence δ change in time t is zero. Thus, mathematically we define

$$\delta r_i = (dr_i)_{dt=0}.$$

Thus the work done by the system of forces in causing imaginary displacement is called virtual work. It is an amount of work that would have been done if the actual displacement had been caused. Hence the expression for the virtual work done by the forces is given by

$$\delta W = \text{Virtual work} = \sum_i F_i \, \delta r_i. \qquad \qquad ...(46)$$

Principle of Virtual Work

If the forces are in equilibrium then the resultant is zero. Hence the algebraic sum of the virtual work is zero. Conversely, if the algebraic sum of the virtual work is zero then the forces are in equilibrium.

Note that this principle is applicable in statics - a branch of mechanics in which we study particles at rest under the action of forces. However, an analogous principle in dynamics was put forward by D'Alembert and is referred as D'Alembert's principle. It is given below.

D'Alembert's Principle

D'Alembert principle is the fundamental principle in mechanics from which any equation of motion can be derived. We will make use of this principle to derive Lagrange's equations of motion in the sequel, and the Hamilton's equations of motion in the Chapter 3. Furthermore, we will derive from this principle the Hamilton's principles for conservative and non-conservative systems in the Chapter 3.

D'Alembert started with the equation of motion of a particle of the system as $F_i = \dot{p}_i$, where p_i is the linear momentum of the ith particle. This can be written as $F_i - \dot{p}_i = 0$. Hence the equation of motion of the system becomes $\sum_i (F_i - \dot{p}_i) = 0$, implying that the system of particles is in equilibrium. This equation states that the dynamical system appears to be in equilibrium under the action of applied forces F_i and an equal and opposite 'effective forces' $\dot{p}_i$. In this way dynamics reduces to static. Thus we have

$$\sum_i (F_i - \dot{p}_i) = 0 \Leftrightarrow \text{the system is in equilibrium (the resultant is zero).}$$

Hence the virtual work done by the forces is zero. This implies that

$$\sum_i (F_i - \dot{p}_i)\delta r_i = 0. \qquad \qquad ...(47)$$

Equation (47) is known as the mathematical form of D'Alembert principle. This equation states that "a system of particles moves in such a way that the total virtual work done by the applied forces and reverse effective forces is zero".

Now we use the D'Alembert's principle and solve some examples.

WORKED EXAMPLES

Example 3: A particle is constrained to move in a circle in a vertical plane xy. Apply the D'Alembert's principle to show that for equilibrium we must have $\ddot{x}y - \ddot{y}x - gx = 0$.

Solution: Let a particle of mass m be moving along a circle of radius r in xy plane. Let (x, y) be the position of the particle at any instant t with respect to the fixed-point 0. The constraint on the motion of the particle is that the position co-ordinates of the particle always lie on the circle. Hence the equation of the constraint is

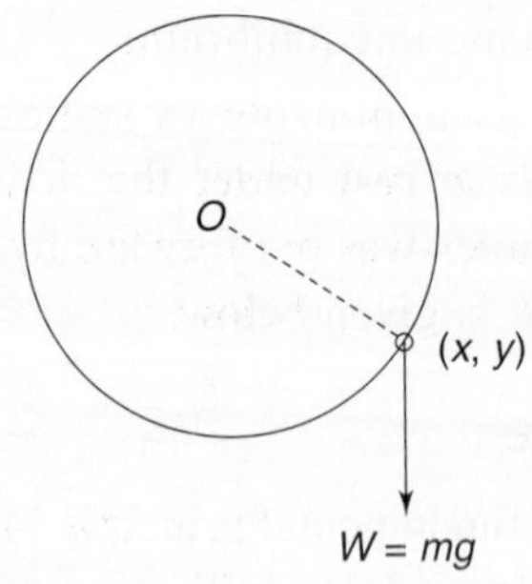

$$x^2 + y^2 = r^2 \qquad \qquad ...(48)$$

$$\Rightarrow \qquad 2x\delta x + 2y\delta y = 0$$

$$\text{or} \qquad \delta x = -\frac{y}{x}\,\delta y \qquad \qquad ...(49)$$

where δx and δy are displacement in x and y respectively. Now from D'Alembert's principle (47) we have for the single particle

$$(F - m\ddot{r})\delta r = 0.$$

In terms of components we have

$$(F_x - m\ddot{x})\,\delta x + (F_y - m\ddot{y})\,\delta y = 0. \qquad \ldots(50)$$

However, the only force acting on the particle at any instant t is its weight mg in the downward direction. Resolving the force horizontally and vertically, we have $F_x = 0$ and $F_y = -mg$. Therefore equation (50) becomes

$$-m\ddot{x}\delta x - (mg + m\ddot{y})\,\delta y = 0.$$

On using (49) we have

$$m(-\ddot{x}y + \ddot{y}x + gx)\delta x = 0.$$

For $\delta x \neq 0$, and $m \neq 0$ we have

$$\ddot{x}y - \ddot{y}x - gx = 0, \qquad \ldots(51)$$

which is the required equation of motion.

Example 4: Two particles of mass m_1 and m_2 are joined by a rod of fixed length. The particle of mass m_2 is constrained to move along a horizontal axis x. Apply D'Alembert's principle and show that the equation of motion of the system is given by

$$m_1 y\ddot{x} - m_2 x\ddot{y} - m_2 gx = 0.$$

g being the acceleration due to gravity.

Solution: Two masses are joined by a rod of fixed length l and the system is shown in the figure. Let $(x, 0)$ and $(0, y)$ be the positions of the two masses m_1 and m_2 at any instant t. The equation of the constraint is

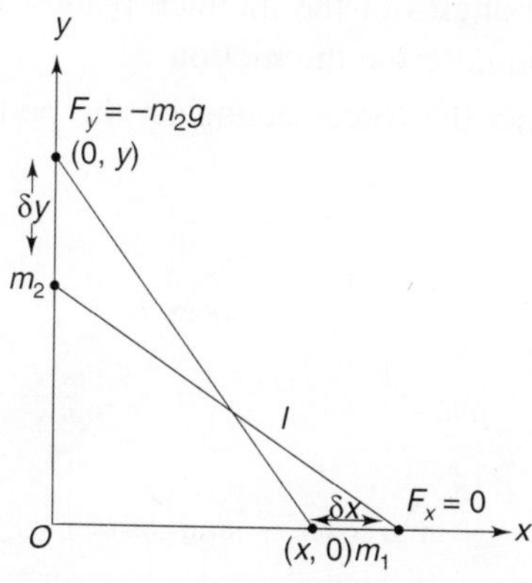

$$x^2 + y^2 = l^2 \qquad \qquad \text{...(52)}$$

$$\Rightarrow \qquad x\delta x + y\delta y = 0$$

$$\text{or} \qquad \delta x = -\frac{y}{x}\,\delta y \qquad \qquad \text{...(53)}$$

where δx and δy are displacements in x and y respectively. The D'Alembert's principle for two particles yields

$$\Rightarrow \qquad (F_1 - m_1\ddot{r}_1)\,\delta r_1 + (F_2 - m_2\ddot{r}_2)\delta r_2 = 0.$$

This is equivalent to

$$(F_x - m_1\ddot{x})\,\delta x + (F_y - m_2\ddot{y})\,\delta y = 0,$$

where

$$F_x = 0,\ F_y = -m_2 g,$$

$$\Rightarrow \qquad -m_1\ddot{x}\delta x - (m_2 g + m_2\ddot{y})\,\delta y = 0.$$

On using the equation (53) in the above equation we have

$$(m_1\ddot{x}y - m_2\ddot{y}x - m_2 gx)\delta y = 0.$$

Since $\delta y \neq 0$, we have

$$m_1\ddot{x}y - m_2\ddot{y}x - m_2 gx = 0. \qquad \qquad \text{...(54)}$$

This is the required equation of motion.

Example 5: Two particles of mass m_1 and m_2 are located on a frictionless double inclined planes and connected by an inextensible massless string passing over a smooth peg. Use the principle of virtual work to show that for equilibrium we must have

$$\frac{\sin \alpha_1}{\sin \alpha_2} = \frac{m_2}{m_1},$$

where α_1 and α_2 are the angles of the inclined planes to the horizontal. Apply D'Alembert's principle to describe the motion.

Solution: The system and the forces acting on the particles of the system are shown in the figure.

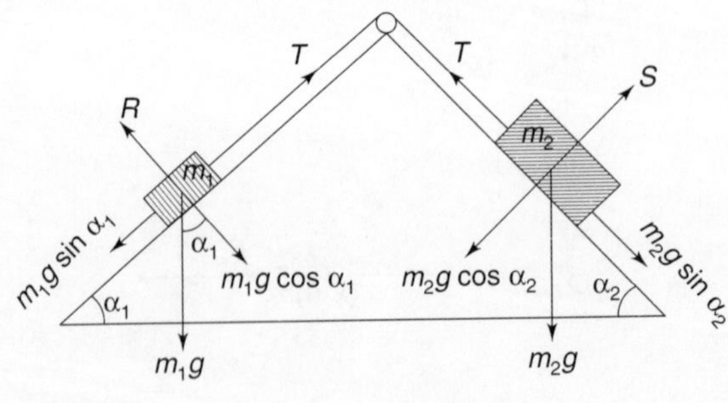

According to the principle of virtual work we have

$$\sum_{i=1}^{2} F_i \, \delta r_i = 0,$$

$$F_1 \delta r_1 + F_2 \delta r_2 = 0,$$

where r_1, r_2 are position vectors of m_1, m_2 with respect to the pulley. The constraint on the motion of the system is that the total length of the string is constant *i.e.*, $r_1 + r_2$ constant.

$$\Rightarrow \qquad \delta r_1 + \delta r_2 = 0, \Rightarrow \delta r_1 = -\delta r_2.$$

For equilibrium we have

$$F_1 \delta r_1 = F_2 \delta r_1 \qquad \qquad \ldots(55)$$

$$\Rightarrow \qquad (m_1 g \sin \alpha_1 - T) \, \delta r_1 = (m_2 g \sin \alpha_2 - T) \, \delta r_1,$$

where T is the tension in the string, $m_1 g \sin \alpha_1$ and $m_2 g \sin \alpha_2$ are the components of the gravitational forces on each of the particles along the respective inclined planes.

$$\Rightarrow \qquad (m_1 g \sin \alpha_1 - m_2 g \sin \alpha_2)\delta r_1 = 0.$$

Since $\delta r_1 \neq 0$, we have

$$m_1 g \sin \alpha_1 - m_2 g \sin \alpha_2 = 0.$$

This readily gives

$$\frac{\sin \alpha_1}{\sin \alpha_2} = \frac{m_2}{m_1}. \qquad \qquad \ldots(56)$$

Now by D'Alembert's principle we have the system of particles moves in such a way that the virtual work done is zero. This yield

$$(F_1 - m_1 \ddot{r}_1) \, \delta r_1 + (F_2 - m_2 \ddot{r}_2) \, \delta r_2 = 0,$$

$$\Rightarrow \qquad (m_1 g \sin \alpha_1 - T - m_1 \ddot{r}_1) \, \delta r_1 + (m_2 g \sin \alpha_2 - T - m_2 \ddot{r}_2) \, \delta r_2 = 0. \qquad \ldots(57)$$

Since from the equation of the constraint we have

$$\ddot{r}_2 = -\ddot{r}_1$$

$$\Rightarrow \qquad (m_1 g \sin \alpha_1 - m_1 \ddot{r}_1 - m_2 g \sin \alpha_2 - m_2 \ddot{r}_1) \, \delta r_1 = 0. \qquad \ldots(58)$$

Since $\delta r_1 \neq 0$, we have therefore,

$$\ddot{r}_1 = \frac{m_1 g \sin \alpha_1 - m_2 g \sin \alpha_2}{m_1 + m_2} \, g. \qquad \qquad \ldots(59)$$

If $m_1 g \sin \alpha_1 > m_2 g \sin \alpha_2$ then the particle m_1 goes down and m_2 goes up.

However, when the frictional force is present during motion and if μ is the coefficient of friction between particle and each of the inclined plane then from equation (55) we have

$$(m_1 g \sin \alpha_1 - \mu R)\, \delta r_1 = (m_2 g \sin \alpha_2 + \mu S)\, \delta r_1,$$

where R and S are the reactions of the inclined planes respectively and are given by

$$R = m_1 g \cos \alpha_1, \; S = m_2 g \cos \alpha_2.$$

Hence from above equation we have readily

$$\mu = \frac{m_1 g \sin \alpha_1 - m_2 g \sin \alpha_2}{m_1 g \cos \alpha_1 + m_2 g \cos \alpha_2}.$$

Example 6: Use D'Alembert's principle to determine the equation of motion of a simple pendulum.

Solution: Consider a particle of mass m attached to one end of the string and other end is fastened to a fixed point 0. Let l be the length of the pendulum and θ the angular displacement of the pendulum shown in the figure. The D'Alembert's principle (47) for single particle gives

$$(F - m\ddot{r})\, \delta r = 0,$$

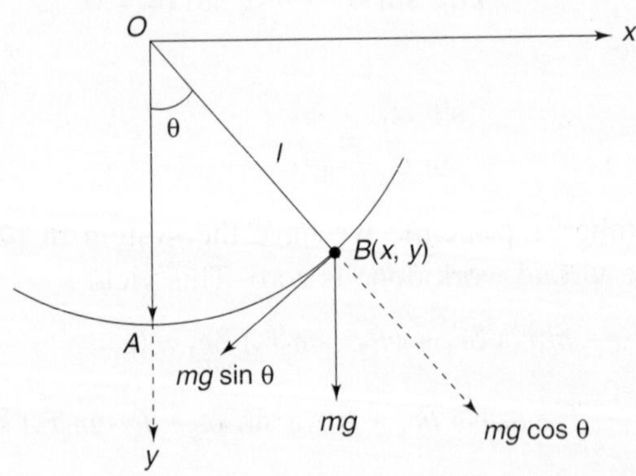

where r is the distance of the particle from starting point along the arc. Resolving the force acting on the particle along the direction of motion and perpendicular to the direction of motion we have

$$(-mg \sin \theta - m\ddot{r})\, \delta r = 0,$$

where the negative sign indicates the force is opposite to the direction of motion. Since $\delta r \neq 0$, we have

$$\ddot{r} = -g \sin \theta.$$

From the figure we have $r = $ arc $AB \Rightarrow r = l\theta \Rightarrow \ddot{r} = l\ddot{\theta}$. Hence the above equation becomes

$$\ddot{\theta} = -\frac{g}{l} \sin \theta. \qquad \ldots(60)$$

For small angle, we have $\sin \theta = \theta \Rightarrow \ddot{\theta} = -\frac{g}{l}\theta$.

Generalized Velocities

Differentiation of the transformation equations (38) with respect to t yields

$$\dot{r} = \sum_j \frac{\partial r_i}{\partial q_j} \dot{q}_j + \frac{\partial r_i}{\partial t} \qquad \ldots(61)$$

where $\dot{q}_j,\ j = 1, 2, 3,\ldots n$ are called generalized velocities.

Virtual Displacement

The δ variation (change) in the transformation equation (38) gives the virtual displacement. Its expression is given by

$$\delta r_i = \sum \frac{\partial r_i}{\partial q_j} \delta q_j. \qquad \ldots(62)$$

Note that δt term is absent because virtual displacement is assumed to take place at a fixed instant t, hence $\delta t = 0$.

Generalized Force

If F_i are forces acting on a dynamical system with position vectors r_i then virtual work done by these forces is given by

$$\delta W = \sum_i F_i \delta r_i,$$

$$= \sum_i \sum_j F_i \frac{\partial r_i}{\partial q_j} \delta q_j,$$

$$= \sum_j \left(\sum_i F_i \frac{\partial r_i}{\partial q_j} \right) \delta q_j,$$

$$\delta W = \sum_j Q_j \delta q_j, \qquad \ldots(63)$$

where $\qquad Q_j = \sum_i F_i \frac{\partial r_i}{\partial q_j} \qquad \ldots(64)$

are called the components of generalized forces.

Note 1. If forces on a system of n particles are conservative then they are derived from potential V. They are given by the equation

$$F_i = -\nabla_i V = -\frac{\partial V}{\partial r_i}. \qquad \qquad \text{...(65)}$$

Consequently, the generalized forces become

$$Q_j = -\frac{\partial V}{\partial q_j}.$$

2. If the forces are non-conservative, the scalar potential U may be function of position, velocity and time. *i.e.*, $U = U(q_j, \dot{q}_j, t)$. This is called velocity dependent potential or generalized potential. Such a potential exits in the case of a motion of a particle of charge q moving in an electromagnetic field. We will see latter in example (8) that how the generalized potential can be determined in the case of a particle moving in an electromagnetic field. In this case generalized forces are given by

$$Q_j = -\frac{\partial U}{\partial q_j} + \frac{d}{dt}\left(\frac{\partial U}{\partial \dot{q}_j}\right). \qquad \qquad \text{...(66)}$$

3. If however, the system is acted upon by some conservative forces F_i and some non-conservative forces $F_i^{(d)}$, in this case generalized forces Q_j are given by

$$Q_j = \Sigma\left(F_i + F_i^{(d)}\right)\frac{\partial r_i}{\partial q_j} \Rightarrow Q_j = -\frac{\partial V}{\partial q_j} + Q_j^{(d)},$$

where $Q_j^{(d)} = \sum\limits_i F_i^{(d)} \dfrac{\partial r_i}{\partial q_j}$ are non-conservative forces which are not derivable from the potential V. Such a situation often arises when frictional forces or dissipative forces are present in the system. It is found by experiment that in general the dissipative or frictional forces are proportional to the velocity of the particle.

$$\Rightarrow \qquad \qquad F_i^{(d)} = -\lambda_i \dot{r}_i,$$

where λ_i are constants. In such cases the generalized forces are obtained as

$$Q_j^{(d)} = \sum\limits_i F_i^{(d)} \frac{\partial r_i}{\partial q_j}$$

$$= -\sum\limits_i \lambda_i \dot{r}_i \frac{\partial r_i}{\partial q_j}$$

However, from transformation equations (38) we obtain

$$\frac{\partial r_i}{\partial q_j} = \frac{\partial \dot{r}_i}{\partial \dot{q}_j}. \qquad \qquad \ldots(67)$$

Thus we write

$$Q_j^{(d)} = \Sigma \, \frac{\partial}{\partial \dot{q}_j} \left(-\frac{1}{2} \lambda_i \dot{r}_i^2 \right),$$

$$Q_j^{(d)} = -\frac{\partial R}{\partial \dot{q}_j},$$

where
$$R = \frac{1}{2} \Sigma \, \lambda_i \dot{r}_i^2 \qquad \qquad \ldots(68)$$

is called the Rayleigh's dissipative function.

Lagrange's Equations of Motion from D'Alembert's Principle

Newtonian approach for the description of a particle involves vector quantities. We now introduce another formulation called the Lagrangian formulation for the description of mechanics of a particle or a system of particles in terms of generalized coordinates, generalized velocities with time t as a parameter. This formulation involves scalar quantities such as kinetic energy and the potential energy and hence proves to be easier than the Newtonian approach, because to deal with scalars is easy than to deal with vectors. In the following theorem we derive the Lagrange's equations of motion for conservative, non-conservative, partially conservative and partially non-conservative systems.

Theorem 3   Obtain Lagrange's equations of motion from D'Alembert's principle.

Proof: We consider a system of n particles of masses m_i and position vectors r_i. The position vectors r_i are expressed as the functions of n generalized co-ordinates $q_1, q_2, q_3, \ldots, q_n$ and time t by the transformation equations (38). *i.e.,*

$$r_i = r_i \, (q_1, q_2, q_3, \ldots, q_n, t).$$

If F_i are the forces acting on the system, then we start from the D'Alembert's principle

$$\sum_i (F_i - \dot{p}_i) \, \delta r_i = 0, \qquad \qquad \ldots(69)$$

where, $\dot{p}_i = m_i \ddot{r}_i$, and the virtual displacement δr_i from equation (63) is given by

$$\delta r_i = \sum_j \frac{\partial r_i}{\partial q_j} \delta q_j.$$

Here the term δt is absent because the virtual displacement is assumed to take place only in the co-ordinates and at the particular instant. Hence the equation (69) becomes

$$\sum_i \sum_j F_i \frac{\partial r_i}{\partial q_j} \delta q_j = \sum_i \sum_j m_i \ddot{r}_i \frac{\partial r_i}{\partial q_j} \delta q_j.$$

$$\sum_j \left(\sum_i F_i \frac{\partial r_i}{\partial q_j} \right) \delta q_j = \sum_{i,j} m_i \ddot{r}_i \frac{\partial r_i}{\partial q_j} \delta q_j,$$

or

$$\sum_j Q_j \, \delta q_j = \sum_{i,j} m_i \ddot{r}_i \frac{\partial r_i}{\partial q_j} \delta q_j, \qquad \qquad \ldots(70)$$

where

$$Q_j = \sum_i F_i \frac{\partial r_i}{\partial q_j} \qquad \qquad \ldots(71)$$

are called the components of generalized forces.

Consider

$$\frac{d}{dt} \left(\dot{r}_i \frac{\partial r_i}{\partial q_j} \right) = \ddot{r}_i \frac{\partial r_i}{\partial q_j} + \dot{r}_i \frac{d}{dt} \left(\frac{\partial r_i}{\partial q_j} \right).$$

Substituting this in equation (70), we get

$$\sum_j Q_j \delta q_j = \sum_{i,j} m_i \left[\frac{d}{dt} \left(\dot{r}_i \frac{\partial r_i}{\partial q_j} \right) - \dot{r}_i \frac{d}{dt} \left(\frac{\partial r_i}{\partial q_j} \right) \right] \delta q_j. \qquad \ldots(72)$$

Now from the transformation equation (38) we have

$$\dot{r}_i = \sum_k \frac{\partial r_i}{\partial q_k} \dot{q}_k + \frac{\partial r_i}{\partial t}. \qquad \qquad \ldots(73)$$

Differentiating this with respect to $\dot{q}_j$ we get

$$\frac{\partial \dot{r}_i}{\partial \dot{q}_j} = \frac{\partial r_i}{\partial q_j}. \qquad \qquad \ldots(74)$$

Further, differentiating equation (73) with respect to q_j, we get

$$\frac{\partial \dot{r}_i}{\partial q_j} = \sum_k \frac{\partial^2 r_i}{\partial q_k \partial q_j} \dot{q}_k + \frac{\partial^2 r_i}{\partial t \partial q_j}. \qquad \qquad \ldots(75)$$

Also we have

$$\frac{d}{dt}\left(\frac{\partial r_i}{\partial q_j}\right) = \sum_k \frac{\partial^2 r_i}{\partial q_j \partial q_k}\, \dot{q}_k + \frac{\partial^2 r_i}{\partial q_j \partial t}. \qquad \ldots(76)$$

We notice from equations (75) and (76) that

$$\frac{\partial \dot{r}_i}{\partial q_j} = \frac{d}{dt}\left(\frac{\partial r_i}{\partial q_j}\right).$$

In general we have

$$\frac{\partial}{\partial q_j}\left(\frac{d}{dt}\right) = \frac{d}{dt}\left(\frac{\partial}{\partial q_j}\right). \qquad \ldots(77)$$

On using equation (77) in equation (72), we get

$$\sum_j Q_j \delta q_j = \sum_{i,j}\left[\frac{d}{dt}\left(m_i v_i \frac{\partial v_i}{\partial \dot{q}_j}\right) - m_i v_i \frac{\partial v_i}{\partial q_j}\right]\delta q_j.$$

We write this as

$$\sum_j Q_j \delta q_j = \sum_j\left[\frac{d}{dt}\left(\frac{\partial}{\partial \dot{q}_j}\left(\sum_i \frac{1}{2} m_i v_i^2\right)\right) - \frac{\partial}{\partial q_j}\left(\sum_i \frac{1}{2} m_i v_i^2\right)\right]\delta q_j$$

or $$\sum_j Q_j \delta q_j = \sum_j\left[\frac{d}{dt}\left(\frac{\partial T}{\partial \dot{q}_j}\right) - \frac{\partial T}{\partial q_j}\right]\delta q_j,$$

where $$T = \frac{1}{2}\sum_i m_i v_i^2$$

is the total kinetic energy of the system of particles.

$$\Rightarrow \qquad \sum_j\left[\frac{d}{dt}\left(\frac{\partial T}{\partial \dot{q}_j}\right) - \frac{\partial T}{\partial q_j} - Q_j\right]\delta q_j = 0. \qquad \ldots(78)$$

If the constraints on the motion of particles in the system are holonomic then δq_j are independent. In this case we infer from equation (78) that

$$\frac{d}{dt}\left(\frac{\partial T}{\partial \dot{q}_j}\right) - \frac{\partial T}{\partial q_j} - Q_j = 0.$$

$$\Rightarrow \qquad \frac{d}{dt}\left(\frac{\partial T}{\partial \dot{q}_j}\right) - \frac{\partial T}{\partial q_j} = Q_j, \quad j = 1, 2, 3, \ldots, n. \qquad \ldots(79)$$

These are called the Lagrange's equations of motion for non-conservative system. We see that, to derive these equations of motion the knowledge of forces acting on the system of particles was not necessary. We note here that if the constraints are non-holonomic then the generalized co-ordinates are not all independent of each other. Hence we can't conclude equation (79) from equation (78). We also note that in deriving Lagrange's equations of motion the requirement of holonomic constraints does not appear until the last step. Now we will derive in the following the form of the Lagrange's equations of motion for conservative, non-conservative and partially conservative and partially non-conservative systems.

Case (1): Conservative system

If the system is conservative so that particles move under the influence of a potential V which is dependent on co-ordinates only, then the forces derived from the potential V are given by equation (65). In this case the components of generalized forces become

$$Q_j = \frac{\partial V}{\partial q_j}, \quad \text{and} \quad V \neq V(\dot{q}_j).$$

Hence equation (79) becomes

$$\frac{d}{dt}\left(\frac{\partial(T-V)}{\partial \dot{q}_j}\right) - \frac{\partial(T-V)}{\partial q_j} = 0.$$

Define a new function

$$L = T - V \qquad \qquad \text{...(80)}$$

where L is a function of $q_1, q_2, q_3, \ldots, q_n, \dot{q}_1, \dot{q}_2, \dot{q}_3, \ldots \dot{q}_n$ and time t is called a Lagrangian function of the system of particles. Then the equations of motion become

$$\frac{d}{dt}\left(\frac{\partial L}{\partial \dot{q}_j}\right) - \frac{\partial L}{\partial q_j} = 0. \qquad \qquad \text{...(81)}$$

These are called the Lagrange's equations of motion for conservative holonomic system.

Note: The Lagrangian L satisfying equation (81) is not unique. Refer Example (13).

Case (2): Non-conservative system

In the case of non-conservative system the scalar potential U may be function of both position and velocity. *i.e.*, $U = U(q_j, \dot{q}_j, t)$. Such a potential is called as velocity dependent potential or generalized potential. In this case the associated generalized forces are given by the equations (66). Using these equations in the equation (79), we get

$$\frac{d}{dt}\left(\frac{\partial L}{\partial \dot{q}_j}\right) - \frac{\partial L}{\partial q_j} = 0, j = 1, 2, ..., n. \qquad ...(82)$$

These are the Lagrange's equations of motion for non-conservative forces. We see from equations (81) and (82) that the same Lagrange's equations hold good for both the conservative and the non-conservative systems but the Lagrangian L in equation (82) contains the potential U of non-conservative system, while the Lagrangian L in equation (81) contains the potential V of the conservative forces.

Case (3): *Partially conservative and partially non-conservative system*

Consider the system is acted upon by conservative forces F_i and non-conservative forces $F_i^{(d)}$. Such a situation often occurs when frictional forces or dissipative forces are present in the system. In this case the components of generalized force are given by

$$Q_j = \sum_i \left(F_i + F_i^{(d)}\right)\frac{\partial r_i}{\partial q_j},$$

$$\Rightarrow \qquad Q_j = -\frac{\partial V}{\partial q_j} + Q_j^{(d)},$$

where the non-conservative forces which are not derivable from potential function V are represented in $Q_j^{(d)}$. Substituting this in equation (79) we readily obtain

$$\frac{d}{dt}\left(\frac{\partial L}{\partial \dot{q}_j}\right) - \frac{\partial L}{\partial q_j} = Q_j^{(d)}, j = 1, 2, ..., n \qquad ...(83)$$

where the Lagrangian L contains the potential of the conservative forces, and $Q_j^{(d)}$ represents the forces not arising from the potential V. However, it is found by experiment that, in general the dissipative or frictional forces are proportional to the velocity of the particles.

$$F_i^{(d)} = -\lambda_i \dot{r}_i, \, \lambda_i \text{ are constants.}$$

Hence we have

$$Q_j^{(d)} = \sum_i F_i^{(d)} \frac{\partial r_i}{\partial q_j},$$

$$Q_j^{(d)} = -\sum_i \lambda_i \dot{r}_i \frac{\partial r_i}{\partial q_j}.$$

But we know from the equation (67) that

$$\frac{\partial \dot{r}_i}{\partial \dot{q}_j} = \frac{\partial r_i}{\partial q_j}.$$

Hence

$$Q_j^{(d)} = \sum_i \frac{\partial}{\partial \dot{q}_j} \left(-\frac{1}{2} \lambda_i \, \dot{r}_i^2 \right) = -\frac{\partial R}{\partial \dot{q}_j},$$

where

$$R = \frac{1}{2} \sum_i \lambda_i \, \dot{r}_i^2$$

is called Rayleigh's dissipation function. Hence the Lagrange's equations of motion become

$$\frac{d}{dt} \left(\frac{\partial L}{\partial \dot{q}_j} \right) - \frac{\partial L}{\partial q_j} + \frac{\partial R}{\partial \dot{q}_j} = 0. \qquad \qquad ...(84)$$

These are Lagrange's equation of motion for partially conservative and partially non-conservative system.

One should remember that the Lagrange's formulation is not the result of new theory, but it can be derived from Newton's theory. To see it, consider a single particle of mass m moving under the action of a conservative force $\overline{F}$. Let $\bar{r}$ be the position vector of the particle then from Newton's second law of motion we have

$$m\ddot{\bar{r}} = \overline{F},$$

where $\overline{F}$ is conservative, therefore, we have $\overline{F} = -\dfrac{\partial V}{\partial r}$. Hence the above equation becomes

$$m\ddot{\bar{r}} + \frac{\partial V}{\partial r} = 0.$$

This equation can be written as $\dfrac{d}{dt}(m\dot{\bar{r}}) + \dfrac{\partial V}{\partial r} = 0,$

$$\frac{d}{dt} \left[\frac{\partial}{\partial \dot{r}} \left(\frac{1}{2} m\dot{\bar{r}}^2 \right) \right] + \frac{\partial V}{\partial r} = 0,$$

$$\frac{d}{dt} \left(\frac{\partial L}{\partial \dot{r}} \right) - \frac{\partial L}{\partial r} = 0,$$

where $L = T - V$, $T = \dfrac{1}{2} m\dot{r}^2$, $V = V(r)$. Conversely, Newton's equation of motion can also be derived in the same way from the Lagrange's equation of motion. Similarly, the respective Lagrange's equations of motion can also be derived from the corresponding Newton's equations of motion when dissipative forces are present in the system and vice versa.

WORKED EXAMPLES

Conservation of Energy

Example 7: Show that the total energy of a particle moving in a conservative force field remains constant, if the potential energy is not an explicit function of time.

Solution: Let a particle of mass m be moving in the conservative field of force $\overline{F}$. Let r be the position vector of the particle at any instant. The total energy of the particle is

$$E = T + V, \qquad\qquad \ldots(85)$$

where T is the kinetic energy $= \dfrac{1}{2} mv^2$ and V is the potential energy of the particle.

Differentiating (85) with respect to t we get

$$\frac{dE}{dt} = \frac{dT}{dt} + \frac{dV}{dt}, \qquad\qquad \ldots(86)$$

where the force is given by $F = m \dfrac{dv}{dt}$. Therefore, we write

$$F dr = m \frac{dv}{dt} \frac{dr}{dt} dt$$

$$F dr = mv\,dv,$$

$$F dr = d\left(\frac{1}{2} mv^2 \right)$$

$$\Rightarrow \qquad F dr = dT,$$

$$\Rightarrow \qquad F \frac{dr}{dt} = \frac{dT}{dt}. \qquad\qquad \ldots(87)$$

Similarly, we have the potential energy $V = V(r, t)$ therefore,

$$dV = \frac{\partial V}{\partial x} dx + \frac{\partial V}{\partial y} dy + \frac{\partial V}{\partial z} dz + \frac{\partial V}{\partial t} dt,$$

$$\Rightarrow \qquad dV = \nabla V \cdot dr + \frac{\partial V}{\partial t} dt,$$

$$\Rightarrow \qquad \frac{dV}{dt} = \nabla V \cdot \frac{dr}{dt} + \frac{\partial V}{\partial t}. \qquad\qquad \ldots(88)$$

Substituting this in equation (86), we get

$$\frac{dE}{dt} = \overline{F}\,\frac{dr}{dt} + \nabla V\,\frac{dr}{dt} + \frac{\partial V}{\partial t},$$

$$\frac{dE}{dt} = (\overline{F} + \nabla V)\,\frac{dr}{dt} + \frac{\partial V}{\partial t}.$$

The force $\overline{F}$ is conservative means $\overline{F} + \nabla V = 0$,

$$\Rightarrow \qquad \frac{dE}{dt} = \frac{\partial V}{\partial t}.$$

Now if the potential energy V is independent of time t then $\dfrac{dE}{dt} = 0$. This implies that E is conserved.

Theorem 4   If the force acting on a particle is conservative then the total energy is conserved.

Proof: If the particle is acted upon by the force $\overline{F}$, and if it moves from position P_1 to P_2, then the work done by the force is given by

$$W = \int_{P_1}^{P_2} \overline{F} \cdot dr \qquad\qquad \ldots(89)$$

where

$$\overline{F} = \dot{p} = m\,\frac{dv}{dt},$$

Therefore,

$$W = \int_{P_1}^{P_2} m\,\frac{dv}{dt} \cdot \frac{dr}{dt}\,dt = \int_{P_1}^{P_2} m\,\frac{dv}{dt} \cdot v\,dt$$

$$= \int_{P_1}^{P_2} \frac{d}{dt}\left(\frac{1}{2}\,mv^2\right)dt,$$

$$= \left(\frac{1}{2}\,mv^2\right)\Bigg|_{P_1}^{P_2},$$

$$W = \frac{1}{2}\,mv_2^2 - \frac{1}{2}\,mv_1^2.$$

Thus

$$W = T_2 - T_1. \qquad\qquad \ldots(90)$$

Now, if the force $\overline{F}$ is conservative then it is derivable from a scalar potential function V, which is a function of position only. Therefore, we have $\overline{F} = -\nabla V = -\dfrac{\partial V}{\partial r}$, where V is the potential energy. Substituting this value in the equation (89), we get

$$W = -\int_{P_1}^{P_2} \frac{\partial V}{\partial r}\,dr$$

$$= -\int_{P_1}^{P_2} dV$$

$$= -(V)_{P_1}^{P_2}$$

$$W = V_1 - V_2. \qquad \qquad \ldots(91)$$

From equations (90) and (91), we have

$$T_2 - T_1 = V_1 - V_2,$$

$$T_1 + V_1 = T_2 + V_2 = \text{constant}$$

$$\Rightarrow \qquad \qquad T + V = \text{constant}.$$

This shows that the total energy of the particle is conserved.

Aliter: The force field is conservative. This implies that

$$\overline{F} = -\nabla V, \qquad \qquad \ldots(92)$$

where V is the potential energy. Newton's second law of motion defines the force by

$$\overline{F} = m\ddot{r}. \qquad \qquad \ldots(93)$$

Thus we have

$$m\ddot{r} = -\frac{\partial V}{\partial r}.$$

Multiplying this equation by $\dot{r}$, we get

$$m\ddot{r}\,\dot{r} = -\frac{\partial V}{\partial r}\,\dot{r}.$$

This we write as

$$\frac{d}{dt}\left(\frac{1}{2}\,m\dot{r}^2 + V\right) = 0.$$

Integrating we get

$$\frac{1}{2}\,m\dot{r}^2 + V = \text{const.}$$

This shows that the total energy of the particle moving in the conservative field of force is constant.

Theorem 5   If the external and internal forces are both conservative, then show that the total potential energy V of the system is given by

$$V = \sum_i V_i^{(e)} + \frac{1}{2}\sum_{i,j} V_{ij}^{(\text{int})},$$

where $V_i^{(e)}$ is the potential energy arises due to the external forces $\overline{F}_i^{(e)}$ and $V_{ij}^{(int)}$ is the internal energy arises due to internal forces $\overline{F}_{ji}^{(int)}$. Further show that the total energy of the system is conserved.

Proof: Two types of forces viz., external and internal forces are acting on the system of particles. To find the total energy of the system, we find the work done by all the forces external as well as internal in moving the system from initial configuration 1 to the final configuration 2. It is given by

$$W = \sum_i \int_1^2 \overline{F}_i d\overline{r}_i,$$

where

$$\overline{F}_i = \overline{F}_i^{(e)} + \sum_j \overline{F}_{ji}^{(int)}$$

$$\Rightarrow \qquad W = \sum_i \int_1^2 \overline{F}_i^{(e)} \, d\overline{r}_i + \sum_{i,j,i\neq j} \int_1^2 \overline{F}_{ji}^{(int)} \, d\overline{r}_i. \qquad \ldots(94)$$

Let $\overline{F}_i^{(e)}$ be conservative, then there exists a potential $V_i^{(e)}$ such that

$$\overline{F}_i^{(e)} = -\nabla_i V_i^{(e)} = -\frac{\partial V_i^{(e)}}{\partial r_i}$$

$$\sum_i \int_1^2 \overline{F}_i^{(e)} \, d\overline{r}_i = -\sum_i \int_1^2 \frac{\partial V_i^{(e)}}{\partial r_i} \, d\overline{r}_i = -\sum_i \int_1^2 dV_i^{(e)},$$

$$\sum_i \int_1^2 \overline{F}_i^{(e)} d\overline{r}_i = -\left[\sum_i V_i^{(e)}\right]_1^2. \qquad \ldots(95)$$

Now consider the second term on the right hand side of equation (94)

$$\sum_{i,j,i\neq j} \int_1^2 \overline{F}_{ji}^{(int)} \, d\overline{r}_i = \sum_{i,j,i\neq j} \int_1^2 \overline{F}_{ji}^{(int)} \, d\overline{r}_i. \qquad \ldots(96)$$

Interchanging i and j on the right hand side of equation (96), we get

$$\sum_{i,j,i\neq j} \int_1^2 \overline{F}_{ji}^{(int)} \, d\overline{r}_i = \sum_{i,j,i\neq j} \int_1^2 \overline{F}_{ij}^{(int)} \, d\overline{r}_j.$$

Since the internal forces obey the Newton's third law of motion, we have therefore $\overline{F}_{ji}^{(int)} = -\overline{F}_{ij}^{(int)}$.

$$\Rightarrow \qquad \sum_{i,j,i\neq j} \int_1^2 \overline{F}_{ji}^{(int)} \, d\overline{r}_i = -\sum_{i,j,i\neq j} \int_1^2 \overline{F}_{ji}^{(int)} \, d\overline{r}_j, \qquad \ldots(97)$$

Adding equations (96) and (97), we get

$$\sum_{i,j,i\neq j} \int_1^2 \overline{F}_{ji}^{(\text{int})} \, d\overline{r}_i = \frac{1}{2} \sum_{i,j,i\neq j} \int_1^2 \overline{F}_{ji}^{(\text{int})} \, (d\overline{r}_i - d\overline{r}_j),$$

$$\sum_{i,j,i\neq j} \int_1^2 \overline{F}_{ji}^{(\text{int})} \, d\overline{r}_i = \frac{1}{2} \sum_{i,j,i\neq j} \int_1^2 \overline{F}_{ji}^{(\text{int})} \, d\overline{r}_{ij}, \quad \text{for} \quad d\overline{r}_{ij} = d\overline{r}_i - d\overline{r}_j.$$

Now if the internal forces $\overline{F}_{ji}^{(\text{int})}$ are conservative, there exists a potential $V_{ij}^{(\text{int})}$ such that

$$\overline{F}_{ji}^{(\text{int})} = -\nabla_{ji} V_{ji}^{(\text{int})} = -\frac{\partial V_{ij}^{(\text{int})}}{\partial r_{ij}},$$

where ∇_{ji} is the gradient with respect to r_{ji}. Thus the above equation becomes

$$\sum_{i,j,i\neq j} \int_1^2 \overline{F}_{ji}^{(\text{int})} \, d\overline{r}_i = -\frac{1}{2} \sum_{i,j,i\neq j} \int_1^2 \frac{\partial V_{ij}^{(\text{int})}}{\partial r_{ij}} \, d\overline{r}_{ij},$$

$$\sum_{i,j,i\neq j} \int_1^2 \overline{F}_{ji}^{(\text{int})} \, d\overline{r}_i = -\frac{1}{2} \sum_{i,j,i\neq j} \int_1^2 \, dV_{ij}^{(\text{int})},$$

$$\sum_{i,j,i\neq j} \int_1^2 \overline{F}_{ji}^{(\text{int})} d\overline{r}_i = -\left[\frac{1}{2} \sum_{i,j} V_{ij}^{(\text{int})}\right]_1^2 \qquad \qquad \ldots(98)$$

Substituting the values from equations (95) and (98), in the equation (94) we get,

$$W = -[V]_1^2 = V_1 - V_2, \qquad \qquad \ldots(99)$$

where

$$V = \sum_i V_i^{(e)} + \frac{1}{2} \sum_{i,j} V_{ij}^{(\text{int})} \qquad \qquad \ldots(100)$$

represents the total potential energy of the system of particles. Similarly the total work done by the forces on the system in terms of kinetic energy is given by

$$W = \sum_i \int_1^2 \overline{F}_i \, dr_i$$

$$W = \sum_i \int_1^2 \frac{d}{dt} (m_i v_i) \frac{d\overline{r}_i}{dt} \, dt$$

$$= \sum_i \int_1^2 m_i \frac{dv_i}{dt} \, v_i \, dt,$$

$$W = \sum_i \int_1^2 m_i \frac{d}{dt}\left(\frac{1}{2} v_i^2\right) dt$$

$$W = \sum_i \int_1^2 d\left(\frac{1}{2} m_i v_i^2\right)$$

$$W = \sum_i \left(\frac{1}{2} m_i v_i^2\right)_1^2 = T_2 - T_1. \qquad \ldots(101)$$

From equations (99) and (101) we have

$$T_1 + V_1 = T_2 + V_2.$$

This shows that the total energy of the system is conserved.

Example 8: Find the velocity dependent potential and hence the Lagrangian for a particle of charge q moving in an electromagnetic field.

Solution: Consider a charge particle of charge q moving with velocity v in an electric field $\overline{E}$ and magnetic field $\overline{B}$. The force acting on the particle is called Lorentz force and is given by

$$\overline{F} = q(\overline{E} + \overline{v} \times \overline{B}) \qquad \ldots(102)$$

where $\overline{E}$ and $\overline{B}$ satisfy the Maxwell's field equations

$$\nabla \cdot \overline{B} = 0,$$

$$\nabla \times \overline{E} = -\frac{\partial \overline{B}}{\partial t}. \qquad \ldots(103)$$

The vector identity $\nabla \cdot \nabla \times \overline{A} = 0$ gives $\overline{B} = \nabla \times \overline{A}$, where $\overline{A}$ is the vector potential function of co-ordinates and velocities. Substituting this in the second Maxwell equation we get

$$\nabla \times \overline{E} + \frac{\partial}{\partial t}(\nabla \times \overline{A}) = 0$$

$$\Rightarrow \qquad \nabla \times \overline{E} + \left(\nabla \times \frac{\partial \overline{A}}{\partial t}\right) = 0$$

$$\Rightarrow \qquad \nabla \times \left(\overline{E} + \frac{\partial \overline{A}}{\partial t}\right) = 0. \qquad \ldots(104)$$

Comparing the equation (104) with the vector identity $\nabla \times \nabla \phi = 0$, we see that, there exists a scalar potential ϕ function of co-ordinates such that

$$\overline{E} + \frac{\partial \overline{A}}{\partial t} = -\nabla \phi$$

$\Rightarrow$
$$\overline{E} = -\nabla\phi - \frac{\partial\overline{A}}{\partial t}. \qquad\qquad ...(105)$$

Consequently the equation (102) becomes

$$\overline{F} = q\left[-\nabla\phi - \frac{\partial\overline{A}}{\partial t} + \overline{v} \times \nabla \times \overline{A}\right], \qquad\qquad ...(106)$$

where we have

$$\nabla\phi = i\frac{\partial\phi}{\partial x} + j\frac{\partial\phi}{\partial y} + k\frac{\partial\phi}{\partial z},$$

$$\frac{\partial\overline{A}}{\partial t} = i\frac{\partial A_x}{\partial t} + j\frac{\partial A_y}{\partial t} + k\frac{\partial A_z}{\partial t},$$

$$\nabla \times \overline{A} = \begin{vmatrix} i & j & k \\ \dfrac{\partial}{\partial x} & \dfrac{\partial}{\partial y} & \dfrac{\partial}{\partial z} \\ A_x & A_y & A_z \end{vmatrix} = i\left(\frac{\partial A_z}{\partial y} - \frac{\partial A_y}{\partial z}\right) + j\left(\frac{\partial A_x}{\partial z} - \frac{\partial A_z}{\partial x}\right) + k\left(\frac{\partial A_y}{\partial x} - \frac{\partial A_x}{\partial y}\right),$$

and

$$\overline{v} \times \nabla \times \overline{A} = i\left[v_y\left(\frac{\partial A_y}{\partial x} - \frac{\partial A_x}{\partial y}\right) - v_z\left(\frac{\partial A_x}{\partial z} - \frac{\partial A_z}{\partial x}\right)\right] +$$

$$+ j\left[v_x\left(\frac{\partial A_y}{\partial x} - \frac{\partial A_x}{\partial y}\right) - v_z\left(\frac{\partial A_z}{\partial y} - \frac{\partial A_y}{\partial z}\right)\right] +$$

$$+ k\left[v_x\left(\frac{\partial A_x}{\partial z} - \frac{\partial A_z}{\partial x}\right) - v_y\left(\frac{\partial A_z}{\partial y} - \frac{\partial A_y}{\partial z}\right)\right].$$

Therefore the *x*-component of the Lorentz force defined in the equation (106) becomes

$$F_x = q\left[-\frac{\partial\phi}{\partial x} - \frac{\partial A_x}{\partial t} + v_y\left(\frac{\partial A_y}{\partial x} - \frac{\partial A_x}{\partial y}\right) - v_z\left(\frac{\partial A_x}{\partial z} - \frac{\partial A_z}{\partial x}\right)\right]. \qquad ...(107)$$

Now consider

$$v_y\left(\frac{\partial A_y}{\partial x} - \frac{\partial A_x}{\partial y}\right) - v_z\left(\frac{\partial A_x}{\partial z} - \frac{\partial A_z}{\partial x}\right) = v_x\frac{\partial A_x}{\partial x} + v_y\frac{\partial A_y}{\partial x} +$$

$$+ v_z \frac{\partial A_z}{\partial x} - \left(v_x \frac{\partial A_x}{\partial x} + v_y \frac{\partial A_x}{\partial y} + v_z \frac{\partial A_x}{\partial z} \right). \qquad \ldots(108)$$

Also we have

$$\frac{dA_x}{dt} = v_x \frac{\partial A_x}{\partial x} + v_y \frac{\partial A_x}{\partial y} + v_z \frac{\partial A_x}{\partial z} + \frac{\partial A_x}{\partial t},$$

$$v_x \frac{\partial A_x}{\partial x} + v_y \frac{\partial A_x}{\partial y} + v_z \frac{\partial A_x}{\partial z} = \frac{dA_x}{dt} - \frac{\partial A_x}{\partial t}. \qquad \ldots(109)$$

Also

$$\frac{\partial}{\partial x} (\bar{v} \cdot \bar{A}) = \frac{\partial}{\partial x} (v_x A_x + v_y A_y + v_z A_z),$$

$$\frac{\partial}{\partial x} (\bar{v} \cdot \bar{A}) = v_x \frac{\partial A_x}{\partial x} + v_y \frac{\partial A_y}{\partial x} + v_z \frac{\partial A_z}{\partial x}. \qquad \ldots(110)$$

Substituting from equations (109) and (110) in equation (108), we get

$$v_y \left(\frac{\partial A_y}{\partial x} - \frac{\partial A_x}{\partial y} \right) - v_z \left(\frac{\partial A_x}{\partial z} - \frac{\partial A_z}{\partial x} \right) = \frac{\partial}{\partial x} (\bar{v} \cdot \bar{A}) - \frac{dA_x}{dt} + \frac{\partial A_x}{\partial t}. \qquad \ldots(111)$$

Hence equation (107) becomes

$$F_x = q \left[-\frac{\partial}{\partial x} (\phi - \bar{v} \cdot \bar{A}) - \frac{dA_x}{dt} \right]. \qquad \ldots(112)$$

Also

$$\frac{\partial}{\partial v_x} (\bar{v} \cdot \bar{A}) = \frac{\partial}{\partial v_x} (v_x A_x + v_y A_y + v_z A_z) = A_x.$$

As ϕ is independent of v_x, therefore we write

$$\frac{\partial}{\partial v_x} (\phi - \bar{v} \cdot \bar{A}) = -A_x.$$

$$\Rightarrow \qquad \frac{d}{dt} \cdot \frac{\partial}{\partial v_x} (\phi - \bar{v} \cdot \bar{A}) = -\frac{dA_x}{dt}. \qquad \ldots(113)$$

Substituting this in equation (112), we get

$$F_x = q \left[-\frac{\partial}{\partial x} (\phi - \bar{v} \cdot \bar{A}) - \frac{d}{dt} \left\{ \frac{\partial}{\partial v_x} (\phi - \bar{v} \cdot \bar{A}) \right\} \right]. \qquad \ldots(114)$$

Define the generalized potential

$$U = q(\phi - \overline{v} \cdot \overline{A}).$$
...(115)

Hence we write equation (114) as

$$F_x = \left[-\frac{\partial U}{\partial x} - \frac{d}{dt}\left\{\frac{\partial U}{\partial \dot{x}}\right\} \right]. \quad \text{for } \dot{x} = v_x$$
...(116)

Hence the x-Lagrange's equation of motion $\dfrac{d}{dt}\left(\dfrac{\partial T}{\partial \dot{x}}\right) - \dfrac{\partial T}{\partial x} = F_x$ becomes

$$\frac{d}{dt}\left(\frac{\partial T}{\partial \dot{x}}\right) - \frac{\partial T}{\partial x} = -\frac{\partial U}{\partial x} + \frac{d}{dt}\left(\frac{\partial U}{\partial \dot{x}}\right).$$

$$\frac{d}{dt}\left(\frac{\partial (T - U)}{\partial \dot{x}}\right) - \frac{\partial (T - U)}{\partial x} = 0.$$

$$\frac{d}{dt}\left(\frac{\partial L}{\partial \dot{x}}\right) - \frac{\partial L}{\partial x} = 0.$$

Similarly, y and z-Lagrange's equations of motion are determined.

The Lagrangian of the particle $L = T - U$ becomes

$$L = \frac{1}{2} m \left(\dot{x}^2 + \dot{y}^2 + \dot{z}^2\right) - q\phi + q\overline{v} \cdot \overline{A}.$$
...(117)

Example 9:   Show that the Lagrange's equation

$$\frac{d}{dt}\left(\frac{\partial T}{\partial \dot{q}_j}\right) - \frac{\partial T}{\partial q_j} = Q_j.$$

can also be written in the form

$$\frac{\partial \dot{T}}{\partial \dot{q}_j} - 2\frac{\partial T}{\partial q_j} = Q_j.$$

Solution: The kinetic energy T of a particle is in general a function of generalized co-ordinates, generalized velocities and time. Thus we have

$$T = T(q_j, \dot{q}_j, t).$$
...(118)

Differentiating this with respect to time t, we get

$$\frac{dT}{dt} = \dot{T} = \sum_k \frac{\partial T}{\partial q_k}\dot{q}_k + \sum_k \frac{\partial T}{\partial \dot{q}_k}\ddot{q}_k + \frac{\partial T}{\partial t}.$$
...(119)

Differentiating equation (119) partially with respect to $\dot{q}_j$ we get

$$\frac{\partial \dot{T}}{\partial \dot{q}_j} = \sum_k \left(\frac{\partial^2 T}{\partial \dot{q}_j \partial q_k} \dot{q}_k + \frac{\partial T}{\partial q_k} \delta_k^j \right) + \sum_k \frac{\partial^2 T}{\partial \dot{q}_j \partial \dot{q}_k} \ddot{q}_k + \frac{\partial^2 T}{\partial \dot{q}_j \partial t}$$

$$\frac{\partial \dot{T}}{\partial \dot{q}_j} = \sum_k \frac{\partial^2 T}{\partial \dot{q}_j \partial q_k} \dot{q}_k + \frac{\partial T}{\partial q_j} + \sum_k \frac{\partial^2 T}{\partial \dot{q}_j \partial \dot{q}_k} \ddot{q}_k + \frac{\partial^2 T}{\partial \dot{q}_j \partial t}. \qquad \text{...(120)}$$

Also we find the expression

$$\frac{d}{dt} \left(\frac{\partial T}{\partial \dot{q}_j} \right) = \sum_k \frac{\partial^2 T}{\partial q_k \partial \dot{q}_j} \dot{q}_k + \sum_k \frac{\partial^2 T}{\partial \dot{q}_k \partial \dot{q}_j} \ddot{q}_k + \frac{\partial^2 T}{\partial t \partial \dot{q}_j}. \qquad \text{...(121)}$$

From equations (120) and (121), we have

$$\frac{\partial \dot{T}}{\partial \dot{q}_j} - \frac{d}{dt} \left(\frac{\partial T}{\partial \dot{q}_j} \right) = \frac{\partial T}{\partial q_j}. \qquad \text{...(122)}$$

But it is given that

$$\frac{d}{dt} \left(\frac{\partial T}{\partial \dot{q}_j} \right) = \frac{\partial T}{\partial q_j} + Q_j.$$

Consequently equation (122) becomes

$$\frac{\partial \dot{T}}{\partial \dot{q}_j} - \left(\frac{\partial T}{\partial q_j} + Q_j \right) = \frac{\partial T}{\partial q_j},$$

$$\Rightarrow \qquad \frac{\partial \dot{T}}{\partial \dot{q}_j} - 2 \frac{\partial T}{\partial q_j} = Q_j.$$

Example 10:   A particle of mass M moves in a plane in the field of force given by $F = -\hat{i}_r kr \cos \theta$, where k is constant and $\hat{i}_r$ is the radial unit vector. Show that angular momentum of the particle about the origin is conserved and obtain the differential equation of the orbit of the particle.

Solution: Let (x, y) and (r, θ) be the Cartesian and polar co-ordinates of a particle of mass M moving in a plane under the action of the given field of force

$$F = -\hat{i}_r kr \cos \theta. \qquad \text{...(123)}$$

Since the force is explicitly given, hence the Lagrange's equations of motion corresponding to the generalized coordinates r and θ are given by

$$\frac{d}{dt}\left(\frac{\partial T}{\partial \dot{r}}\right) - \frac{\partial T}{\partial r} = Q_r, \qquad \ldots(124)$$

and
$$\frac{d}{dt}\left(\frac{\partial T}{\partial \dot{\theta}}\right) - \frac{\partial T}{\partial \theta} = Q_\theta \qquad \ldots(125)$$

where T is the kinetic energy of the particle and is given by

$$T = \frac{1}{2} M(\dot{x}^2 + \dot{y}^2) = \frac{1}{2} M (\dot{r}^2 + r^2 \dot{\theta}^2). \qquad \ldots(126)$$

The components of generalized force along the radial direction and in the direction of θ are given by

$$Q_r = -\hat{i}_r kr \cos \theta,$$

$$Q_\theta = 0.$$

Hence equations (124) and (125) become

$$M\ddot{r} - Mr\dot{\theta}^2 + kr \cos \theta = 0,$$

and
$$\frac{d}{dt}(Mr^2\dot{\theta}) = 0.$$

$$\Rightarrow \qquad Mr^2\dot{\theta} = \text{const} = h.$$

This proves that angular momentum of the particle is conserved.

Eliminating $\dot{\theta}$ from the above equation we get

$$M\ddot{r} - \frac{h^2}{Mr^3} + kr \cos \theta = 0.$$

This is the equation of motion of the orbit of the particle.

Example 11: Show that the Lagrange's equation of motion can also be written as

$$\frac{\partial L}{\partial t} - \frac{d}{dt}\left(L - \Sigma \dot{q}_j \frac{\partial L}{\partial \dot{q}_j}\right) = 0.$$

Solution: A Lagrangian of a particle is

$$L = L (q_j, \dot{q}_j, t).$$

Differentiating this with respect to time t, we obtain

$$\frac{dL}{dt} = \Sigma_j \frac{\partial L}{\partial q_j} \dot{q}_j + \Sigma_j \frac{\partial L}{\partial \dot{q}_j} \ddot{q}_j + \frac{\partial L}{\partial t}. \qquad \ldots(127)$$

Also we write the expression

$$\frac{d}{dt}\left(\sum_{j} \dot{q}_j \frac{\partial L}{\partial \dot{q}_j}\right) = \sum_{j} \dot{q}_j \frac{d}{dt}\left(\frac{\partial L}{\partial \dot{q}_j}\right) + \sum_{j} \frac{\partial L}{\partial \dot{q}_j} \ddot{q}_j. \qquad \ldots(128)$$

Subtracting equation (128) from (127), we get

$$\frac{d}{dt}\left(L - \sum_{j} \dot{q}_j \frac{\partial L}{\partial \dot{q}_j}\right) = \frac{\partial L}{\partial t} + \sum_{j} \left(\frac{\partial L}{\partial q_j} - \frac{d}{dt}\left(\frac{\partial L}{\partial \dot{q}_j}\right)\right) \dot{q}_j. \qquad \ldots(129)$$

Using the Lagrange's equation of motion (81) in the equation (129) we obtain

$$\frac{d}{dt}\left(L - \sum_{j} \dot{q}_j \frac{\partial L}{\partial \dot{q}_j}\right) - \frac{\partial L}{\partial t} = 0. \qquad \ldots(130)$$

This is the required form.

Example 12: A particle of mass m moves in a plane under the action of a conservative force F with components $F_x = -k^2(2x + y)$, $F_y = -k^2(x + 2y)$, k is a constant. Find the total energy of the motion, the Lagrangian, and the equations of motion of the particle.

Solution: A particle is moving in a plane. Let (x, y) be the co-ordinates of the particle at any instant t. If T and V are the kinetic and potential energies of the particle then we have

$$T = \frac{1}{2} m(\dot{x}^2 + \dot{y}^2), \qquad \ldots(131)$$

and
$$V = V(x, y). \qquad \ldots(132)$$

Since the force is given by

$$\overline{F} = -\nabla V,$$

$$\Rightarrow \qquad iF_x + jF_y = -\left(i\frac{\partial V}{\partial x} + j\frac{\partial V}{\partial y} + k\frac{\partial V}{\partial z}\right)$$

$$-k^2(2x + y)\, i - k^2(x + 2y)\, j = -\left(i\frac{\partial V}{\partial x} + j\frac{\partial V}{\partial y}\right).$$

$$\Rightarrow \qquad \frac{\partial V}{\partial x} = k^2(2x + y),\ \frac{\partial V}{\partial y} = k^2(x + 2y).$$

Using these expressions in the equation $dV = \dfrac{\partial V}{\partial x}\, dx + \dfrac{\partial V}{\partial y}\, dy$, we get

$$dV = k^2(2x + y)\, dx + k^2(x + 2y)\, dy.$$

$$dV = k^2(2xdx + d(xy) + 2ydy).$$

On integrating we get,

$$V = k^2 \left(x^2 + xy + y^2\right). \qquad \ldots(133)$$

The total energy of motion of the particle is therefore given by

$$E = T + V.$$

While the Lagrangian of the motion is given by

$$L = \frac{1}{2} m(\dot{x}^2 + \dot{y}^2) - k^2(x^2 + xy + y^2). \qquad \ldots(134)$$

The Lagrange's equations of motion corresponding to the generalized co-ordinates x and y respectively yield

$$m\ddot{x} + k^2 (2x + y) = 0,$$

$$m\ddot{y} + k^2 (x + 2y) = 0.$$

Kinetic Energy as a Homogeneous Quadratic Function of Generalized Velocities

Theorem 6   Find the expression for the kinetic energy as the quadratic function of generalized velocities. Further show that

   (i) when the constraints are scleronomic and the kinetic energy T is a homogeneous quadratic function of generalized velocities then $\sum\limits_{j} \dot{q}_j \dfrac{\partial T}{\partial \dot{q}_j} = 2T,$

   (ii) when the constraints are rheonomic then $\sum\limits_{j} \dot{q}_j \dfrac{\partial T}{\partial \dot{q}_j} = 2T_2 + T_1$, where T_1, T_2 are homogeneous functions of generalized velocities of degree one and two respectively.

Proof: Consider a system of particles of masses m_i and position vectors r_i. Generally, for a rheonomic system we have $r_i = r_i(q_1, q_2, q_3,\ldots q_n, t)$. The time derivative of this equation gives

$$\dot{r}_i = \sum_{k} \frac{\partial r_i}{\partial q_k} \dot{q}_k + \frac{\partial r_i}{\partial t}. \qquad \ldots(135)$$

In this case the kinetic energy $T = \dfrac{1}{2} \sum\limits_{i} m_i \dot{r}_i^2$ of the system becomes

$$T = \frac{1}{2} \sum_{i} m_i \left[\sum_{j} \left(\frac{\partial r_i}{\partial q_j} \dot{q}_j + \frac{\partial r_i}{\partial t} \right) \sum_{k} \left(\frac{\partial r_i}{\partial q_k} \dot{q}_k + \frac{\partial r_i}{\partial t} \right) \right],$$

$$T = \frac{1}{2} \sum_{i} m_i \left[\sum_{j,k} \frac{\partial r_i}{\partial q_j} \cdot \frac{\partial r_i}{\partial q_k} \dot{q}_j \dot{q}_k + 2\sum_{j} \frac{\partial r_i}{\partial q_j} \frac{\partial r_i}{\partial t} \dot{q}_j + \left(\frac{\partial r_i}{\partial t} \right)^2 \right].$$

Since finite summations can be interchanged, hence we write the above expression as

$$T = \sum_{j,k}\left[\frac{1}{2}\sum_i m_i \frac{\partial r_i}{\partial q_j}\frac{\partial r_i}{\partial q_k}\right]\dot{q}_j\dot{q}_k + \sum_j\left[\sum_i m_i \frac{\partial r_i}{\partial q_j}\frac{\partial r_i}{\partial t}\right]\dot{q}_j + \sum_i \frac{1}{2}m_i\left(\frac{\partial r_i}{\partial t}\right)^2$$

or
$$T = \sum_{j,k} a_{jk}\dot{q}_j\dot{q}_k + \sum_j a_j\dot{q}_j + a, \qquad \qquad \ldots(136)$$

where $a_{jk} = \sum_i \frac{1}{2}m_i \frac{\partial r_i}{\partial q_j}\frac{\partial r_i}{\partial q_k}$, $a_j = \sum_i m_i \frac{\partial r_i}{\partial q_j}\frac{\partial r_i}{\partial t}$, $a = \sum_i \frac{1}{2}m_i\left(\frac{\partial r_i}{\partial t}\right)^2$, $\qquad \ldots(137)$

are definite functions of r and t and hence functions of $q's$ and t. Thus for rheonomic system the kinetic energy obtained in the equation (136) is a quadratic function of the generalized velocities.

Case 1: If the constraints are scleronomic, then in general the transformation equations do not contain time t explicitly, and then we have $\dfrac{\partial r_i}{\partial t} = 0$, and consequently a and a_j vanish. Therefore, equation (136) reduces to

$$T = \sum_{j,k} a_{jk}\dot{q}_j\dot{q}_k. \qquad \qquad \ldots(138)$$

This shows that for scleronomic system the kinetic energy is a homogeneous quadratic function of generalized velocities. Now applying Euler's theorem for the kinetic energy function T defined in the equation (138), we get

$$\sum_j \dot{q}_j \frac{\partial T}{\partial \dot{q}_j} = 2T. \qquad \qquad \ldots(139)$$

However, it is important to note here that the result (139) is not always true even if the constraints are scleronomic. It may happen that even if the constraints are scleronomic, the transformation equations either involve time t or do not involve time t explicitly, because of the rotating (non-inertial) coordinate system or due to some other reasons. Consequently, the kinetic energy T will not be a homogeneous quadratic function of generalized velocities as it contains additional rotational terms involving or not involving generalized velocities. In this case the equation (139) reads

$$\sum_j \dot{q}_j \frac{\partial T}{\partial \dot{q}_j} = 2T + \text{terms containing } T_1 \text{ and / or } T_0, \qquad \ldots(139a)$$

where T_1 is a homogeneous function of generalized velocities of degree one and T_0 is a term not involving generalized velocities. (refer Examples 46 and 56). Nevertheless, if a dynamical system only in translational motion (inertial coordinate system *i.e.,* rotation of any kind is absent) is scleronomic, then obviously kinetic energy T is always a homogeneous quadratic function of generalized velocities and consequently the result (139) holds.

Case 2: If the constraints are rheonomic then none of the functions a_{jk}, a_j and a defined in the equation (137) is zero. In this case we write equation (136) as

$$T = T_2 + T_1 + T_0, \qquad \qquad ...(140)$$

where

$$T_2 = \sum_{j,k} a_{jk}\,\dot{q}_j\,\dot{q}_k,\ T_1 = \sum_j a_j\dot{q}_j,\ T_0 = a = \sum_i \frac{1}{2}\,m_i\left(\frac{\partial r_i}{\partial t}\right)^2, \qquad ...(141)$$

are homogeneous functions of generalized velocities of degree two, one and zero respectively. Differentiating the equation (140) with respect to $\dot{q}_j$, multiplying it by $\dot{q}_j$ and then summing over j we obtain

$$\sum_j \dot{q}_j\,\frac{\partial T}{\partial \dot{q}_j} = \sum_j \dot{q}_j\,\frac{\partial T_2}{\partial \dot{q}_j} + \sum_j \dot{q}_j\,\frac{\partial T_1}{\partial \dot{q}_j} + \sum_j \dot{q}_j\,\frac{\partial T_0}{\partial \dot{q}_j}.$$

Now on applying Euler's theorem for the homogeneous function to each term on the right hand side of the above equation we readily get

$$\sum_j \dot{q}_j\,\frac{\partial T}{\partial \dot{q}_j} = 2T_2 + T_1. \qquad \qquad ...(142)$$

This completes the proof.

Note: (1) It may also happen that though the constraints are rheonomic yet the kinetic energy expression of the system leads to the result $\sum_j \dot{q}_j\,\dfrac{\partial T}{\partial \dot{q}_j} = 2T - 2T_0$ due to the rotating (non-inertial) coordinate system. (refer the example (45) and the example (43) of the Chapter 3).

(2) The result (142) can also be obtained by direct differentiating equation (136) with respect to $\dot{q}_j$. Thus

$$\frac{\partial T}{\partial \dot{q}_j} = 2 \sum_k a_{jk}\dot{q}_k + a_j.$$

Next multiplying this equation by $\dot{q}_j$ and summing over j we get

$$\sum_j \dot{q}_j\,\frac{\partial T}{\partial \dot{q}_j} = 2 \sum_{j,k} a_{jk}\dot{q}_j\dot{q}_k + \sum_j a_j\dot{q}_j$$

$$\sum_j \dot{q}_j\,\frac{\partial T}{\partial \dot{q}_j} = 2T_2 + T_1.$$

The result (139) can similarly be derived by direct differentiating equation (138).

Kinetic Energy in Polar Form

Consider a single particle moving in a plane. If the generalized co-ordinates are r, θ then we have $q_1 = r$, $q_2 = \theta$. In this case the expression for kinetic energy (136) becomes

$$T = a_{rr}\dot{r}\dot{r} + 2a_{r\theta}\dot{r}\dot{\theta} + a_{\theta\theta}\dot{\theta}\dot{\theta} + a_r\dot{r} + a_\theta\dot{\theta} + a, \qquad \ldots(143)$$

where from equation (137) we have

$$a_{rr} = \frac{1}{2}m\,\frac{\partial\bar{r}}{\partial r}\frac{\partial\bar{r}}{\partial r}, \qquad\qquad a_{r\theta} = m\,\frac{\partial\bar{r}}{\partial r}\frac{\partial\bar{r}}{\partial\theta},$$

$$a_{\theta\theta} = \frac{1}{2}m\,\frac{\partial\bar{r}}{\partial\theta}\frac{\partial\bar{r}}{\partial\theta}, \qquad\qquad a_r = m\,\frac{\partial\bar{r}}{\partial r}\frac{\partial\bar{r}}{\partial t},$$

$$a_\theta = m\,\frac{\partial\bar{r}}{\partial\theta}\frac{\partial\bar{r}}{\partial t}, \qquad\qquad a = \frac{1}{2}m\left(\frac{\partial\bar{r}}{\partial t}\right)^2. \qquad \ldots(144)$$

Now for $\bar{r} = xi + yj$, and $x = r\cos\theta$, $y = r\sin\theta$, we find all the coefficients of equation (143) from equations (144) and obtain the expression

$$T = \frac{1}{2}m\big[(i\cos\theta + j\sin\theta)^2\big]\dot{r}\dot{r} + m\big[(i\cos\theta + j\sin\theta)(-ir\sin\theta + jr\cos\theta)\big]\dot{r} +$$

$$+ \frac{1}{2}m(-ir\sin\theta + jr\cos\theta)^2\,\dot{\theta}\dot{\theta}.$$

Which on simplifying we get

$$T = \frac{1}{2}m(\dot{r}^2 + r^2\dot{\theta}^2). \qquad \ldots(145)$$

Another way of Proving Conservation Theorem for Energy

Theorem 7　If the Lagrangian does not contain time t explicitly, the total energy of the conservative system is conserved.

Proof:　Consider a conservative system, in which the forces are derivable from a potential V which is dependent on position only. The Lagrangian of the system is defined as $L = T - V$, where

$$L = L(q_j, \dot{q}_j, t) \qquad \ldots(146)$$

satisfies the Lagrange's equation (81). Differentiating equation (146) with respect to time t we obtain

$$\frac{dL}{dt} = \sum_j \left[\frac{\partial L}{\partial q_j}\dot{q}_j + \frac{\partial L}{\partial\dot{q}_j}\ddot{q}_j\right] + \frac{\partial L}{\partial t}.$$

Since L does not contain time t explicitly implies $\dfrac{\partial L}{\partial t} = 0$.

$$\frac{dL}{dt} = \sum_j \left[\frac{\partial L}{\partial q_j} \dot{q}_j + \frac{\partial L}{\partial \dot{q}_j} \ddot{q}_j \right].$$

On using Lagrange's equation motion (81) we write this equation as

$$\frac{dL}{dt} = \sum_j \frac{d}{dt}\left(\frac{\partial L}{\partial \dot{q}_j} \right) \dot{q}_j + \sum_j \frac{\partial L}{\partial \dot{q}_j} \ddot{q}_j.$$

$$\frac{dL}{dt} = \frac{d}{dt}\left(\sum_j \frac{\partial L}{\partial \dot{q}_j} \dot{q}_j \right).$$

$$\Rightarrow \qquad \frac{d}{dt}\left[L - \sum_j \dot{q}_j \frac{\partial L}{\partial \dot{q}_j} \right] = 0.$$

Integrating we get

$$L - \sum_j \dot{q}_j \frac{\partial L}{\partial \dot{q}_j} = \text{const.} \qquad\qquad \dots(147)$$

Since the potential energy V for the conservative system depends upon the position co-ordinates only and does not involve generalized velocities. Hence we have $\dfrac{\partial L}{\partial \dot{q}_j} = \dfrac{\partial T}{\partial \dot{q}_j}$.

The generalized momentum is defined as (refer the definition on page 1.116).

$$p_j = \frac{\partial T}{\partial \dot{q}_j}.$$

Thus we have from the equation (147) that

$$\sum_j p_j \dot{q}_j - L = \text{const } (H). \qquad\qquad \dots(148)$$

The Lagrangian L does not contain time t means neither the kinetic energy nor the potential energy of the particle involves time t. In this case the constraints may be scleronomic and hence the kinetic energy T may be a homogeneous quadratic function of generalized velocities, then we have from equation (139)

$$\sum_j \dot{q}_j \frac{\partial T}{\partial \dot{q}_j} = 2T.$$

Consequently, equation (148) becomes

$$2T - L = H$$

$$\Rightarrow \qquad\qquad 2T - T + V = H$$

$$\Rightarrow \qquad\qquad T + V = H,$$

$$\Rightarrow \qquad\qquad\qquad E = H \text{ (const)}.$$

This proves the total energy E of a conservative system is conserved. The constant of total energy is chosen as H because we will show in the Chapter 3 that this energy constant usually represents the Hamiltonian of the system. One should note here that when the constraints are scleronomic and the kinetic energy is not homogeneous quadratic function of generalized velocities due to rotation of the system, then energy is not conserved as it is evident from the equation (139a).

The Theorem 7 can also be proved from the Lagrange's equations of motion. We give its proof below:

Consider a conservative system described by the Lagrangian L which is independent of time t. Consequently, neither the kinetic energy nor the potential energy involves time t explicitly. The Lagrange's equations of motion in this case reduces to

$$\frac{d}{dt}\left(\frac{\partial T}{\partial \dot{q}_j}\right) - \frac{\partial T}{\partial q_j} = -\frac{\partial V}{\partial q_j}. \qquad\qquad \ldots(149)$$

Multiplying the equation (149) by $\dot{q}_j$ and summing over j we obtain, after adding and subtracting the term $\sum\limits_j \ddot{q}_j \dfrac{\partial T}{\partial \dot{q}_j}$ to it, the equation

$$\frac{d}{dt}\left[\sum_j \dot{q}_j \frac{\partial T}{\partial \dot{q}_j}\right] - \sum_j \ddot{q}_j \frac{\partial T}{\partial \dot{q}_j} - \sum_j \dot{q}_j \frac{\partial T}{\partial q_j} = -\sum_j \dot{q}_j \frac{\partial V}{\partial q_j} \qquad \ldots(150)$$

Since the system is conservative and the Lagrangian L is independent of time t, then constraints may be scleronomic. We consider the case when the kinetic energy is homogeneous quadratic function of generalized velocities, then from the equation (139), we have

$$\sum_j \dot{q}_j \frac{\partial T}{\partial \dot{q}_j} = 2T.$$

On using this equation in the equation (150) we obtain

$$\frac{d}{dt}(2T) - \frac{dT}{dt} = -\frac{dV}{dt},$$

where
$$\frac{dT}{dt} = \sum_j \frac{\partial T}{\partial q_j}\dot{q}_j + \sum_j \frac{\partial T}{\partial \dot{q}_j}\ddot{q}_j,$$

and
$$\frac{dV}{dt} = \sum_j \dot{q}_j \frac{\partial V}{\partial q_j}.$$

$\Rightarrow \qquad \frac{d}{dt}(T + V) = 0$

$\Rightarrow \qquad T + V = \text{const.}$

This proves that the total energy is conserved.

Note: It is important to note here that the conditions that a dynamical system is conservative and scleronomic are not enough for the total energy of the system to be conserved. The same is illustrated in the Theorem 9 below.

Theorem on Total Energy

Theorem 8 Show that non-conservation of total energy is directly associated with the existence of non-conservative forces even if the transformation equation does not contain time t.

Proof: Consider a partially conservative and partially non-conservative system described by the Lagrangian L which is a function of generalized coordinates q_j, generalized velocities $\dot{q}_j$ and time t. i.e., $L = L(q_j, \dot{q}_j, t)$. Differentiating this with respect to time t, we get

$$\frac{dL}{dt} = \sum_j \left[\frac{\partial L}{\partial q_j} \dot{q}_j + \frac{\partial L}{\partial \dot{q}_j} \ddot{q}_j \right] + \frac{\partial L}{\partial t}. \qquad \ldots(151)$$

The Lagrange's equations of motion for such system are given in the equations (83). These are given by

$$\frac{\partial L}{\partial q_j} = \frac{d}{dt}\left(\frac{\partial L}{\partial \dot{q}_j} \right) - Q_j^{(d)},$$

where the Lagrangian L contains the potential of the conservative forces and the forces which are not arising from potential V are represented by $Q_j^{(d)}$. Therefore, using this equation in the equation (151) we obtain

$$\frac{dL}{dt} = \sum_j \frac{d}{dt}\left(\frac{\partial L}{\partial \dot{q}_j} \right) \dot{q}_j + \sum_j \frac{\partial L}{\partial \dot{q}_j} \ddot{q}_j - \sum_j Q_j^{(d)} \dot{q}_j + \frac{\partial L}{\partial t}.$$

$$\frac{dL}{dt} = \frac{d}{dt}\left(\sum_j \frac{\partial L}{\partial \dot{q}_j} \dot{q}_j \right) - \sum_j Q_j^{(d)} \dot{q}_j + \frac{\partial L}{\partial t}. \qquad \ldots(152)$$

Since L contains the potential of the conservative forces, then we have

$$\frac{\partial L}{\partial \dot{q}_j} = \frac{\partial T}{\partial \dot{q}_j},$$

$$\frac{dL}{dt} = \frac{d}{dt}\left(\sum_j \frac{\partial T}{\partial \dot{q}_j}\dot{q}_j\right) - \sum_j Q_j^{(d)}\dot{q}_j + \frac{\partial L}{\partial t}. \qquad \ldots(153)$$

In general the kinetic energy T of a system is a quadratic function of generalized velocities.

Therefore, using the equation (142) in the equation (153) we obtain

$$\frac{dL}{dt} = \frac{d}{dt}(2T_2 + T_1) - \sum_j Q_j^{(d)}\dot{q}_j + \frac{\partial L}{\partial t}.$$

Since $\qquad\qquad T = T_2 + T_1 + T_0,$

$\Rightarrow \qquad\qquad 2T_2 + T_1 = 2T - (2T_0 + T_1).$

Hence $\qquad \dfrac{dL}{dt} = 2\dfrac{dT}{dt} - \dfrac{d}{dt}(2T_0 + T_1) - \sum_j Q_j^{(d)}\dot{q}_j + \dfrac{\partial L}{\partial t}.$

$$\frac{d}{dt}(T - V) = 2\frac{dT}{dt} - \frac{d}{dt}(2T_0 + T_1) - \sum_j Q_j^{(d)}\dot{q}_j + \frac{\partial L}{\partial t},$$

$\Rightarrow \qquad \dfrac{dE}{dt} = \dfrac{d}{dt}(2T_0 + T_1) + \sum_j Q_j^{(d)}\dot{q}_j - \dfrac{\partial L}{\partial t}. \qquad \ldots(154)$

If the transformation equations do not contain time t explicitly, then we have $T_0 = 0$ and $T_1 = 0$ hence the kinetic energy does not contain time t. This means that $\dfrac{\partial T}{\partial t} = 0$. Since the system is partially conservative, and Lagrangian contains the potential of conservative forces, we have therefore, $V = V(q_j)$ and hence $\dfrac{\partial V}{\partial t} = 0$. Consequently, we have $\dfrac{\partial L}{\partial t} = 0$. Hence equation (154) becomes

$$\frac{dE}{dt} = \sum_j Q_j^{(d)}\dot{q}_j \qquad \ldots(155)$$

This shows that the non-conservation of total energy is directly associated with the existence of non-conservative forces $Q_j^{(d)}$ in the system.

Note:

(i) It is apparent from the equation (155) that if the system is conservative and the transformation equations do not contain time t explicitly, then the total energy is always conserved.

(ii) If the constraints are independent of time, the coordinate transformation equations may involve time t explicitly because of the rotation of coordinate axes or due to some other reasons. In such cases the kinetic energy is no more homogeneous quadratic function of generalized velocities. Hence it is evident from the equation (154) that the total energy is not conserved even if the system is conservative.

(iii) In the Example 7 and in the Theorem 4 it was tacitly assumed that the kinetic energy is a homogeneous quadratic function of velocity if not explicitly mentioned.

Theorem 9   Prove the conditions that a dynamical system is conservative and scleronomic are not enough for the total energy of the system is to be conserved.

Proof: Let $T = T(q_j, \dot{q}_j, t)$ be the kinetic energy of a dynamical system. Then its time derivative gives

$$\frac{dT}{dt} = \sum_j \frac{\partial T}{\partial q_j}\dot{q}_j + \sum_j \frac{\partial T}{\partial \dot{q}_j}\ddot{q}_j + \frac{\partial T}{\partial t}. \qquad \ldots(156)$$

On using the Lagrange's equations of motion (79) we find

$$\frac{dT}{dt} = \frac{d}{dt}\left(\sum_j \frac{\partial T}{\partial \dot{q}_j}\dot{q}_j\right) + \frac{\partial T}{\partial t} - \sum_j Q_j\dot{q}_j. \qquad \ldots(157)$$

Since the kinetic energy is a quadratic function of generalized velocities, hence on using the equation (142) in the equation (157) we find

$$\frac{dT}{dt} = \frac{d}{dt}(2T_2 + T_1) + \frac{\partial T}{\partial t} - \sum_j Q_j\dot{q}_j,$$

where from the equation (140) we obtain

$$2\frac{dT}{dt} = \frac{d}{dt}(2T_2 + T_1) + \frac{d}{dt}(T_1 + 2T_0).$$

Using this in the above equation we get

$$\frac{dT}{dt} = \frac{d}{dt}(T_1 + 2T_0) - \frac{\partial T}{\partial t} + \sum_j Q_j\dot{q}_j, \qquad \ldots(158)$$

where T_2 contains the terms quadratic in generalized velocities, T_1 contains the terms linear in generalized velocities and T_0 is independent of generalized velocities. So far in deriving the equation (158) neither we use the knowledge of conservative force nor the knowledge of scleronomic constraint. Now we assume the system is conservative, and then the forces acting on the system are derivable from the potential energy V which is just a function of generalized coordinates. In this case the components of generalized force are given by $Q_j = -\dfrac{\partial V}{\partial q_j}$. Consequently, the equation (158) becomes

$$\frac{dT}{dt} = \frac{d}{dt}(T_1 + 2T_0) - \frac{\partial T}{\partial t} - \sum_j \frac{\partial V}{\partial q_j}\dot{q}_j,$$

$$\frac{dT}{dt} = \frac{d}{dt}(T_1 + 2T_0) - \frac{\partial T}{\partial t} - \frac{dV}{dt},$$

$$\frac{dE}{dt} = \frac{d}{dt}(T_1 + 2T_0) - \frac{\partial T}{\partial t}. \qquad \qquad ...(159)$$

We also assume that the system is scleronomic. In this case the constraints are independent of time t but it does not mean that the transformation equations are independent of time t. The coordinate transformations may involve time t explicitly even if the constraints are independent of time because of the rotation of the system. Though the transformation equations involve time t explicitly, in vast majority of the cases, the kinetic energy T is independent of time t but not necessarily a homogeneous quadratic function of generalized velocities as it involves some rotational terms. In such cases $\frac{\partial T}{\partial t} = 0$ and $\frac{\partial r_i}{\partial t} \neq 0 \Rightarrow T_1 \neq 0, T_0 \neq 0$. Consequently, the equation (159) yields

$$\frac{dE}{dt} = \frac{d}{dt}(T_1 + 2T_0) \neq 0.$$

This implies that the total energy of the system is not conserved due to the presence of the terms T_1, T_0 in the expression of kinetic energy. These terms usually enter in to the picture due to the rotation of the system. This proves that the conditions that the system is conservative and scleronomic are not enough for total energy is conserved.

Note:

(i) Nevertheless, if a dynamical system only in translational motion (rotation of any kind is absent) is conservative and scleronomic, then the kinetic energy of the system does not involve time t and it is always a homogeneous quadratic function of generalized velocities. Hence $\frac{\partial T}{\partial t} = 0$ and $\frac{\partial r_i}{\partial t} = 0 \Rightarrow T_1 = 0, T_0 = 0$, proving $\frac{dE}{dt} = 0$, and hence the total energy is conserved.

(ii) In the case of rheonomic system, the constraint equations involve time t explicitly even then the kinetic energy T may not involve time t. Hence in such cases $\frac{\partial T}{\partial t} = 0$ and $T_1 \neq 0, T_0 \neq 0$. Hence from equation (159) we see that the total energy is not conserved.

(iii) If the system is partially conservative and partially non-conservative, then in this case the components of the generalized force are given by $Q_j = -\frac{\partial V}{\partial q_j} + Q_j^{(d)}$, where V is the potential energy arises from conservative forces and $Q_j^{(d)}$ represents non conservative forces which are not arising from potential V. In this case the equation (158) becomes

$$\frac{dE}{dt} = \frac{d}{dt}(T_1 + 2T_0) - \frac{\partial T}{\partial t} + \sum_j Q_j^{(d)} \dot{q}_j. \qquad \ldots(160)$$

We see from equation (160) that even if a dynamical system in only translational motion is scleronomic, then as discussed earlier, we have $\frac{\partial T}{\partial t} = 0$ and $\frac{\partial r_i}{\partial t} = 0 \Rightarrow T_1 = 0, T_0 = 0$, the total energy is not conserved due to the presence of dissipative forces in the system.

Example 13: Show that the new Lagrangian L' defined by

$$L' = L + \frac{d}{dt} f(q_j, t), j = 1, 2, \ldots, n$$

satisfies Lagrange's equation of motion, where f is an arbitrary differentiable functions of q_j and t, and L is a Lagrangian for a system of n degrees of freedom.

Solution: Given that $L' = L + \frac{d}{dt} f(q_j, t)$, where L satisfies the Lagrange's equations (81).

We prove that

$$\frac{d}{dt}\left(\frac{\partial L'}{\partial \dot{q}_j}\right) - \frac{\partial L'}{\partial q_j} = 0.$$

Since

$$f = f(q_j, t).$$

Differentiation of this equation with respect to time t gives

$$\frac{df}{dt} = \sum_k \frac{\partial f}{\partial q_k} \dot{q}_k + \frac{\partial f}{\partial t}. \qquad \ldots(161)$$

Differentiating the equation (161) partially with respect to q_j we get

$$\frac{\partial}{\partial q_j}\left(\frac{df}{dt}\right) = \sum_k \frac{\partial^2 f}{\partial q_k \partial q_j} \dot{q}_k + \frac{\partial^2 f}{\partial t \partial q_j}. \qquad \ldots(162)$$

Now differentiating the equation (161) with respect to $\dot{q}_j$ we have

$$\frac{\partial}{\partial \dot{q}_j}\left(\frac{df}{dt}\right) = \frac{\partial f}{\partial q_j}.$$

Differentiating this with respect to time t, we get

$$\frac{d}{dt}\left[\frac{\partial}{\partial \dot{q}_j}\left(\frac{df}{dt}\right)\right] = \sum_k \frac{\partial^2 f}{\partial q_k \partial q_j} \dot{q}_k + \frac{\partial^2 f}{\partial t \partial q_j}. \qquad \ldots(163)$$

Subtracting equations (162) from (163), we get

$$\frac{d}{dt}\left[\frac{\partial}{\partial \dot{q}_j}\left(\frac{df}{dt}\right)\right] - \frac{\partial}{\partial q_j}\left(\frac{df}{dt}\right) = 0.$$

i.e.,
$$\frac{d}{dt}\left[\frac{\partial L'}{\partial \dot{q}_j}\right] - \frac{\partial L'}{\partial q_j} = 0.$$

This proves that the Lagrangian of the system is not unique.

2. Lagrange's Equations for Non-holonomic Constraints

We have seen that a system is also said to be non-holonomic, if it corresponds to non-integrable differential equations of constraints. Such constraints cannot be expressed in the form of equation of the type

$$f_l(q_j,\, t) = 0,\ l = 1,\, 2,\, 3,\dots,\, m, \qquad\qquad \dots(164)$$

where l represents the number of constraints. Hence such constraints are called non-holonomic constraints. Obviously, holonomic system has integrable differential equations of constraints expressible in the form of equation. Consider non-integrable differential constraints of the type

$$\sum_{k=1}^{n} a_{lk}dq_k + a_{lt}dt = 0. \qquad\qquad \dots(165)$$

where a_{lk} and a_{lt} are functions of q_j and t. Constraints of this type will be holonomic only if, an integrating factor can be found that turns it in to an exact differential, and hence the constraints can be reduced to the form of equations.

However, neither equations (165) can be integrated nor one can find an integrating factor that will turn either of the equations in to perfect differentials. Hence the constraints cannot be reduced to the form (164). Hence the constraints of the type (165) are therefore non-holonomic.

Note also that non-integrable differential constraints of the type (165) are not the only type of non-holonomic constraints. The non-holonomic constraint conditions may involve higher order derivatives or may appear in the form of inequalities.

There is no general way of attacking non-holonomic problems. However, the constraints are not integrable, the differential equations of the constraint can be introduced in to the problem along with the differential equations of motion and the dependent equations are eliminated by the method of Lagrange's multipliers. The method is illustrated in the following theorem.

Theorem 10 Explain the method of Lagrange's undetermined multipliers to construct equations of motion of the system with non-holonomic constraints.

Proof: Consider a conservative non-holonomic system, where the equations of the non-holonomic constraints are given by equation (165). Since the constraints are non-holonomic, hence the equations expressing the constraints (165) cannot be used to eliminate the dependent co-ordinates and hence all the generalized co-ordinates are not independent, but are related by constraint relations.

In the variational (Hamilton's) principle, (refer the Chapter 3) the time for each path is held fixed ($\delta t = 0$). Hence the virtual displacement δq_k must satisfy the following equations of constraints.

$$\sum_{k=1}^{n} a_{lk} \delta q_k = 0, \; l = 1, 2, 3, \ldots, m. \qquad \ldots(166)$$

We can use these m equations (166) to eliminate the dependent virtual displacement and reduce the number of virtual displacement to $n - m$ independent one by the method of Lagrange's multipliers. Hence we multiply equations (166) by $\lambda_1, \lambda_2, \lambda_3, \ldots, \lambda_m$ respectively and summing over l and integrating it between the limits t_0 to t_1 we get

$$\int_{t_0}^{t_1} \sum_{l=1}^{m} \sum_{k=1}^{n} \lambda_l a_{lk} \, \delta q_k \, dt = 0, \qquad \ldots(167)$$

Hamilton's principle is assumed to hold for non-holonomic system, (refer Theorem 2 of Chapter 3) we therefore have

$$\int_{t_0}^{t_1} \delta \, L dt = 0.$$

$$\Rightarrow \qquad \int_{t_0}^{t_1} \sum_{k=1}^{n} \left[\frac{\partial L}{\partial q_k} - \frac{d}{dt} \left(\frac{\partial L}{\partial \dot{q}_k} \right) \right] \delta q_k dt = 0. \qquad \ldots(168)$$

Adding equations (167) and (168), we get

$$\int_{t_0}^{t_1} \left\{ \sum_{k=1}^{n} \left[\frac{\partial L}{\partial q_k} - \frac{d}{dt} \left(\frac{\partial L}{\partial \dot{q}_k} \right) + \sum_{l=1}^{m} \lambda_l a_{lk} \right] \delta q_k \right\} dt = 0. \qquad \ldots(169)$$

Note all the virtual displacement δq_k, $k = 1, 2, \ldots, n$ are not independent but connected by m equations (166). Now to eliminate the extra dependent virtual displacements we choose the multipliers $\lambda_1, \lambda_2, \lambda_3, \ldots, \lambda_m$, such that the coefficients of m-dependent virtual displacements in equation (169) are zero. *i.e.,*

$$\frac{\partial L}{\partial q_k} - \frac{d}{dt} \left(\frac{\partial L}{\partial \dot{q}_k} \right) + \sum_{l=1}^{m} \lambda_l a_{lk} = 0, \quad \text{for} \quad k = n - (m-1), \ldots, (n-1), n. \quad \ldots(170)$$

Hence from equation (170), we have

$$\int_{t_0}^{t_1} \left\{ \sum_{k=1}^{n-m} \left[\frac{\partial L}{\partial q_k} - \frac{d}{dt} \left(\frac{\partial L}{\partial \dot{q}_k} \right) + \sum_{l=1}^{m} \lambda_l a_{lk} \right] \delta q_k \right\} dt = 0. \qquad \ldots(171)$$

where δq_1, δq_2, δq_3,..., δq_{n-m} are all independent. Hence it follows that

$$\frac{\partial L}{\partial q_k} - \frac{d}{dt} \left(\frac{\partial L}{\partial \dot{q}_k} \right) + \sum_{l=1}^{m} \lambda_l a_{lk} = 0, \quad \text{for} \quad k = 1, 2,..., n - m \qquad \ldots(172)$$

Combining equations (170) and (172), we have finally the complete set of Lagrange's equations of motion for non-holonomic system

$$\frac{d}{dt} \left(\frac{\partial L}{\partial \dot{q}_k} \right) - \frac{\partial L}{\partial q_k} = \sum_{l=1}^{m} \lambda_l a_{lk}, \; k = 1, 2,..., n - m,..., n. \qquad \ldots(173)$$

Remarks:

1. The n-equations in (173) together with m-equations of constraints (165) are sufficient to determine $n + m$ unknowns viz., the n – generalized co-ordinates q_j and m Lagrange's multipliers λ_l.

2. Lagrange's multiplier method can also be used for holonomic constraints, when it is inconvenient to reduce all the $q's$ to independent co-ordinates, and then obtain the forces of constraints.

WORKED EXAMPLES

Example 14: Use Lagrange's undetermined multipliers to construct the equation of motion of simple pendulum and obtain the force of constraint.

Solution: Consider a simple pendulum of mass m and of constant length l. Let $P\,(x, y)$ be the position co-ordinates of the pendulum. Then the equation of the constraint is

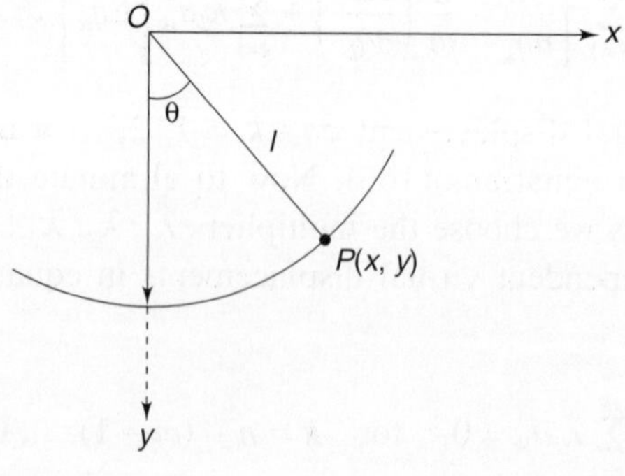

$$x^2 + y^2 = l^2. \qquad \ldots(174)$$

This shows that x, y are not the generalized co-ordinates. If (r, θ) are the polar co-ordinates of the pendulum, then the equation of constraint is (say)

$$f_1 \equiv r - l = 0. \qquad \ldots(175)$$

If this constraint is not used to eliminate the dependent variable r, then r, θ are the generalized coordinates. Hence the kinetic energy and potential energy of the pendulum are respectively given by

$$T = \frac{1}{2} mr^2 \dot\theta^2,$$

$$V = -mgr \cos \theta.$$

The Lagrangian of the pendulum $L = T - V$ becomes

$$L = \frac{1}{2} mr^2 \dot\theta^2 + mgr \cos \theta. \qquad \ldots(176)$$

Differentiating the equation of the constraint (175), we get

$$dr = 0.$$

Comparing this equation with the standard equation of constraint $a_{1r} dr + a_{1\theta} d\theta = 0$, we get

$$a_{1r} = 1, \, a_{1\theta} = 0.$$

The Lagrange's equations of motion (173) become

$$\frac{d}{dt}\left(\frac{\partial L}{\partial \dot\theta}\right) - \frac{\partial L}{\partial \theta} = \lambda_1 a_{1\theta}$$

and
$$\frac{d}{dt}\left(\frac{\partial L}{\partial \dot r}\right) - \frac{\partial L}{\partial r} = \lambda_1 a_{1r}.$$

These equations after solving become

$$\ddot\theta + \frac{g}{r} \sin \theta = 0, \qquad \ldots(177)$$

$$mr\dot\theta^2 + mg \cos \theta = -\lambda_1, \qquad \ldots(178)$$

where λ_1 is the force of constraint, in this case it is the tension in the string. Equation (177) determines the motion of the pendulum under the constraint force given in (178).

Example 15: Use Lagrange's undetermined multipliers to construct the equations of motion of spherical pendulum. (See the defination on page 1.89)

Solution: Let a particle of mass m move on a frictionless surface of radius r under the action of gravity. Let $P(x, y, z)$ be the position co-ordinates of the pendulum. If (r, θ, ϕ) are the spherical polar co-ordinates of the pendulum, then we have the relations

$$x = r \sin \theta \cos \phi,$$
$$y = r \sin \theta \sin \phi, \qquad \qquad \text{...(179)}$$
$$z = r \cos \theta,$$

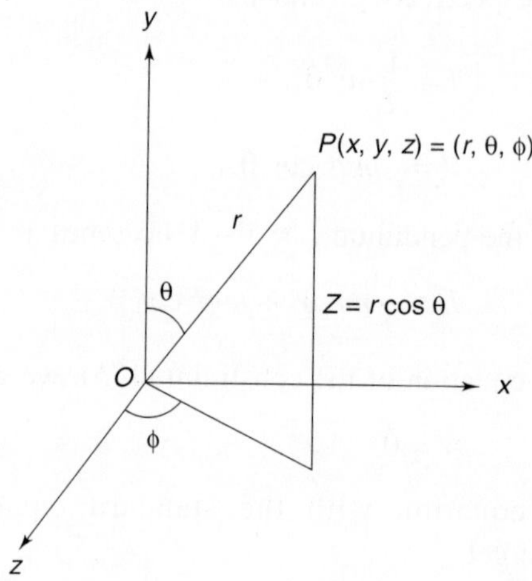

where $\qquad x^2 + y^2 + z^2 = r^2.$

This shows that x, y, z are not the generalized co-ordinates. The kinetic and potential energies of the spherical pendulum are given by respectively

$$T = \frac{1}{2} m(\dot{r}^2 + r^2\dot{\theta}^2 + r^2\sin^2 \theta \dot{\phi}^2),$$

$$V = mgr \cos \theta.$$

Hence the Lagrangian of the system becomes

$$L = \frac{1}{2} m\left(\dot{r}^2 + r^2\dot{\theta}^2 + r^2 \sin^2 \theta \dot{\phi}^2\right) - mgr \cos \theta.$$

The equation of the constraint on the motion of the particle moving on the sphere is

$$f_1 \equiv r - l = 0. \qquad \qquad \text{...(180)}$$

If this constraint is not used to eliminate the dependent variable r, then the generalized co-ordinates are (r, θ, ϕ). Differentiating equation (180) we get $dr = 0$. Comparing this with the standard equation $a_{1r}dr + a_{1\theta}d\theta = 0$, we get $a_{1r} = 1$, $a_{1\theta} = 0$ $a_{1\phi} = 0$.

Hence the, Lagrangian function become

$$L = \frac{1}{2} mr^2(\dot{\theta}^2 + \sin^2 \theta \dot{\phi}^2) - mgr \cos \theta. \qquad \dots(181)$$

In this case the Lagrange's equations of motion (173) become

$$\frac{d}{dt}\left(\frac{\partial L}{\partial \dot{r}}\right) - \frac{\partial L}{\partial r} = \lambda_1 a_{1r}, \qquad \dots(182)$$

$$\frac{d}{dt}\left(\frac{\partial L}{\partial \dot{\theta}}\right) - \frac{\partial L}{\partial \theta} = \lambda_1 a_{1\theta}, \qquad \dots(183)$$

and
$$\frac{d}{dt}\left(\frac{\partial L}{\partial \dot{\phi}}\right) - \frac{\partial L}{\partial \phi} = \lambda_1 a_{1\phi}. \qquad \dots(184)$$

Consequently, these equations reduce to

$$mr\left(\dot{\theta}^2 + \sin^2 \theta \dot{\phi}^2 - \frac{g}{r} \cos \theta\right) = -\lambda_1. \qquad \dots(185)$$

This equation determines the constraint force. Similarly, from equations (183) and (184) we obtain

$$\ddot{\theta} - \sin \theta \cos \theta \dot{\phi}^2 - \frac{g}{r} \sin \theta = 0, \qquad \dots(186)$$

and
$$mr^2 \sin^2 \theta \dot{\phi}^2 = p_\phi \text{ (const)}. \qquad \dots(187)$$

Eliminating $\dot{\phi}$ between equations (186) and (187), we get

$$\ddot{\theta} - \frac{p_\phi^2}{mr^4\sin^3 \theta} \cos \theta - \frac{g}{r} \sin \theta = 0. \qquad \dots(188)$$

Equations (185) and (188) determine the motion of the spherical pendulum.

Example 16: Use Lagrange's undetermined multipliers to construct the equations of motion of a cylinder of mass m rolling down an inclined plane without slipping. Determine the force of constraint.

Solution: Consider a cylinder of radius 'a' starts rolling from point O on an inclined plane of length l and making an angle θ with the horizontal. The equation of the constraint is

$$ad\phi = dx$$

$$\Rightarrow \qquad ad\phi - dx = 0. \qquad \dots(189)$$

If this constraint is not used to eliminate ϕ in terms of x then the system has two degrees of freedom and hence two generalized co-ordinates ϕ and x. Thus the kinetic energy of the rolling cylinder has linear kinetic energy and rotational kinetic energy and is given by

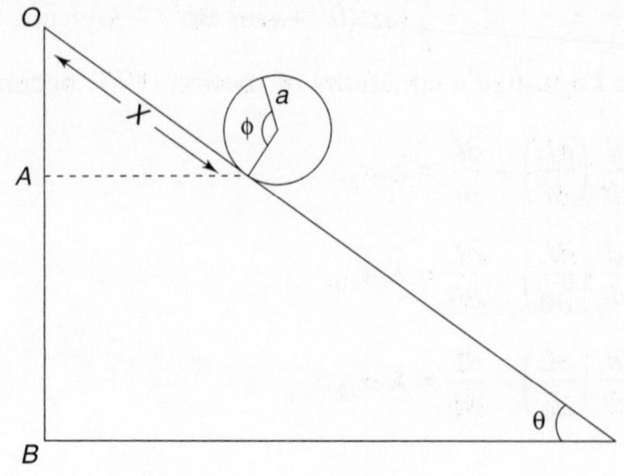

$$T = \frac{1}{2} m \dot{x}^2 + \frac{1}{2} I \omega^2,$$

where I is the moment of inertia of the cylinder and is given by

$$I = \frac{1}{2} m a^2.$$

Hence

$$T = \frac{1}{2} m \dot{x}^2 + \frac{1}{4} m a^2 \omega^2.$$

Since $\omega = \dot{\phi}$, hence the expression for the kinetic energy becomes

$$T = \frac{1}{2} m \dot{x}^2 + \frac{1}{4} m a^2 \dot{\phi}^2. \qquad \ldots(190)$$

The potential energy of the cylinder is

$$V = mgAB = mg\,(OB - OA),$$

$$\Rightarrow \qquad V = mg(l - x) \sin\theta. \qquad \ldots(191)$$

Hence the Lagrangian of the cylinder becomes

$$L = \frac{1}{2} m \dot{x}^2 + \frac{1}{4} m a^2 \dot{\phi}^2 - mg \sin\theta \,(l - x). \qquad \ldots(192)$$

The equations of the non-holonomic constraints (165) in this case become

$$a_{1x} dx + a_{1\phi} d\phi = 0. \qquad \ldots(193)$$

Comparing equations (193) and (189), we get

$$a_{1x} = -1, \; a_{1\phi} = a. \hspace{3cm} \text{...(194)}$$

Hence the x-Lagrange's and ϕ-Lagrange's equations of motion give

$$m\ddot{x} - mg \sin \theta + \lambda_1 = 0, \hspace{2cm} \text{...(195)}$$

$$\frac{1}{2} ma\ddot{\phi} = \lambda_1. \hspace{3cm} \text{...(196)}$$

From the constraint equation (189), we have

$$a\dot{\phi} = \dot{x} \Rightarrow a\ddot{\phi} = \ddot{x}. \hspace{2.5cm} \text{...(197)}$$

Hence the equations (196) and (197) give

$$\frac{1}{2} m\ddot{x} = \lambda_1. \hspace{3cm} \text{...(198)}$$

This gives the force of constraint. Substituting this value of λ_1 in the equation (195) we obtain

$$\ddot{x} = \frac{2}{3}g \sin \theta, \hspace{3cm} \text{...(199)}$$

and from equation (197), we have

$$\ddot{\phi} = \frac{2}{3a}g \sin \theta. \hspace{3cm} \text{...(200)}$$

Hence the force of constraint is

$$\lambda_1 = \frac{1}{3}mg \sin \theta. \hspace{3cm} \text{...(201)}$$

This in fact is the frictional force.

Example 17:   A particle is constrained to move on the plane curve $xy = c$, where c is a constant, under gravity. Obtain the Lagrangian and hence the equation of motion.

Solution: Given that a particle is constrained to move on the plane curve

$$xy = c. \hspace{3cm} \text{...(202)}$$

The kinetic energy of the particle is given by

$$T = \frac{1}{2} m \left(\dot{x}^2 + \dot{y}^2 \right). \hspace{2.5cm} \text{...(203)}$$

The potential energy is given by

$$V = mgy, \quad y \text{ is vertical.} \qquad \qquad \text{...(204)}$$

We see that x and y are not linearly independent as they are related by the equation of constraint (202) and hence they are not the generalized co-ordinates. However, we eliminate the variable y by putting $y = \dfrac{c}{x}$ and hence $\dot{y} = -\dfrac{c}{x^2}\dot{x}$ in equations (203) and (204), we get

$$T = \frac{1}{2} m \dot{x}^2 \left(1 + \frac{c^2}{x^4}\right), \quad V = mg \frac{c}{x}.$$

Here x is the generalized co-ordinate. Hence the Lagrangian of the particle becomes

$$L = \frac{1}{2} m \dot{x}^2 \left(1 + \frac{c^2}{x^4}\right) - \frac{mgc}{x}. \qquad \qquad \text{...(205)}$$

The x-Lagrange's equation of motion (81) becomes

$$m\ddot{x}\left(1 + \frac{c^2}{x^4}\right) - 2\frac{c^2 m}{x^5}\dot{x}^2 - \frac{mgc}{x^2} = 0. \qquad \qquad \text{...(206)}$$

Example 18: A particle is constrained to move on the surface of a cylinder of fixed radius. Obtain the Lagrange's equation of motion.

Solution: The surface of the cylinder is characterized by the parametric equations given by

$$x = r \cos\theta, \quad y = r \sin\theta, \quad z = z. \qquad \qquad \text{...(207)}$$

However, x, y, z are not the generalized co-ordinates as x and y are related by the equation of constraint $x^2 + y^2 = r^2$, r is a constant radius of the circle. Hence the generalized co-ordinates are θ and z. In terms of these generalized co-ordinates the kinetic and potential energies become

$$T = \frac{1}{2} m \left(r^2\dot{\theta}^2 + \dot{z}^2\right),$$

$$V = mgz.$$

Hence the Lagrangian is given by

$$L = \frac{1}{2} m \left(r^2\dot{\theta}^2 + \dot{z}^2\right) - mgz. \qquad \qquad \text{...(208)}$$

Therefore, the θ – Lagrange's equation yields

$$\frac{d}{dt}\left(mr^2\dot{\theta}\right) = 0 \Rightarrow mr^2\dot{\theta} = \text{const}(l).$$

Integrating we get

$$\theta = \frac{l}{mr^2}\, t + \theta_0, \qquad\qquad \ldots(209)$$

where θ_0 is a constant of integration. Similarly, z-Lagrange's equation of motion gives

$$z = ut - \frac{1}{2}\, gt^2, \qquad\qquad \ldots(210)$$

where $\qquad \dot{z} = u$ at $t = 0$.

Example 19: A particle of mass m is projected with initial velocity u at an angle α with the horizontal. Use Lagrange's equation to describe the motion of the projectile.

Solution: Let a particle of mass m be projected from O with an initial velocity u unit making an angle α with the horizontal line referred as x-axis. Let $P(x, y)$ be the position of the particle at any instant t. Since x and y are independent and hence the generalized co-ordinates. The kinetic energy and the potential energy of the projectile are respectively given by

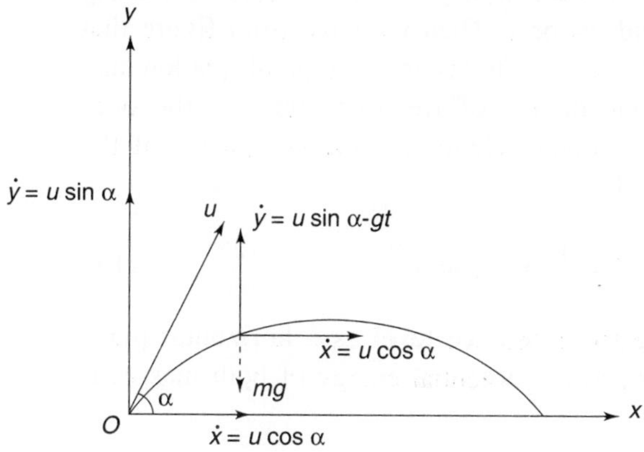

$$T = \frac{1}{2}\, m \left(\dot{x}^2 + \dot{y}^2 \right) \quad \text{and} \quad V = mgy.$$

Thus the Lagrangian function of the projectile is

$$L = \frac{1}{2}\, m \left(\dot{x}^2 + \dot{y}^2 \right) - mgy. \qquad\qquad \ldots(211)$$

The x-Lagrange's equation of motion and y-Lagrange's equation of motion respectively gives

$$\ddot{x} = 0 \quad \text{and} \quad \ddot{y} + g = 0. \qquad\qquad \ldots(212)$$

To find the velocity of the projectile and its path at any instant we integrate equations (212) and using initial conditions we readily obtain

$$\dot{x} = u \cos \alpha, \quad \dot{y} = u \sin \alpha - gt. \qquad \ldots(213)$$

These equations determine velocity at any time t. Integrating (213) once again and using initial conditions we get

$$x = u \cos \alpha.t \text{ and } y = u \sin \alpha.t - \frac{1}{2} gt^2. \qquad \ldots(214)$$

Eliminating t between equations (214), we get

$$y = x \tan \alpha - \frac{1}{2} g \frac{x^2}{u^2 \cos^2 \alpha}. \qquad \ldots(215)$$

This represents the path of the projectile and it is a parabola.

Atwood's Machine

Example 20: Explain Atwood Machine and discuss it's motion.

Solution: Atwood machine consists of two masses m_1 and m_2 attached at the ends of a massless, inextensible string going over a frictionless pulley. Let the length of the string between m_1 and m_2 be l. Then we have from figure that $PA = x$ and $QA = l - x$. This is an example of a holonomic system with one degree of freedom and x is the only generalized coordiante. Hence the kinetic energy of the system is given by

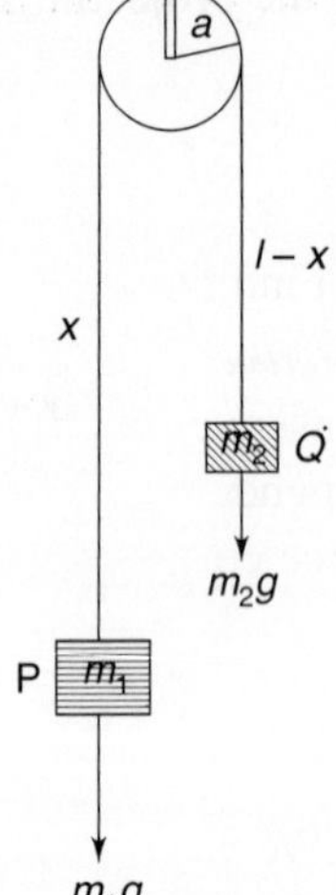

$$T = \frac{1}{2} (m_1 + m_2)\dot{x}^2. \qquad \ldots(216)$$

Considering the reference level as a horizontal plane passing through A, the potential energy of both masses is given by

$$V = -m_1 gx - m_2 g(l - x). \qquad \ldots(217)$$

Hence the Lagrangian of the system becomes

$$L = \frac{1}{2} (m_1 + m_2)\dot{x}^2 + (m_1 - m_2)gx + m_2 gl. \qquad \ldots(218)$$

The corresponding Lagrange's equation of motion gives

$$\ddot{x} = \frac{(m_1 - m_2)}{m_1 + m_2} g. \qquad \ldots(219)$$

The solution of this equation gives

$$x = \frac{1}{2} \frac{(m_1 - m_2)}{m_1 + m_2} gt^2 + x_0 t + y_0, \qquad \ldots(220)$$

where x_0, y_0 are constants of integration.

Example 21: A particle of mass m moves in one dimension such that it has the Lagrangian

$$L = \frac{m^2 \dot{x}^4}{12} + m\dot{x}^2 V(x) - V^2(x),$$

where V is some differentiable function of x. Find equation of motion for $x(t)$.

Solution: We see from the given Lagrangian function that x is the only generalized co-ordinate. Therefore the corresponding Lagrange's equation of motion gives

$$m\ddot{x} + \frac{\partial V}{\partial x} = 0. \qquad \qquad ...(221)$$

This equation of motion shows that the particle moves in a straight line under the action of a force $\overline{F} = -\dfrac{\partial V}{\partial x}$.

Example 22: Let a particle be moving in a field of force given by

$$F = \frac{1}{r^2}\left(1 - \frac{\dot{r}^2 - 2r\ddot{r}}{c^2}\right).$$

Find the Lagrangian of motion and hence the equation of motion.

Solution: One can check that $\nabla \times \overline{F} \neq 0$, hence the force is non-conservative; consequently, the corresponding potential is generalized potential or velocity dependent potential. From the expression (64) the component of generalized force corresponding to the generalized coordinate r becomes

$$Q_r = F = \frac{1}{r^2}\left(1 - \frac{\dot{r}^2 - 2r\ddot{r}}{c^2}\right).$$

We write this force as

$$Q_r = \frac{1}{r^2} - \frac{\dot{r}^2}{c^2 r^2} + \frac{2\ddot{r}}{c^2 r},$$

$$Q_r = \frac{1}{r^2} + \frac{\dot{r}^2}{c^2 r^2} + \frac{2\ddot{r}}{c^2 r} - \frac{2\dot{r}^2}{c^2 r^2},$$

$$Q_r = -\frac{\partial}{\partial r}\left(\frac{1}{r} + \frac{\dot{r}^2}{c^2 r}\right) + \frac{d}{dt}\frac{\partial}{\partial \dot{r}}\left(\frac{1}{r} + \frac{\dot{r}^2}{c^2 r}\right),$$

$$Q_r = -\frac{\partial U}{\partial r} + \frac{d}{dt}\frac{\partial U}{\partial \dot{r}},$$

$$U = \frac{1}{r}\left(1 + \frac{\dot{r}^2}{c^2}\right). \qquad \qquad ...(222)$$

We notice that the potential energy U is the velocity dependent potential. The kinetic energy of the particle is given by

$$T = \frac{1}{2} m\dot{r}^2. \qquad \qquad ...(223)$$

Hence the Lagrangian of the particle becomes

$$L = \frac{1}{2} m\dot{r}^2 - \frac{1}{r}\left(1 + \frac{\dot{r}^2}{c^2}\right). \qquad \qquad ...(224)$$

We see that r is the only generalized co-ordinate; hence the corresponding Lagrange's equation yields the equation of motion in the form

$$\ddot{r}\left(m - \frac{2}{rc^2}\right) + \frac{\dot{r}^2}{r^2 c^2} - \frac{1}{r^2} = 0. \qquad \qquad ...(225)$$

Example 23: Derive the equation of motion of a particle falling vertically under the influence of gravity, when frictional forces obtainable from dissipation function $\frac{1}{2} K v^2$ are present. Integrate the equation to obtain the velocity as a function of time. Show also that the maximum possible velocity for fall from rest is $v = \dfrac{mg}{K}$.

Solution: Let a particle of mass m be falling vertically under the influence of gravity. Let z be the height of the particle at any instant t. Therefore the only generalized co-ordinate is z. Thus the Kinetic energy and the potential energy of the particle are given by

$$T = \frac{1}{2} m\dot{z}^2 \text{ and } V = -mgz.$$

The Lagrangian function is given by

$$L = \frac{1}{2} m\dot{z}^2 + mgz. \qquad \qquad ...(226)$$

Some frictional forces obtainable from dissipation function $\frac{1}{2} K\dot{z}^2$ are also present in the system. Consequently, the system is partially conservative and partially non-conservative. In this case the Lagrange's equation of motion for a system is given by the equation (84). Consequently, the equation (84) gives

$$m\ddot{z} + K\dot{z} - mg = 0. \qquad \qquad ...(227)$$

On integrating the equation (227), we get

$$m\dot{z} + Kz - mgt + c_1 = 0, \qquad \qquad ...(228)$$

where c_1 is a constant of integration and is to be determined. Using the initial conditions we obtained at $t = 0$, $\dot{z} = v = 0$, $z = 0 \Rightarrow c_1 = 0$.

We have therefore

$$\dot{z} + \left(\frac{K}{m}\right) z = gt. \qquad \qquad \dots(229)$$

This is a linear differential equation of first order whose solution is given by

$$z = \left(\frac{mg}{K}\right) t - \left(\frac{m}{K}\right)^2 g + c_2 \, e^{\frac{-Kt}{m}}.$$

As

$$t = 0 \Rightarrow z = 0 \Rightarrow c_2 = \left(\frac{m}{K}\right)^2 g,$$

hence

$$z = \left(\frac{mg}{K}\right) t - \left(\frac{m}{K}\right)^2 g + \left(\frac{m}{K}\right)^2 g \, e^{\frac{-Kt}{m}}. \qquad \dots(230)$$

Differentiating the equation (230) we obtain

$$\dot{z} = \left(\frac{mg}{K}\right) - \left(\frac{m}{K}\right) g \, e^{\frac{-Kt}{m}}. \qquad \qquad \dots(231)$$

This shows that the velocity $\dot{z}$ is the function of time only. For maximum velocity we have

$$\frac{d\dot{z}}{dt} = 0.$$

Hence the maximum velocity is obtained from the equation (227) by putting $\ddot{z} = 0$ and is given by

$$\dot{z} = \frac{mg}{K}.$$

Example 24: Two mass points of mass m_1 and m_2 are connected by a string passing through a hole in a smooth table so that m_1 rests on the table surface and m_2 hangs suspended. Assuming m_2 moves only in a vertical line, what are the generalized co-ordinates for the system? Write down the Lagrangian for the system. Reduce the problem to a single second order differential equation and obtain the first integral of the equation.

Solution: Let the two mass points m_1 and m_2 be connected by a string passing through a hole in a smooth table so that m_1 rests on the table surface and m_2 hangs suspended. We assume that m_2 moves only in a vertical line. The system is shown in the figure.

Let l be the length of a string. Consider OX as an initial line. Let (r, θ) be the position of the particle of mass m_1.

$\Rightarrow \qquad \qquad Om_2 = l - r.$

Thus the system is specified by two generalized co-ordinates r and θ. The kinetic energy of the system is the sum of the kinetic energies of the two masses and is given by

$$T = \frac{1}{2} m_1 \left(\dot{r}^2 + r^2\dot{\theta}^2\right) + \frac{1}{2} m_2\dot{r}^2. \qquad \ldots(232)$$

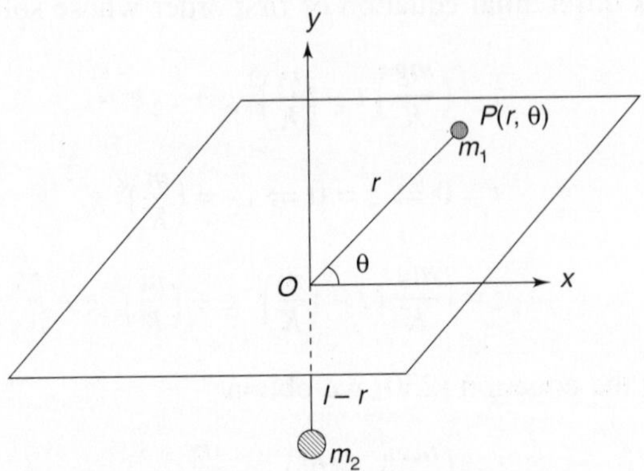

Potential energy of mass m_1 is zero, while that of mass m_2 is $-m_2 g(l - r)$. Hence the Lagrangian of the system becomes

$$L = \frac{1}{2} m_1 \left(\dot{r}^2 + r^2\dot{\theta}^2\right) + \frac{1}{2} m_2\dot{r}^2 + m_2 g(l - r). \qquad \ldots(233)$$

The Lagrange's equations corresponding to the generalized co-ordinates r and θ respectively give the equations of motion

$$(m_1 + m_2)\,\ddot{r} - m_1 r\dot{\theta}^2 + m_2 g = 0, \qquad \ldots(234)$$

and $$m_1 r^2\,\dot{\theta} = \text{const. } h(\text{say}) \qquad \ldots(235)$$

Now eliminating $\dot{\theta}$ between (234) and (235), we obtain

$$(m_1 + m_2)\,\ddot{r} - \frac{h^2}{m_1 r^3} + m_2 g = 0. \qquad \ldots(236)$$

This is the required single second order differential equation of motion. Now to find the first integral of motion, multiply equation (236) by $2\dot{r}$ and integrating it with respect to time t, we get

$$(m_1 + m_2)\int 2\dot{r}\ddot{r}\,dt - \frac{2h^2}{m_1}\int\left(\frac{\dot{r}}{r^3}\right)dt + 2m_2 g\int \dot{r}\,dt = \text{const.}$$

$$(m_1 + m_2)\int d(\dot{r}^2) - \frac{2h^2}{m_1}\int d\left(-\frac{1}{2r^2}\right) + 2m_2 g\int dr = \text{const.}$$

$$(m_1 + m_2)\,\dot{r}^2 + \frac{h^2}{m_1 r^2} + 2m_2 gr = \text{const.} \qquad \ldots(237)$$

This is required first integral of motion which represents the total energy of the system.

Example 25: A body of mass m is thrown up an inclined plane which is moving horizontally with a constant velocity v. Use Lagrangian equation to find the locus of the position of the body at any time t after the motion sets in.

Solution: Let AB be an inclined plane moving horizontally with constant velocity v. Therefore, at some instant t the distance moved by the plane AB is given by

$$OA = v \cdot t \qquad \qquad \ldots(238)$$

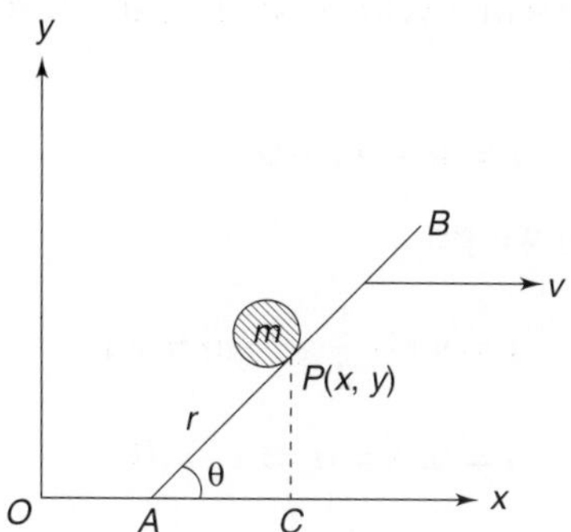

Let at $t = 0$, a body of mass m be thrown up an inclined plane AB. Let P be the position of the particle at any instant t, where $AP = r$. If (x, y) are the co-ordinates of the particle at P, then we have from figure

$$x = OA + AP \cos\theta,$$

$\Rightarrow \qquad \qquad x = vt + r\cos\theta, \qquad \qquad \ldots(239)$

and $\qquad \qquad y = r\sin\theta. \qquad \qquad \ldots(240)$

Note that θ is a fixed angle. The kinetic energy of the particle is given by

$$T = \frac{1}{2} m \left(\dot{x}^2 + \dot{y}^2 \right).$$

We notice that the position coordinates x and y are related by the rheonomic equation of constraint $y = (x - vt)\tan\theta$, and hence x and y are not the generalized co-ordinates. The only generalized co-ordinate is r. Hence using equations (239) and (240) we obtain the expression for the kinetic energy in terms of generalized co-ordinate r in the form

$$T = \frac{1}{2} m \left(v^2 + \dot{r}^2 + 2\dot{r}v\cos\theta \right). \qquad \qquad \ldots(241)$$

The potential energy of the particle is given by

$$V = mgr \sin \theta. \qquad \qquad \qquad \text{...(242)}$$

Hence the Lagrangian of the system becomes

$$L = \frac{1}{2} m \left(v^2 + \dot{r}^2 + 2\dot{r}v \cos \theta\right) - mgr \sin \theta. \quad \text{...(243)}$$

The corresponding r – Lagrange's equation of motion yields

$$\ddot{r} = -g \sin \theta. \qquad \qquad \qquad \text{...(244)}$$

Integrating we get

$$\dot{r} = -g \sin \theta t + c_1.$$

At $t = 0$ let $\dot{r} = u$ be the initial velocity of the particle with which it is projected. This gives $c_1 = u$, hence

$$\dot{r} = u - g \sin \theta t. \qquad \qquad \qquad \text{...(245)}$$

Integrating once again we get

$$r = u.t - \frac{1}{2} gt^2 \sin \theta + c_2.$$

At $\qquad \qquad \qquad t = 0, r = 0 \Rightarrow c_2 = 0.$

$\Rightarrow \qquad \qquad \qquad r = ut - \frac{1}{2} gt^2 \sin \theta.$

Hence the locus of the position of the particle is given by

$$r^2 = \left(ut - \frac{1}{2} gt^2 \sin \theta\right)^2 = (x - vt)^2 + y^2. \qquad \text{...(246)}$$

Example 26: Set up the Lagrangian and the Lagrange's equation of motion for simple pendulum.

Solution: Consider a simple pendulum of point mass m attached to one end of an inextensible light string of length l and other end is fixed at point O. The system is shown in figure. If $B(x, y)$ are the position co-ordinates of the pendulum at any instant t, then the equation of the constraint is given by $x^2 + y^2 + l^2$, where $x = l \sin \theta$, $y = l \cos \theta$, θ is the angle made by the pendulum with the vertical. This shows that x and y are not the generalized co-ordinates. We see that the angle θ determines the position of pendulum at any given time; hence it is a generalized co-ordinate. Hence the kinetic and potential energies of the pendulum become

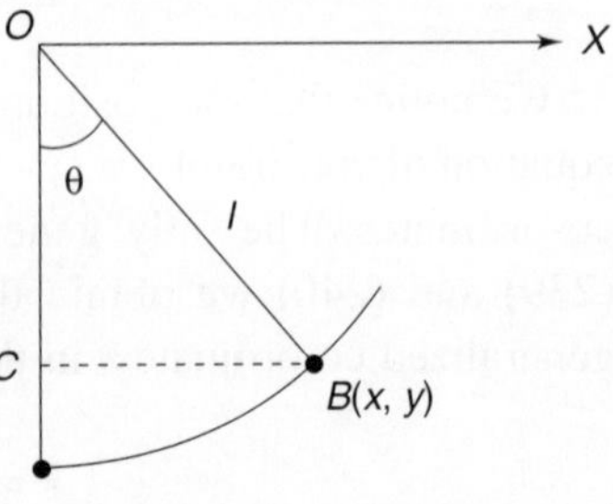

$$T = \frac{1}{2} m (l\dot{\theta})^2, \quad V = mgl(1 - \cos \theta).$$

Hence the Lagrangian of the motion becomes

$$L = \frac{1}{2}\, m(l\dot{\theta})^2 - mgl(1 - \cos\theta). \qquad \ldots(247)$$

The θ-Lagrange's equation of motion gives

$$\ddot{\theta} + \frac{g}{l}\sin\theta = 0. \qquad \ldots(248)$$

This is the second order differential equation that determines the motion of the simple pendulum.

Example 27: The length of a simple pendulum changes with time such that $l = a + bt$, where a and b are constants. Find the Lagrangian equation of motion.

Solution: Consider a simple pendulum of point mass m. Let l be the length of the pendulum which is not constant but a function of time t, given by

$$l = a + bt, \; a, \, b \text{ are constants.}$$

If (x, y) are the position co-ordinates of the pendulum at any time t then we have

$$x = l\sin\theta, \; y = l\cos\theta,$$

where θ is the angle made by the pendulum with vertical and determines the position of the pendulum at any instant, hence it is the generalized co-ordinate. The constraint in this case is holonomic and rheonomic given by the equation

$$x^2 + y^2 = l^2 = (a + bt)^2. \qquad \ldots(249)$$

The kinetic energy of the particle is given by

$$T = \frac{1}{2}\, m \left(\dot{l}^2 + l^2\dot{\theta}^2 \right),$$

and the potential energy is given by

$$V = mg\, (a + bt)\, (1 - \cos\theta).$$

Hence the Lagrangian function becomes

$$L = \frac{1}{2}\, m \left(\dot{l}^2 + l^2\dot{\theta}^2 \right) - mg(a + bt)\, (1 - \cos\theta). \qquad \ldots(250)$$

The Lagrange's equation corresponding to the generalized co-ordinate θ is

$$\frac{d}{dt}\left(\frac{\partial L}{\partial \dot{\theta}} \right) - \frac{\partial L}{\partial \theta} = 0. \qquad \ldots(251)$$

Solving this equation we obtain the required equation of motion in the form

$$(a + bt)\, \ddot{\theta} + 2b\dot{\theta} + g \sin \theta = 0. \qquad \ldots(252)$$

Example 28: A pendulum bob of radius r rolling on a circular track of radius $R(> r)$. Construct the Lagrangian and derive the equation of motion.

Solution: Let a pendulum bob (solid sphere) of radius r be rolling on a circular track of radius $R(> r)$. The rolling body has kinetic energy as well as rotational kinetic energy. If (x, y) are the position co-ordinates of the bob, then from the figure.

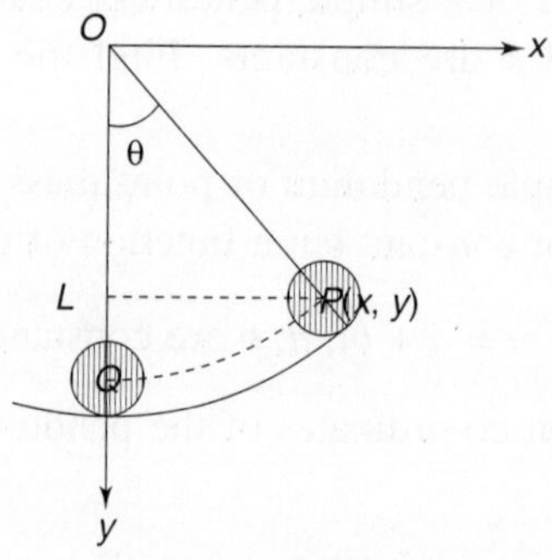

we have
$$x = (R - r) \sin \theta, \quad y = (R - r) \cos \theta, \qquad \ldots(253)$$

where θ determines the position of the pendulum, hence it is the generalized co-ordinate. The total kinetic energy of the system is

$$T = T_1 + T_2, \qquad \ldots(254)$$

where

$$T_1 = \frac{1}{2}\, m\, (\dot{x}^2 + \dot{y}^2) \text{ and } T_2 = \frac{1}{2}\, I\omega^2. \qquad \ldots(255)$$

Here ω is the angular velocity of the bob and I is its moment of inertia about the fixed point O and is given by

$$I = \frac{2}{5}\, mr^2. \qquad \ldots(256)$$

Using equations (253) and (255), in the equation (254), the kinetic energy becomes

$$T = \frac{1}{2}\, m\, (R - r)^2\, \dot{\theta}^2 + \frac{1}{5}\, mr^2\, \omega^2. \qquad \ldots(257)$$

If v is the linear velocity of the bob, then its angular velocity is given by $\omega = \dfrac{v}{r}$, where

$$v^2 = (\dot{x}^2 + \dot{y}^2) = (R - r)^2\, \dot{\theta}^2.$$

Hence equation (257) becomes

$$T = \frac{7}{10} m (R - r)^2 \dot{\theta}^2. \qquad \qquad ...(258)$$

The potential energy of the bob is given by

$$V = mg \cdot QL,$$

$$V = mg (OQ - OL),$$

$$V = mg [(R - r)) - (R - r) \cos \theta],$$

$$V = mg (R - r)) (1 - \cos \theta). \qquad \qquad ...(259)$$

Hence the Lagrangian of the bob becomes

$$L = \frac{7}{10} m (R - r)^2 \dot{\theta}^2 - mg(R - r) (1 - \cos \theta). \qquad \qquad ...(260)$$

Since θ is the only generalized co-ordinate, hence the θ-Lagrange's equation gives the differential equation of motion of the bob in the form

$$\ddot{\theta} + \frac{5}{7} \left(\frac{g}{R - r} \right) \sin \theta = 0. \qquad \qquad ...(261)$$

Example 29: A solid homogeneous cylinder of radius r rolls without slipping on the inside of a stationary large cylinder of radius R. Find the Lagrangian and equations of motion.

Solution: Consider a solid homogeneous cylinder of radius r and mass m rolls without slipping on the inside of a stationary cylinder of radius $R > r$. Let (x, y) be the position of the solid cylinder at any instant t. Let θ be the angular distance that the solid cylinder rolled and ϕ- the angle that the radius vector of the cylinder makes with the vertical. From the figure, we have $OO' = R - r$. If $O'(x, y)$ is the position of the solid cylinder, then $x = (R - r) \sin \phi$, $y = (R - r) \cos \phi$.

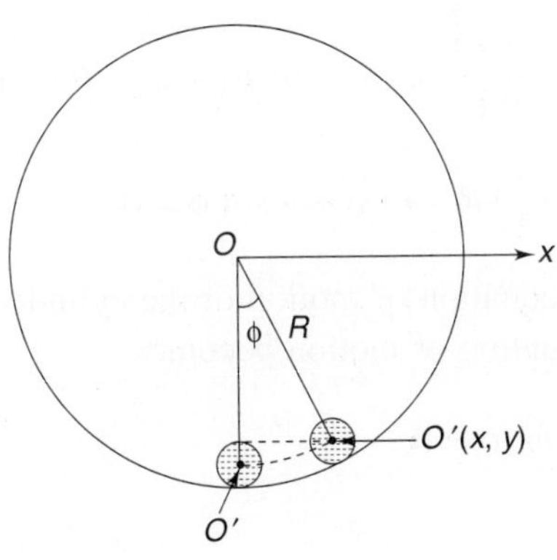

Since the cylinder rolls without slipping, the equation of the constraint is

$$r\theta = (R - r)\phi$$

$$\Rightarrow \quad r\dot{\theta} = (R - r)\dot{\phi}. \qquad\qquad \ldots(262)$$

Kinetic energy of the rolling cylinder is the sum of the kinetic energies of translation of the centre of mass and the rotation about the centre of mass and is given by

$$T = \frac{1}{2} mv^2 + \frac{1}{2} I\omega^2, \qquad\qquad \ldots(263)$$

where moment of inertia of solid cylinder is given by $I = \frac{1}{2} mr^2$, and the magnitude of the angular velocity is given by $\omega = \frac{v}{r}$. Hence the kinetic energy expression (263) becomes

$$T = \frac{3}{4} mv^2,$$

where

$$v^2 = \dot{x}^2 + \dot{y}^2 = (R - r)^2 \dot{\phi}^2.$$

Hence the kinetic energy becomes

$$T = \frac{3}{4} m(R - r)^2 \dot{\phi}^2. \qquad\qquad \ldots(264)$$

The potential energy of the cylinder is

$$V = -mgy$$

$$V = -mg\,(R - r)\cos\phi. \qquad\qquad \ldots(265)$$

Thus the Lagrangian of the cylinder becomes

$$L = \frac{3}{4} m(R - r)^2 \dot{\phi}^2 + mg\,(R - r)\cos\phi. \qquad\qquad \ldots(266)$$

Hence the corresponding Lagrange's equation of motion becomes

$$\frac{d}{dt}\left[\frac{3}{2} m\,(R - r)^2\,\dot{\phi}\right] + mg\,(R - r)\sin\phi = 0$$

$$\Rightarrow \quad \frac{3}{2}\,(R - r)\,\ddot{\phi} + g\sin\phi = 0. \qquad\qquad \ldots(267)$$

This is the required equation of motion of the cylinder. For small ϕ, we have $\sin\phi = \phi$ and hence equation of motion becomes

$$\ddot{\phi} + \omega^2\,\phi = 0,$$

where $\omega = \sqrt{\dfrac{2g}{3(R-r)}}$ is the frequency of small oscillations.

Example 30:   A pendulum of mass m is attached to a block of mass M. The block slides on a horizontal frictional less surface. Find the Lagrangian and equation of motion of the pendulum. For small amplitude oscillations derive an expression for periodic time.

Solution: Let a pendulum of point mass m be attached to one end of the light and inextensible string of length l and other end is attached to a block of mass M. The system is shown in figure. Let at any instant t the position co-ordinates of the block M and the pendulum m be $(x_1, 0)$ and (x_2, y_2) respectively,

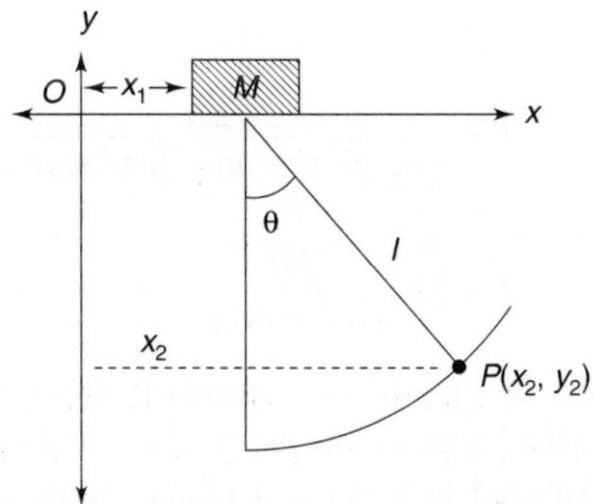

where $x_2 = x_1 + l \sin \theta$, $y_2 = l \cos \theta$.

We see that the position co-ordinates of the pendulum are related by the constraint equation; hence these are not the generalized co-ordinates. The generalized co-ordinates in this case are x_1 and θ. The kinetic energy of the system is the sum of the kinetic energy of the pendulum and the kinetic energy of the block. It is given by

$$T = \frac{1}{2} M \dot{x}_1^2 + \frac{1}{2} m \left(\dot{x}_1^2 + l^2 \dot{\theta}^2 + 2l\dot{x}_1 \dot{\theta} \cos \theta \right).$$

The potential energy of the pendulum is given by

$$V = -mgl \cos \theta.$$

Hence the Lagrangian of the system becomes

$$L = \frac{1}{2} (M + m) \, \dot{x}_1^2 + \frac{1}{2} m \left(l^2 \dot{\theta}^2 + 2l\dot{x}_1 \dot{\theta} \cos \theta \right) + mgl \cos \theta. \qquad ...(268)$$

Since θ and x_1 are the generalized co-ordinates, hence on simplifying the corresponding Lagrange's equations of motion yield the equations

$$ml^2 \, \ddot{\theta} + ml \cos \theta \ddot{x}_1 + mgl \sin \theta = 0. \qquad ...(269)$$

$$(M + m)\,\ddot{x}_1 + ml\ddot{\theta} = 0. \qquad \ldots(270)$$

However, if θ is small then we have $\sin \theta = \theta$ and $\cos \theta = 1$. Consequently equation (269) becomes

$$\ddot{\theta} + \frac{\ddot{x}_1}{l} + \frac{g}{l}\,\theta = 0. \qquad \ldots(271)$$

Eliminating $\ddot{x}_1$ between equations (270) and (271), we get

$$\ddot{\theta} = \frac{(M + m)g}{Ml}\,\theta. \qquad \ldots(272)$$

This is the required equation of simple harmonic motion. The periodic time T is given by

$$T = \frac{2\pi}{\sqrt{\text{accel}^n\ \text{per unit displacement}}}$$

$$\Rightarrow \qquad T = 2\pi\,\sqrt{\frac{Ml}{(M + m)g}}.$$

Example 31: The point of support of a simple pendulum of length l and mass m is moved along a vertical line according to the equation $y = y(t)$. The motion of the pendulum is restricted to a vertical plane. Show that the kinetic energy of the pendulum is given by $T = \dfrac{1}{2}\,m(l\dot{\theta})^2 + \dfrac{1}{2}\,m\dot{y}^2 + ml\dot{y}\dot{\theta}\,\sin\,\theta$. Derive the equation of motion.

Solution: Let l be the length of the simple pendulum of mass m. The point of support of the pendulum is moved along a vertical line according the equation $y = y(t)$. Let $(0, y)$ and (x_1, y_1) be the position co-ordinates of the point of support and the pendulum at any instant t respectively, where from the figure. we have

$$x_1 = l \sin\,\theta, \quad y_1 = l \cos\,\theta - y.$$

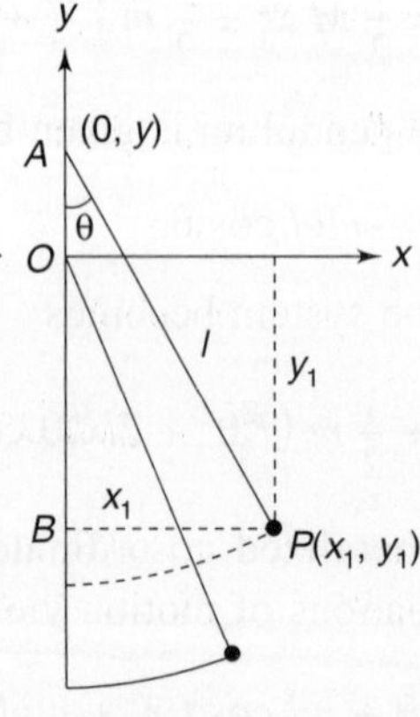

It is clear that x_1, y_1 are not the generalized co-ordinates. The generalized co-ordinates in this case are y and θ. Thus the kinetic energy and the potential energy of the pendulum are respectively given by

$$T = \frac{1}{2} m \, (l\dot{\theta})^2 + \frac{1}{2} m\dot{y}^2 + ml\dot{y}\dot{\theta} \, \sin \theta \text{ and } V = mgy - mgl \cos \theta.$$

Hence the Lagrangian of the particle becomes

$$L = \frac{1}{2} m \, (l\dot{\theta})^2 + \frac{1}{2} m\dot{y}^2 + ml\dot{y}\dot{\theta} \, \sin \theta - mgy + mgl \cos \theta. \qquad \ldots(273)$$

Since θ and y are the generalized co-ordinates, hence the corresponding Lagrange's equations of motion give

$$\ddot{\theta} + \frac{\sin \theta}{l} \, \ddot{y} + \frac{g}{l} \sin \theta = 0, \qquad \ldots(274)$$

$$\ddot{y} + l\ddot{\theta} \sin \theta + l\dot{\theta}^2 \cos \theta + g = 0. \qquad \ldots(275)$$

Eliminating $\ddot{y}$ between equations (274) and (275), we get

$$\ddot{\theta} - \dot{\theta}^2 \tan \theta = 0. \qquad \ldots(276)$$

This equation determines the motion of the pendulum.

Example 32: Consider a simple pendulum of mass m and length l whose point of suspension moves uniformly on a vertical circle of radius R with angular velocity ω. Obtain the Lagrangian of the pendulum in the form

$$L = \frac{1}{2} m \left[l^2\dot{\theta}^2 + 2lR\omega^2 \sin (\theta - \omega t) \right] + mgl \cos \theta,$$

and the equation of motion.

Solution: Consider a simple pendulum of point mass m and length l whose point of suspension moves uniformly on a vertical circle centered at O and of radius R, with angular velocity ω. Let (x_1, y_1) and (x_2, y_2) be the position co-ordinates of the point of support and the pendulum with respect to the co-ordinate system through the center of the circle. We have therefore,

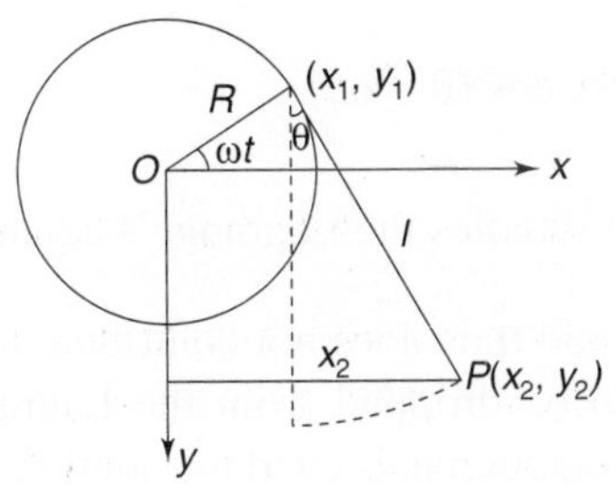

$$x_1 = R \cos \omega t, \; y_1 = R \sin \omega t,$$

$$x_2 = R \cos \omega t + l \sin \theta,$$

$$y_2 = l \cos \theta - R \sin \omega t. \qquad \qquad \ldots(277)$$

where θ is the angle made by the pendulum with the vertical. We see that θ is the only generalized co-ordinate. Therefore the kinetic energy and the potential energy of the pendulum are respectively given by

$$T = \frac{1}{2} m \left[R^2 \omega^2 + l^2 \dot{\theta}^2 + 2 l R \omega \dot{\theta} \sin (\theta - \omega t) \right],$$

and $\qquad \qquad V = -mg \, (l \cos \theta - R \sin \omega t).$

Hence the Lagrangian of the system becomes

$$L = \frac{1}{2} m \left[R^2 \omega^2 + l^2 \dot{\theta}^2 + 2 l R \omega \dot{\theta} \sin (\theta - \omega t) \right] + mg(l \cos \theta - R \sin \omega t) \quad \ldots(278)$$

We see that the terms containing only time and constant do not contribute in the equation of motion. Hence those terms can be dropped from equation of Lagrangian (278). Hence the reduced form of the Lagrangian may be given by

$$L = \frac{1}{2} m \left[l^2 \dot{\theta}^2 + 2 l R \omega \dot{\theta} \sin (\theta - \omega t) \right] + mg \, l \cos \theta \qquad \ldots(279)$$

This Lagrangian can also be written as

$$L = \frac{1}{2} m \left[l^2 \dot{\theta}^2 + 2 l R \omega^2 \sin (\theta - \omega t) \right] + mg \, l \cos \theta - \frac{d}{dt} \left[m l R \omega \cos(\theta - \omega t) \right].$$

$$\ldots(280)$$

This is of the form

$$L = L' + \frac{d}{dt} f(q_j, t),$$

where

$$L' = \frac{1}{2} m \left[l^2 \dot{\theta}^2 + 2 l R \omega^2 \sin (\theta - \omega t) \right] + mg \, l \cos \theta. \qquad \ldots(281)$$

and $\qquad \qquad f = m l R \omega \cos (\theta - \omega t). \qquad \qquad \ldots(282)$

The function $\dfrac{d}{dt} f(q_j, t)$ satisfies the Lagrange's equation of motion identically

(refer the Example (13)) and thus does not contribute in the equation of motion. Hence this term can also be dropped from the Lagrangian given in equation (280). Thus the required Lagrangian is cited in equation (281). The corresponding equation of motion can be obtained either from the Lagrangian (279) or from Lagrangian (281) in the form

$$\ddot{\theta} - \frac{R\omega^2}{l} \cos(\theta - \omega t) + \frac{g}{l} \sin\theta = 0. \qquad \ldots(283)$$

Example 33: Consider a simple pendulum of mass m and length l whose point of suspension oscillates horizontally in the plane of motion of the pendulum according to the law

$$x = R \cos \omega t.$$

Find the Lagrangian of the pendulum in the form

$$L = \frac{1}{2} m \left[l^2 \dot{\theta}^2 + 2 l R \omega^2 \sin\theta \cos\omega t \right] + mgl \cos\theta,$$

and the equation of motion.

Solution: Consider a simple pendulum of point mass m and length l, whose point of support oscillates horizontally in the plane of motion of the pendulum according to the law

$$x = R \cos \omega t \qquad \ldots(284)$$

where R is a constant and ω is uniform velocity. Let $(x, 0)$ and (x_2, y_2) be the co-ordinates of the point of support and the pendulum with respect to the fixed origin O. Therefore, the equation of the constraint on the motion of the particle is $(x_2 - R \cos \omega t)^2 + y_2^2 = l^2$.

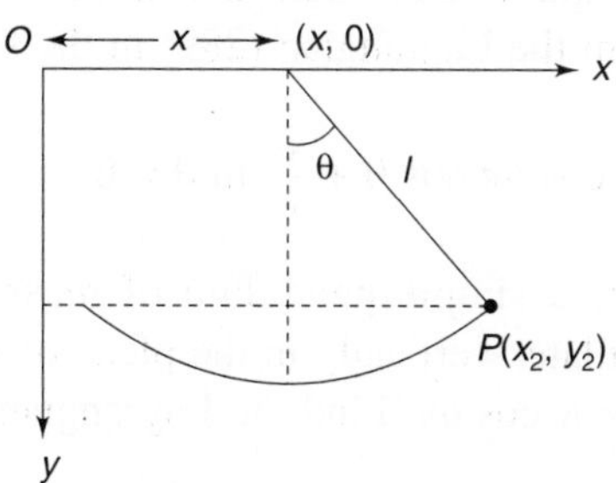

The constraint is rheonomic. The corresponding transformation equations are

$$x_2 = R \cos \omega t + l \sin\theta, \quad y_2 = l \cos\theta, \qquad \ldots(285)$$

where θ is the angle made by the pendulum with the vertical and it is the only generalized co-ordinate. Therefore the kinetic energy and the potential energy of the pendulum are respectively given by

$$T = \frac{1}{2}m\left[R^2\omega^2\sin^2\omega t + l^2\dot{\theta}^2 - 2lR\omega\dot{\theta}\sin\omega t\cos\theta\right],$$

and
$$V = -mgl\cos\theta.$$

Hence the Lagrangian of the system becomes

$$L = \frac{1}{2}m\left[R^2\omega^2\sin^2\omega t + l^2\dot{\theta}^2 - 2lR\omega\dot{\theta}\sin\omega t\cos\theta\right] + mgl\cos\theta. \quad ...(286)$$

We eliminate from the equation (286) the constant term and the terms containing only time as they do not contribute in the equation of motion. Hence the reduced form of the Lagrangian becomes

$$L = \frac{1}{2}m\left[l^2\dot{\theta}^2 - 2lR\omega\dot{\theta}\sin\omega t\cos\theta\right] + mg\,l\cos\theta. \qquad ...(287)$$

We can also write this equation in the form

$$L = \frac{1}{2}m\left[l^2\dot{\theta}^2 + 2lR\omega^2\sin\theta\cos\omega t\right] + mg\,l\cos\theta -$$

$$-\frac{d}{dt}\left[mlR\omega\sin\theta\sin\omega t\right]. \qquad ...(288)$$

On the same arguments as cited in the above example (32), the required Lagrangian is obtained by eliminating the last term of the equation (288) as it satisfies the Lagrange's equation of motion. Thus we have

$$L' = \frac{1}{2}m\left[l^2\dot{\theta}^2 + 2lR\omega^2\sin\theta\cos\omega t\right] + mg\,l\cos\theta. \qquad ...(289)$$

The corresponding equation of motion can be obtained either from the Lagrangian (287) or from the Lagrangian (289) in the form

$$\ddot{\theta} - \frac{R\omega^2}{l}\cos\omega t\cos\theta + \frac{g}{l}\sin\theta = 0. \qquad ...(290)$$

Example 34: Consider a simple pendulum of mass m and length l whose point of suspension oscillates vertically in the plane of motion of the pendulum according to the law $y = R\cos\omega t$. Find the Lagrangian of the pendulum in the form

$$L = \frac{1}{2}m\left[l^2\dot{\theta}^2 - 2lR\omega^2\cos\theta\cos\omega t\right] + mgl\cos\theta,$$

and the equation of motion.

Solution: Consider a simple pendulum of point mass m and length l, whose point of support oscillates vertically according to the law $y = R\cos\omega t$, where R is a constant and ω is uniform velocity. Let $(0, y)$ and (x_2, y_2) be the position co-ordinates of the point of support and the pendulum with respect to the fixed origin O. Therefore, the transformation equations are

$$x_2 = l \sin \theta,$$

$$y_2 = l \cos \theta - R \cos \omega t. \qquad \qquad \text{...(291)}$$

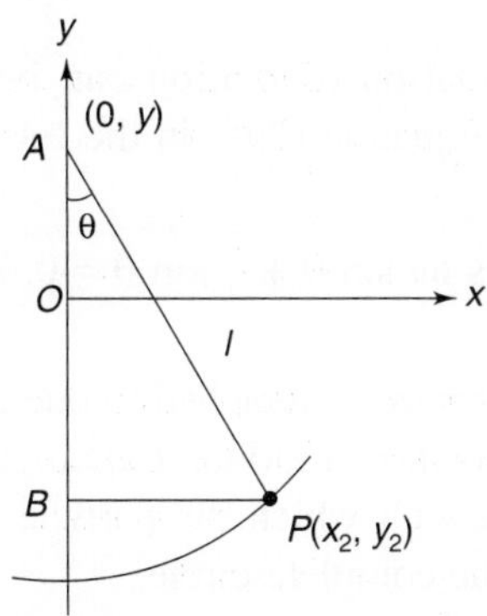

where θ is the angle made by the pendulum with the vertical and it is the only generalized co-ordinate. The equations of the constraint is $x_2^2 + (y_2 + R \cos \omega t)^2 = l^2$ which is holonomic and rheonomic.

$$T = \frac{1}{2} m \left[R^2 \omega^2 \cos^2 \omega t + l^2 \dot{\theta}^2 - 2 l R \omega \dot{\theta} \sin \omega t \sin \theta \right],$$

and $\qquad V = -mg \, (l \cos \theta - R \cos \omega t).$

Hence the Lagrangian of the system becomes

$$L = \frac{1}{2} m \left[R^2 \omega^2 \cos^2 \omega t + l^2 \dot{\theta}^2 - 2 l R \omega \dot{\theta} \sin \omega t \sin \theta \right] +$$

$$+ \, mgl \cos \theta - mgR \cos \omega t. \qquad \qquad \text{...(292)}$$

By neglecting the terms containing only time as these terms do not contribute in the equation of motion, the new Lagrangian given in the equation (292) becomes

$$L = \frac{1}{2} m \left[l^2 \dot{\theta}^2 - 2 l R \omega \dot{\theta} \sin \omega t \sin \theta \right] + mg \, l \cos \theta. \qquad \text{...(293)}$$

We can also write this Lagrangian in the form

$$L = \frac{1}{2} m \left[l^2 \dot{\theta}^2 - 2 l R \omega^2 \cos \theta \cos \omega t \right] + mgl \cos \theta +$$

$$+ \, \frac{d}{dt} \left[m l R \omega \cos \theta \sin \omega t \right]. \qquad \qquad \text{...(294)}$$

The last term of the equation (294) satisfies the Lagrange's equation of motion identically; hence it can be dropped from the equation. Hence the new Lagrangian of the particle is also given by

$$L' = \frac{1}{2} m \left[l^2 \dot{\theta}^2 - 2 l R \omega^2 \cos \theta \cos \omega t \right] + mgl \cos \theta. \qquad ...(295)$$

Now the Lagrange's equation of motion can be obtained either from the equation (293) or from the equation (295) in the form

$$\ddot{\theta} - \frac{R\omega^2}{l} \cos \omega t \sin \theta + \frac{g}{l} \sin \theta = 0.$$

Example 35: A particle of mass m attached to one end of the string starts with velocity u from its lowest position. Find the Lagrangian and equation of motion. Also find the least velocity with which the particle must start from the lowest position so as to describe the complete circle.

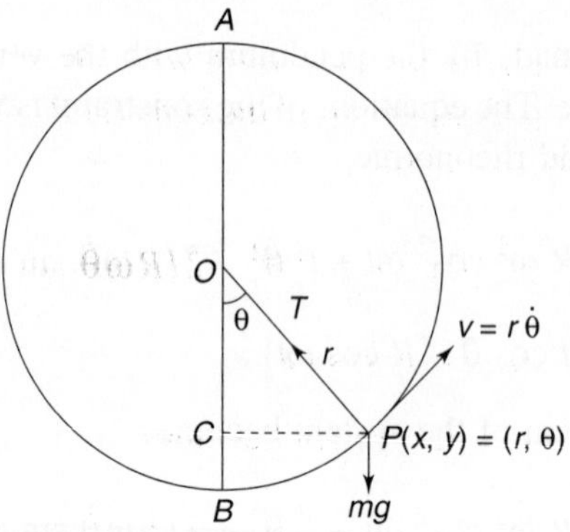

Solution: Let one end of the string be fixed at the point O and to the other end a point mass m is attached. Let the particle start from the lowest position B with velocity u. Let $P(x, y)$ be the position of the particle at some instant. If (r, θ) are the polar co-ordinates of the pendulum, then we have from the figure.

$$x = r \sin \theta, \; y = r \cos \theta.$$

This clearly shows that (x, y) are not the generalized co-ordinates. The linear velocity of the pendulum is v given by $v = r\dot{\theta}$.

Hence the kinetic energy and the potential energy of the pendulum are respectively given by

$$T = \frac{1}{2} m \left(r\dot{\theta} \right)^2 \text{ and } V = mgr \left(1 - \cos \theta \right).$$

Therefore the Lagrangian function of the pendulum becomes

$$L = \frac{1}{2}\, m\, (r\dot{\theta})^2 - mgr\,(1 - \cos\theta). \qquad \ldots(296)$$

Clearly θ is the generalized co-ordinates. Then the corresponding Lagrange's equation of motion gives

$$\ddot{\theta} + \frac{g}{r}\sin\theta = 0. \qquad \ldots(297)$$

Now to find the velocity of the pendulum at any instant, multiply equation (297) by $\dot{\theta}$ we get

$$\ddot{\theta}\dot{\theta} = -\frac{g}{r}\sin\theta\,\dot{\theta},$$

$$\Rightarrow \qquad \frac{d}{dt}\left(\frac{1}{2}\dot{\theta}^2\right) = \frac{g}{r}\frac{d}{dt}(\cos\theta).$$

Integrating we get

$$\frac{1}{2}\dot{\theta}^2 = \frac{g}{r}\cos\theta + c.$$

When $t = 0$, $\theta = 0$ and $r\dot{\theta} = u$ (initial velocity)

$$\Rightarrow \qquad c = \frac{u^2}{2r^2} - \frac{g}{r}.$$

Hence we have

$$v^2 = u^2 - 2gr\,(1 - \cos\theta). \qquad \ldots(298)$$

This equation determines the velocity of the particle at any angular distance θ. At the highest point of ascent, we put $v = 0$ in equation (298), we get

$$u^2 = 2gr\,(1 - \cos\theta). \qquad \ldots(299)$$

Let T be the tension in the string and $\dfrac{v^2}{r}$ the acceleration directed towards the fixed point, then resolving the forces in the direction of the motion and perpendicular to the direction of the motion, we write the equation of motion as

$$m\frac{v^2}{r} = T - mg\cos\theta. \qquad \ldots(300)$$

Eliminating v between equations (298) and (300) we obtain

$$T = m\left[\frac{u^2}{r} - 2g + 3g\cos\theta\right]. \qquad \ldots(301)$$

Equations (298) and (301) determine the motion of the particle for all values of θ. When θ increases from 0 to π, $\cos\theta$ decrease from 1 to -1 and at the highest point A, $\theta = \pi$, $\cos\pi = -1$. Hence equations (298) and (300) become

$$v^2 = u^2 - 4gr, \qquad \qquad \ldots(302)$$

and
$$T = \frac{m}{r}\left(u^2 - 5gr\right). \qquad \qquad \ldots(303)$$

These equations give the velocity and tension at the highest point A.

Condition for tracing complete circle

If $u^2 = 5gr$, we see from equations (302) and (303) that at the highest point A, tension vanishes but not the velocity. In this case the string is momentarily slack at the point A but the velocity is being not zero, the particle will go on tracing the curve. If $u^2 > 5gr$, then we see from equations (302) and (303) that neither the velocity nor the tension vanish at the highest point. In this case the particle will go on tracing the complete circle. Thus the least possible velocity with which the particle must start from the lowest point so as to describe complete circle is given by

$$u^2 = 5gr.$$

Let the velocity vanish when $\theta = \theta_1$. Hence from equation (298) we have

$$\cos\theta_1 = \frac{2gr - u^2}{2gr}. \qquad \qquad \ldots(304)$$

Let the height of this point, where velocity vanishes be h_1 above the point B. Thus we have

$$h_1 = r - r\cos\theta_1,$$

$$h_1 = r\left(1 - \frac{2gr - u^2}{2gr}\right),$$

$$h_1 = \frac{u^2}{2g}. \qquad \qquad \ldots(305)$$

Now let the tension T vanish when $\theta = \theta_2$. Hence from equation (299) we have

$$\cos\theta_2 = \frac{1}{3}\left(2 - \frac{u^2}{gr}\right). \qquad \qquad \ldots(306)$$

Let the height of the point where the tension T vanishes be h_2 above the point B. Thus we have

$$h_2 = r - r \cos \theta_2,$$

$$h_2 = r\left(1 - \frac{2gr - u^2}{3gr}\right),$$

$$h_2 = \frac{u^2 + gr}{3g}. \qquad \qquad ...(307)$$

Condition for velocity vanishing before tension vanishing

If the velocity v vanishes before tension T, then we have

$$h_1 < h_2,$$

$$\Rightarrow \qquad \frac{u^2}{2g} < \frac{u^2 + gr}{3g},$$

$$\Rightarrow \qquad u^2 < 2gr. \qquad \qquad ...(308)$$

This is the condition that the velocity vanishes before the tension. In this case the string remains taut and the particle will trace out its path back and oscillates.

Condition for tension vanishing before velocity vanishing

Let the tension T vanishes before the velocity v vanishes. In this case we have

$$h_2 < h_1,$$

$$\frac{u^2 + gr}{3g} < \frac{u^2}{2g}.$$

$$\Rightarrow \qquad u^2 > 2gr. \qquad \qquad ...(309)$$

This is the necessary and sufficient condition for tension vanishing before velocity vanishes. Thus if $u^2 > 2gr$ and $u^2 < 5gr$ the tension vanishes. The string becomes slack but velocity being not zero, the particle will leave the circular path and trace out a parabolic path.

Let velocity and tension vanish together. In this case we have

$$h_1 = h_2,$$

$$\frac{u^2}{2g} = \frac{u^2 + gr}{3g}$$

$\Rightarrow \qquad u^2 = 2gr.$ ...(310)

In this case if $\cos \theta_1 = 0$ and $\cos \theta_2 = 0$, this gives $\theta_1 = \theta_2 = \dfrac{\pi}{2}$. This shows that the particle will rise up to the horizontal diameter of the circle and oscillates in the semi-circle. Thus the least possible velocity with which the particle must start from the lowest position so as to describe the semi circle is $u^2 = 2gr$.

Example 36: A particle of mass m is projected from the lowest point with velocity u along the inner side of a smooth vertical circle. Find the Lagrangian and the equation of motion. Also find the least velocity with which the particle must be projected from the lowest position so as to describe the complete circle.

Solution: Consider a vertical circle of radius r and centered at O. Let a particle of point mass m be projected from the lowest point B with initial velocity u. Let $P(x, y)$ be the position of the particle at some instant and θ the angular distance covered by the particle at that instant. If (r, θ) are the polar co-ordinates of the pendulum, then we have from figure the transformation equations $x = r \sin \theta$, $y = r \cos \theta$. The constraint equation is $x^2 + y^2 = r^2$. This clearly shows that (x, y) are not the generalized co-ordinates. The linear velocity of the pendulum is v given by $v = r\dot\theta$. Hence the kinetic energy and the potential energy of the pendulum are respectively given by

$$T = \frac{1}{2}\, m\, (r\dot\theta)^2 \text{ and } V = mgr(1 - \cos \theta).$$

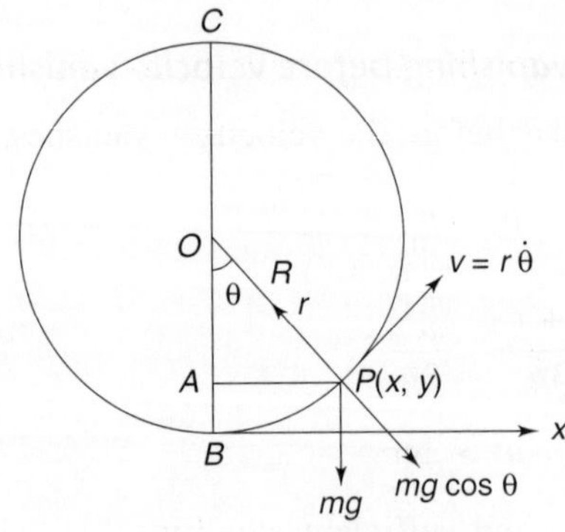

Therefore the Lagrangian function becomes

$$L = \frac{1}{2}\, m\, (r\dot\theta)^2 - mgr(1 - \cos \theta). \qquad \text{...(311)}$$

Clearly θ is the generalized co-ordinates. Then the corresponding Lagrange's equation of motion yields the equation of motion in the form

$$\ddot\theta + \frac{g}{r} \sin \theta = 0. \qquad \text{...(312)}$$

Now to find the velocity of the pendulum at any instant, multiply equation (312) by $\dot{\theta}$, we get

$$\ddot{\theta}\dot{\theta} = -\frac{g}{r}\sin\theta\dot{\theta},$$

$$\Rightarrow \qquad \frac{d}{dt}\left(\frac{1}{2}\dot{\theta}^2\right) = \frac{g}{r}\frac{d}{dt}(\cos\theta).$$

Integrating we get

$$\frac{1}{2}\dot{\theta}^2 = \frac{g}{r}\cos\theta + c.$$

When $t = 0$, $\theta = 0$ and $r\dot{\theta} = u$ (initial velocity)

$$\Rightarrow \qquad c = \frac{u^2}{2r^2} - \frac{g}{r}.$$

Hence we have

$$v^2 = u^2 - 2gr(1 - \cos\theta). \qquad \ldots(313)$$

This equation determines the velocity of the particle at angular distance θ. At the highest point of ascent, we put $v = 0$ in equation (313), we get

$$u^2 = 2gr(1 - \cos\theta). \qquad \ldots(314)$$

Let R be the reaction of the surface of the circle, $\dfrac{v^2}{r}$ is the acceleration directed towards the fixed point, then resolving the forces acting on the particle in the direction of the motion and perpendicular to the direction of the motion, we write the equation of motion as

$$m\frac{v^2}{r} = R - mg\cos\theta. \qquad \ldots(315)$$

Eliminating v between equations (313) and (315) we obtain

$$R = m\left[\frac{u^2}{r} - 2g + 3g\cos\theta\right]. \qquad \ldots(316)$$

Equations (313) and (316) determine the motion of the particle for all values of θ. When θ increases from 0 to π, $\cos\theta$ decrease from 1 to -1 and at the highest point C, $\theta = \pi$, $\cos\pi = -1$. Hence equations (313) and (315) become

$$v^2 = u^2 - 4gr, \qquad \ldots(317)$$

and
$$R = \frac{m}{r}\left(u^2 - 5gr\right). \qquad \qquad \text{...(318)}$$

These equations give the velocity and reaction at the highest point C.

Condition for velocity vanishing before reaction vanishing

From equation (313) the velocity v of the particle vanishes if

$$\cos\theta = 1 - \frac{u^2}{2gr}. \qquad \qquad \text{...(319)}$$

Also from equation (316), reaction R vanishes, if

$$\cos\theta = \frac{1}{3}\left(2 - \frac{u^2}{gr}\right). \qquad \qquad \text{...(320)}$$

The velocity vanishes before reaction vanishes, if the value of θ from Equation (319) is less than the value of θ from equation (320). *i.e.,* v vanishes before R if $\cos\theta$ from (319) is greater than $\cos\theta$ from (320). *i.e.,*

$$1 - \frac{u^2}{2gr} > \frac{1}{3}\left(2 - \frac{u^2}{gr}\right),$$

$$\Rightarrow \qquad \qquad u^2 < 2gr \qquad \qquad \text{...(321)}$$

This is the necessary and sufficient condition for velocity vanishing before reaction R vanishes. In this case the particle does not reach the horizontal position and trace out its path back and will oscillate.

If $u^2 = 2gr$, then we have $\theta = \frac{\pi}{2}$. From equations (313), (316), (319) and (320) we have velocity and reaction vanish simultaneously. In this case, the particle will rise up to the horizontal diameter of the circle and swings through a quadrant on each side of the vertical. If the reaction R vanishes before velocity vanishes, if $\cos\theta$ from the equation (319) is less than $\cos\theta$ from the equation (320). *i.e.,*

$$1 - \frac{u^2}{2gr} < \frac{1}{3}\left(2 - \frac{u^2}{gr}\right),$$

$$\Rightarrow \qquad \qquad u^2 > 2gr. \qquad \qquad \text{...(322)}$$

In this case the particle will leave the circle. We see from equations (317) and (318) that, if $u^2 > 5gr$, neither the velocity nor the tension vanish at the highest point. In this case the particle will go on tracing the complete circle.

Example 37: Set up the Lagrangian and Lagrange's equation of motion of a heavy particle placed on the top of a smooth circle in a vertical plane. If the particle starts from a point whose angular distance is α from the highest point of the circle then find the point when the particle flies off the circle. Show also that if the particle starts from the highest point it will leave the circle at the point whose vertical distance below the point of start is $\frac{1}{3}\,a$.

Solution: Let a particle be initially placed at point P on the circle of radius a. Let OP makes an angle α with the vertical. Let at any time t latter, the position of the particle be at angular distance θ. If v is the linear velocity of the particle, it is given by $v = a\dot{\theta}$. Hence the kinetic energy and the potential energy of the pendulum are respectively given by $T = \frac{1}{2}\,m\,(a\dot{\theta})^2$ and

$$V = -mga\,(\cos\alpha - \cos\theta).$$

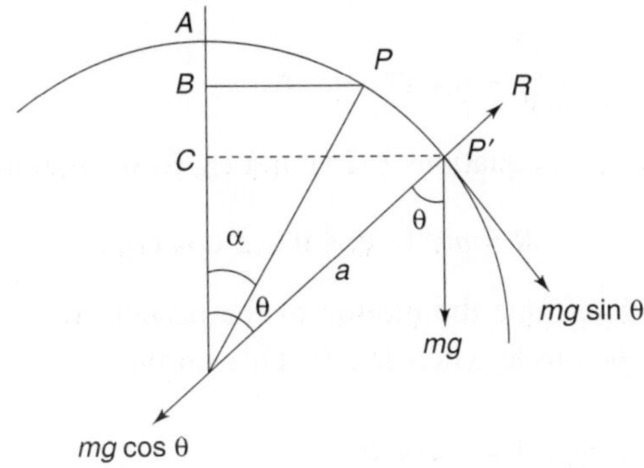

Therefore, the Lagrangian function of the particle becomes

$$L = \frac{1}{2}\,m\,(a\dot{\theta})^2 + mga(\cos\alpha - \cos\theta). \qquad \ldots(323)$$

Clearly θ is the generalized co-ordinates. Then the corresponding Lagrange's equation of motion gives the equation of motion in the form

$$\ddot{\theta} = \frac{g}{a}\sin\theta. \qquad \ldots(324)$$

Now to find the velocity of the pendulum at any instant, multiply equation (324) by $\dot{\theta}$, we get

$$\ddot{\theta}\dot{\theta} = \frac{g}{a}\sin\theta\dot{\theta},$$

$$\Rightarrow \qquad \frac{d}{dt}\left(\frac{1}{2}\dot{\theta}^2\right) = -\frac{g}{a}\frac{d}{dt}(\cos\theta).$$

Integrating we get

$$\frac{1}{2}\dot{\theta}^2 = -\frac{g}{a}\cos\theta + c.$$

When $t = 0$, $\dot{\theta} = 0$ and $\theta = \alpha$ (initial position),

$$\Rightarrow \qquad\qquad c = g \cos \alpha.$$

Hence we have

$$v^2 = 2ga(\cos \alpha - \cos \theta). \qquad\qquad \ldots(325)$$

If R is the reaction of the surface of the circle, $\dfrac{v^2}{a}$ is the acceleration directed towards the fixed point, then resolving the forces acting on the particle in the direction of the motion and perpendicular to the direction of the motion, we write the equation of motion as

$$m\,\frac{v^2}{a} = mg \cos \theta - R. \qquad\qquad \ldots(326)$$

Eliminating v between equations (325) and (326) we obtain

$$R = mg\,[3 \cos \theta - 2 \cos \alpha]. \qquad\qquad \ldots(327)$$

These equations determine the motion of the particle for all values of θ. The particle will fly off the circle when $R = 0$. This implies

$$\cos \theta = \frac{2}{3} \cos \alpha. \qquad\qquad \ldots(328)$$

If however, the particle starts from the highest point then we have $\alpha = 0$. Hence from the equation (328) we have

$$\cos \theta = \frac{2}{3}. \qquad\qquad \ldots(329)$$

This gives the angular distance from the highest point A to the point where the particle leaves the circle. Now the vertical distance of the particle before it leaves the circle is given by AC.

$$AC = OA - OC$$

$$= a - a \cos \theta$$

$$= a\left(1 - \frac{2}{3}\right) \text{ by equation (329)}$$

$$= \frac{1}{3}\,a.$$

This proves the answer.

Spherical Pendulum

A point mass constrained to move on the surface of a sphere is called spherical pendulum.

Example 38: In a spherical pendulum a particle of mass m moves on the surface of a sphere of radius r in a gravitational field. Show that the equation of motion of the particle may be written as

$$\ddot{\theta} - \frac{p_{\phi}^2 \cos \theta}{m^2 r^4 \sin^3 \theta} - \frac{g}{r} \sin \theta = 0,$$

where p_{ϕ} is the constant of angular momentum.

Solution: Let $P\,(x, y, z)$ be the position co-ordinates of the particle moving on the surface of a sphere of radius r. If (r, θ, ϕ) are its spherical polar co-ordinates, then the transformation equations are given by

$$x = r \sin \theta \cos \phi, \quad y = r \sin \theta \sin \phi, \quad z = r \cos \theta.$$

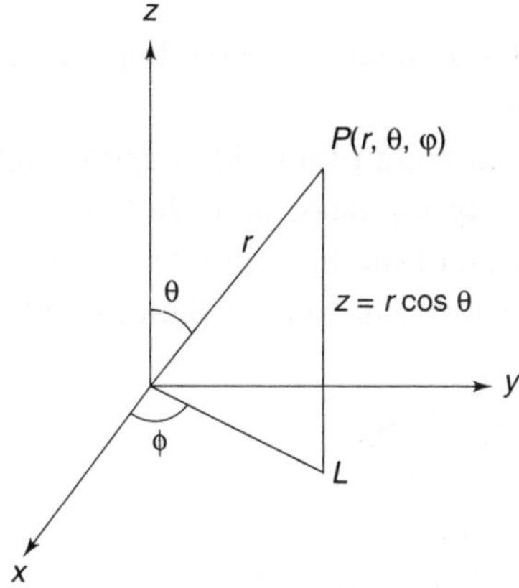

It clearly shows that x, y, z are not the generalized co-ordinates, as they are related by the constraint equations. The generalized co-ordinates are (θ, ϕ). Hence the kinetic and potential energies of particle are respectively given by

$$T = \frac{1}{2} mr^2 \left(\dot{\theta}^2 + \sin^2 \theta \dot{\phi}^2\right), \quad V = mgr \cos \theta.$$

Hence the Lagrangian function becomes

$$L = \frac{1}{2} mr^2 \left(\dot{\theta}^2 + \sin^2 \theta \dot{\phi}^2\right) - mgr \cos \theta. \qquad \ldots(330)$$

The two Lagrange's equations of motion corresponding to two generalized co-ordinates θ and ϕ reduce to

$$mr^2\ \ddot\theta - mr^2 \sin\theta \cos\theta \dot\phi^2 - mgr \sin\theta = 0, \qquad \qquad ...(331)$$

and

$$mr^2 \sin^2\theta \dot\phi = \text{const.} = p_\phi. \qquad \qquad ...(332)$$

Eliminating $\dot\phi$ between equations (331) and (332) we get the single equation of motion as

$$\ddot\theta - \frac{p_\phi^2 \cos\theta}{m^2 r^4 \sin^3\theta} - \frac{g}{r}\sin\theta = 0, \qquad \qquad ...(333)$$

where p_ϕ is a constant of angular momentum.

Compound Pendulum

A rigid body capable of oscillating in a vertical plane about a fixed horizontal axis under the action of gravity is called a compound pendulum.

Example 39:   Set up the Lagrangian and the Lagrange's equation of motion for the compound pendulum.

Solution: Let O be a fixed point of a rigid body through which axis of rotation passes. Let C be the center of mass, and $OC = l$. Let m be the mass of the pendulum and I the moment of inertia about the axis of rotation. If θ is the angle of deflection of the body then the rotational kinetic energy of the pendulum is given by

$$T = \frac{1}{2}\ I\dot\theta^2 \qquad \qquad ...(334)$$

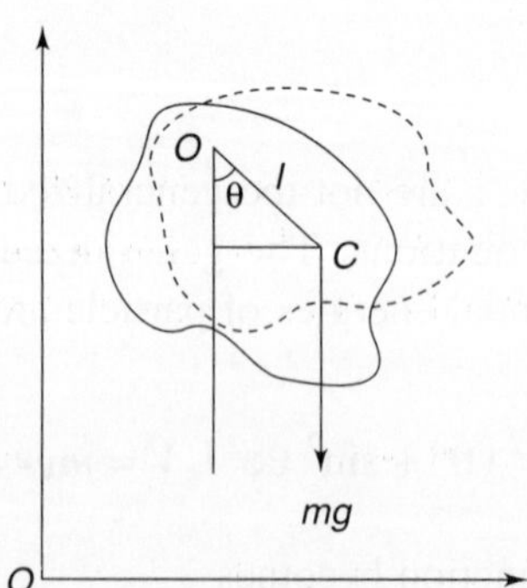

The potential energy relative to the horizontal plane through O is

$$V = -mgl \cos\theta. \qquad \qquad ...(335)$$

Hence the Lagrangian of compound pendulum becomes

$$L = \frac{1}{2} I\dot{\theta}^2 + mgl \cos \theta. \qquad \text{...(336)}$$

Thus the Lagrange's equation of motion corresponding to the generalized co-ordinate θ becomes

$$\ddot{\theta} + \frac{mgl}{I} \sin \theta = 0. \qquad \text{...(337)}$$

The periodic time of oscillation is given by

$$T = 2\pi \sqrt{\frac{I}{mgl}}.$$

Example 40: Obtain the Lagrangian and equations of motion for a double pendulum vibrating in a vertical plane.

Solution: A double pendulum moving in a plane consists of two particles of masses m_1 and m_2 connected by an inextensible string. The system is suspended by another inextensible and weightless string fastened to one of the masses as shown in the figure. Let θ_1 and θ_2 be the deflections of the pendulum from vertical. These are the generalized co-ordinates of the system. Let l_1 and l_2 be the lengths of the strings and (x_1, y_1), (x_2, y_2) be the rectangular position co-ordinates of the masses m_1 and m_2 at any instant t respectively. From the figure, we have

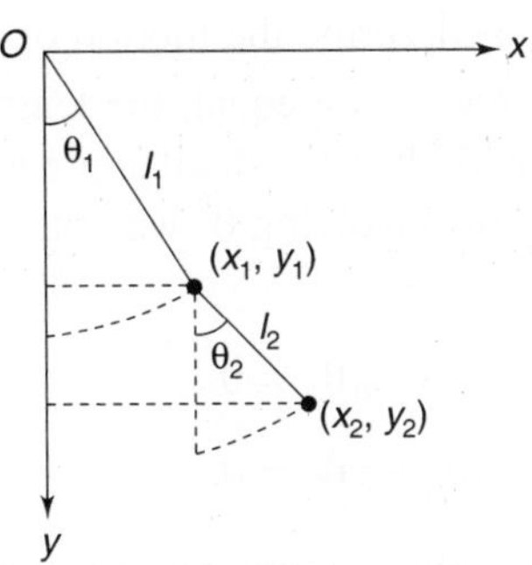

$$x_1 = l_1 \sin \theta_1, \quad y_1 = l_1 \cos \theta_1;$$

$$x_2 = l_1 \sin \theta_1 + l_2 \sin \theta_2. \qquad \text{...(338)}$$

$$y_2 = l_1 \cos \theta_1 + l_2 \cos \theta_2.$$

The total kinetic energy of the system is given by

$$T = \frac{1}{2} m_1 \left(\dot{x}_1^2 + \dot{y}_1^2 \right) + \frac{1}{2} m_2 \left(\dot{x}_2^2 + \dot{y}_2^2 \right).$$

Using equation (338) we obtain the expression for kinetic energy as

$$T = \frac{1}{2} m_1 l_1^2 \, \dot{\theta}_1^2 + \frac{1}{2} m_2 \left[l_1^2 \, \dot{\theta}_1^2 + l_2^2 \, \dot{\theta}_2^2 + 2 l_1 l_2 \, \dot{\theta}_1 \dot{\theta}_2 \cos (\theta_1 - \theta_2) \right]. \quad \ldots(339)$$

Taking the reference level as a horizontal plane through the point of suspension O, the total potential energy of the system is given by

$$V = -m_1 g l_1 \cos \theta_1 - m_2 g \left(l_1 \cos \theta_1 + l_2 \cos \theta_2 \right). \quad \ldots(340)$$

Hence the Lagrangian of the system becomes

$$L = \frac{1}{2} \left(m_1 + m_2 \right) l_1^2 \, \dot{\theta}_1^2 + \frac{1}{2} m_2 \, l_2^2 \, \dot{\theta}_2^2 + m_2 l_1 l_2 \cos (\theta_1 - \theta_2) \dot{\theta}_1 \dot{\theta}_2 +$$

$$+ \, m_1 g l_1 \cos \theta_1 + m_2 g \left(l_1 \cos \theta_1 + l_2 \cos \theta_2 \right). \ \ldots(341)$$

Solving the Lagrange's equations of motion corresponding to the generalized co-ordinates θ_1 and θ_2 we obtain

$$\left(m_1 + m_2 \right) l_1^2 \, \ddot{\theta}_1 + m_2 l_1 l_2 \cos (\theta_1 - \theta_2) \, \ddot{\theta}_2 + m_2 l_1 l_2 \sin (\theta_1 - \theta_2) \, \dot{\theta}_2^2 +$$

$$+ \left(m_1 + m_2 \right) g l_1 \sin \theta_1 = 0, \qquad \ldots(342)$$

and

$$m_2 l_2^2 \, \ddot{\theta}_2 + m_2 l_1 l_2 \cos (\theta_1 - \theta_2) \, \ddot{\theta}_1 - m_2 l_1 l_2 \sin (\theta_1 - \theta_2) \, \dot{\theta}_1^2 +$$

$$+ \, m_2 g l_2 \sin \theta_2 = 0. \qquad (343)$$

Equations (342) and (343) describe the motion of the double pendulum.

Note: If in particular, two masses are equal, the lengths of the pendula are also equal and $\theta_1 - \theta_2$ is very small, then for small angle we have $\sin \theta = \theta$, $\cos \theta = 1$ and hence neglecting the terms involving $\dot{\theta}^2$ we obtain from equations (342) and (343) that

$$2 l \ddot{\theta}_1 + l \ddot{\theta}_2 + 2 g \theta_1 = 0.$$

$$l \ddot{\theta}_2 + l \ddot{\theta}_1 + g \theta_2 = 0. \qquad \ldots(344)$$

Example 41: Find the equation of motion corresponding to the Lagrangian

$$L(x, \dot{x}) = e^{-x^2} \left(e^{-\dot{x}^2} + 2\dot{x} \int_0^{\dot{x}} e^{-\alpha^2} \, d\alpha \right).$$

Find the energy integral for the system.

Solution: The Lagrangian of the system is given in the question. We see that x is the only generalized co-ordinate. We find from the given Lagrangian

$$\frac{\partial L}{\partial x} = -2x e^{-x^2} \left(e^{-\dot{x}^2} + 2\dot{x} \int_0^{\dot{x}} e^{-\alpha^2} \, d\alpha \right),$$

$$\Rightarrow \qquad \frac{\partial L}{\partial x} = -2xL. \qquad \qquad ...(345)$$

Also we find from the Lagrangian function

$$\frac{\partial L}{\partial \dot{x}} = e^{-x^2}\left(-2\dot{x}e^{-\dot{x}^2} + 2\int_0^{\dot{x}} e^{-\alpha^2}\, d\alpha + 2\dot{x}\,\frac{\partial}{\partial \dot{x}}\int_0^{\dot{x}} e^{-\alpha^2}\, d\alpha\right).$$

We recall the formula

$$\frac{d}{dx}\int_{P(x)}^{Q(x)} F(x,y)dy = \int_{P(x)}^{Q(x)} \frac{\partial}{\partial x} F(x,y)dy + F(x,Q)\frac{dQ}{dx} - F(x,P)\frac{dP}{dx}. \qquad ...(346)$$

Using formula (346) we obtain

$$\frac{\partial L}{\partial \dot{x}} = e^{-x^2}\left(-2\dot{x}e^{-\dot{x}^2} + 2\int_0^{\dot{x}} e^{-\alpha^2}\, d\alpha + 2\dot{x}e^{-\dot{x}^2}\right)$$

$$\frac{\partial L}{\partial \dot{x}} = 2e^{-x^2}\int_0^{\dot{x}} e^{-\alpha^2}\, d\alpha. \qquad \qquad ...(347)$$

From equation (347) we find

$$\frac{d}{dt}\left(\frac{\partial L}{\partial \dot{x}}\right) = -4x\dot{x}e^{-x^2}\int_0^{\dot{x}} e^{-\alpha^2}\, d\alpha + 2\ddot{x}e^{-x^2}\, e^{-\dot{x}^2}. \qquad ...(348)$$

Using equations (345) and (348) in the Lagrange's equation of motion, we obtain

$$\ddot{x} + x = 0. \qquad \qquad ...(349)$$

The energy of the system is given by

$$E = \frac{\partial L}{\partial \dot{x}}\,\dot{x} - L. \qquad \qquad ...(350)$$

$$\Rightarrow \qquad E = -\exp\left[-(x^2 + \dot{x}^2)\right].$$

Differentiating the equation (350) with respect to t, we see that

$$\frac{dE}{dt} = 2\dot{x}\exp\left[-(x^2 + \dot{x}^2)\right](\ddot{x} + x)$$

$$\Rightarrow \qquad \frac{dE}{dt} = 0,$$

$$\Rightarrow \qquad E = \text{const.}$$

Example 42:   A cylinder of mass m and radius 'a' rolls down an inclined plane making an angle θ with the horizontal. Set up the Lagrangian and find the equation of motion.

Solution: Consider a cylinder of radius a starts to roll from the point O on an inclined plane making an angle θ with the horizontal. The system is shown in the figure. Let at some instant latter, the distance rolled by the cylinder be $x = a\phi, \Rightarrow \dot{x} = a\dot{\phi}$.

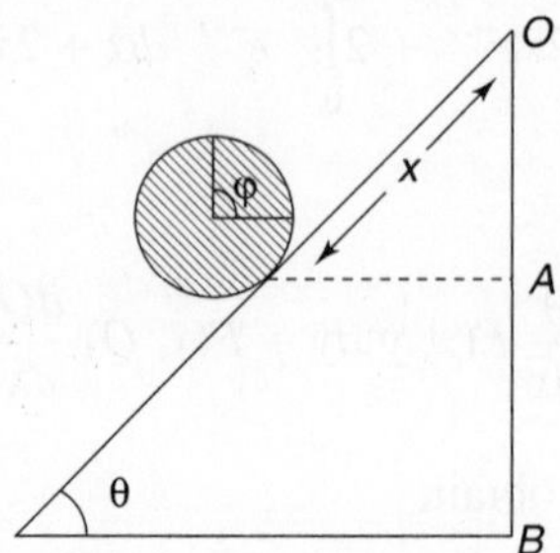

We see that the system has only one degree of freedom and hence only one generalized co-ordinate x. The rolling body has linear kinetic energy as well as rotational kinetic energy. The total energy of the rolling body is given by

$$T = \frac{1}{2}\,m\dot{x}^2 + \frac{1}{2}\,I\omega^2,$$

where the moment of inertia I of the cylinder is given by

$$I = \frac{1}{2}\,ma^2.$$

Hence the kinetic energy of the particle becomes

$$T = \frac{1}{2}\,m\dot{x}^2 + \frac{1}{4}\,ma^2\,\omega^2,$$

where the angular velocity $\omega = \dot{\phi} = \dfrac{\dot{x}}{a}$, hence equation for kinetic energy becomes

$$T = \frac{3}{4}\,m\dot{x}^2. \qquad\qquad ...(351)$$

The potential energy of the cylinder is given by

$$V = mg\,(OA)$$

$$V = mg\,(l - x)\sin\theta. \qquad\qquad ...(352)$$

Hence the Lagrangian of the system becomes

$$L = \frac{3}{4}\,m\dot{x}^2 - mg\,(l - x)\sin\theta. \qquad\qquad ...(353)$$

Hence the equation of motion corresponding to the generalized co-ordinate x gives

$$\frac{3}{2}\, m\ddot{x} - mg\, \sin\theta = 0. \qquad \qquad \text{...(354)}$$

Example 43:   An inclined plane of mass m_1 is sliding on a horizontal smooth surface and a body of mass m_2 is sliding on its smooth inclined surface. Find the Lagrangian of the system. Derive the equations of motion of the body and the inclined plane.

Solution: Consider an inclined plane of mass m_1 slides on the horizontal smooth surface. Take another body of mass m_2 slides on the smooth inclined plane. The system is shown in the figure.

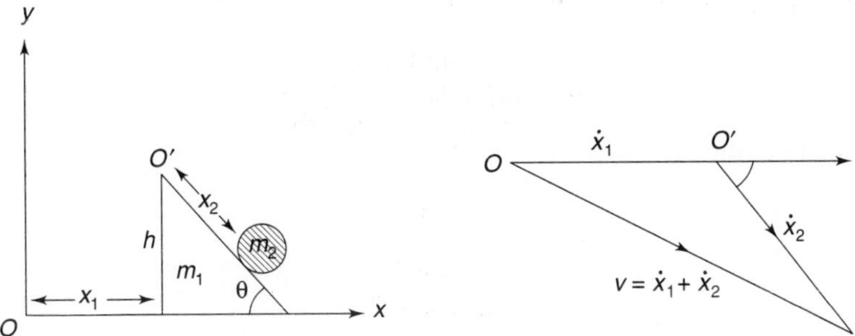

The system has two degrees of freedom and hence two generalized co-ordinates. Let x_1 and x_2 be the displacements of m_1 and m_2 from the points O and O' respectively.

Therefore, velocity of m_1 relative to O is $\dot{x}_1$, while velocity of m_2 relative to O' is $\dot{x}_2$. Hence the velocity of m_2 relative to O is $v = \dot{x}_1 + \dot{x}_2$. This gives

$$v^2 = \dot{x}_1^2 + \dot{x}_2^2 + 2\,\dot{x}_1\dot{x}_2 \cos\theta.$$

Hence the kinetic energy of the whole system is equal to the sum of the kinetic energies of the body of mass m_1 and m_2. Therefore we have

$$T = \frac{1}{2}\, m_1\, \dot{x}_1^2 + \frac{1}{2}\, m_2 v^2,$$

$$T = \frac{1}{2}\, (m_1 + m_2)\, \dot{x}_1^2 + \frac{1}{2}\, m_2 \left(\dot{x}_2^2 + 2\dot{x}_1\dot{x}_2 \cos\theta \right). \qquad \text{...(355)}$$

The potential energy of the system is the potential energy of mass m_2 and is given by

$$V = m_2 g(h - x_2 \sin\theta), \qquad \qquad \text{...(356)}$$

where h is the height of the inclined plane. Hence the Lagrangian of the system becomes

$$L = \frac{1}{2}(m_1 + m_2)\,\dot{x}_1^2 + \frac{1}{2}\,m_2\left(\dot{x}_2^2 + 2\dot{x}_1\dot{x}_2\cos\theta\right) - m_2 g\,(h - x_2 \sin\theta). \quad \text{...(357)}$$

Thus the Lagrange's equations of motion corresponding to the generalized co-ordinates x_1 and x_2 reduce to the forms

$$(m_1 + m_2)\ddot{x}_1 + m_2\ddot{x}_2 \cos\theta = 0, \quad\quad \text{...(358)}$$

and

$$m_2\left(\ddot{x}_2 + \ddot{x}_1 \cos\theta\right) - m_2 g \sin\theta = 0. \quad\quad \text{...(359)}$$

Solving equations (358) and (359) for $\ddot{x}_1$ and $\ddot{x}_2$ we obtain

$$\ddot{x}_1 = \frac{-g\sin\theta\cos\theta}{\dfrac{m_1 + m_2}{m_2} - \cos^2\theta},$$

$$\ddot{x}_2 = \frac{g\sin\theta}{-\dfrac{m_2\cos^2\theta}{m_1 + m_2}}.$$

Example 44: A ladder slides down a smooth wall and smooth floor. Set up the Lagrangian for the system and deduce the equation of motion.

Solution: Let AB be a ladder of length $2l$ and mass m slides down a smooth wall and smooth floor. Let $C\,(x, y)$ be a point through which its weight mg acts in the downward direction. Let θ be the inclination of the ladder with the horizontal floor. Clearly (x, y) are not the generalized co-ordinates but are related by the constraint relations

$$x = l \cos\theta,$$
$$y = l \sin\theta.$$

The generalized co-ordinate is only θ and hence the system has only one degree of freedom. The kinetic energy of the sliding ladder is given by

$$T = \frac{1}{2}\,m\,(\dot{x}^2 + \dot{y}^2) + \frac{1}{2}\,I\omega^2 \quad\quad \text{...(360)}$$

where $I = mk^2$, k is the radius of gyration, and is the distance from C.G. to the axis of rotation, and $\omega = \dot{\theta}$. Thus the kinetic energy of the system becomes

$$T = \frac{1}{2}\,m\,(l\dot{\theta})^2 + \frac{1}{2}\,mk^2\,\dot{\theta}^2 \;\Rightarrow\; T = \frac{1}{2}\,m\,(l^2 + k^2)\,\dot{\theta}^2.$$

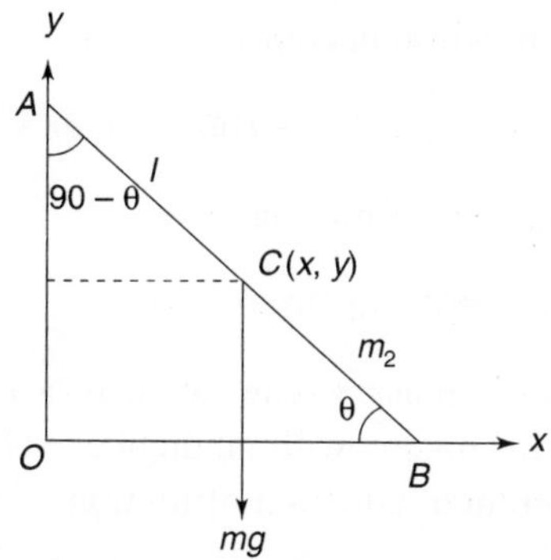

The potential energy of the sliding ladder is given by $V = mgl \sin \theta$. Hence the Lagrangian of the system becomes

$$L = \frac{1}{2} m \left(l^2 + k^2 \right) \dot{\theta}^2 - mgl \sin \theta. \qquad \ldots(361)$$

The Lagrange's equation of motion corresponding to the generalized co-ordinate θ gives

$$\ddot{\theta} = -\left(\frac{gl}{l^2 + k^2} \right) \cos \theta. \qquad \ldots(362)$$

Example 45:   A bead slides on a smooth rod which is rotating about one end in a vertical plane with uniform angular velocity ω. Find the Lagrangian and the Lagrange's equation of motion.

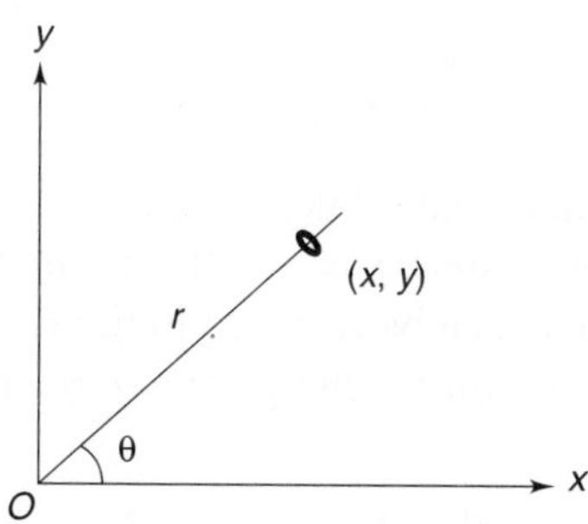

Solution: The system is shown in the figure. Let at any instant t, (x, y) be the co-ordinates of the bead, where $x = r \cos \omega t$, $y = r \sin \omega t$, ω is the angular velocity of the rod. The system is rheonomic and the constraint involved in the system is given by $y = x \tan \omega t$. The system has one degrees of freedom and hence it has one generalized co-ordinates and that is r. Thus the kinetic and potential energies of the particle are given by

$$T = \frac{1}{2} m \left(\dot{r}^2 + r^2 \omega^2 \right), \text{ and } V = mgr \sin \omega t.$$

Hence the Lagrangian function becomes

$$L = \frac{1}{2} m \left(\dot{r}^2 + r^2 \omega^2 \right) - mgr \sin \omega t. \qquad \ldots(363)$$

The r – Lagrange's equation of motion gives

$$\ddot{r} - r\omega^2 + g \sin \omega t = 0. \qquad \ldots(364)$$

Example 46: A particle of mass m can move in a frictionless thin circular wire of radius r. If the wire rotates with an angular velocity ω about a vertical diameter, deduce the differential equation of motion of the particle.

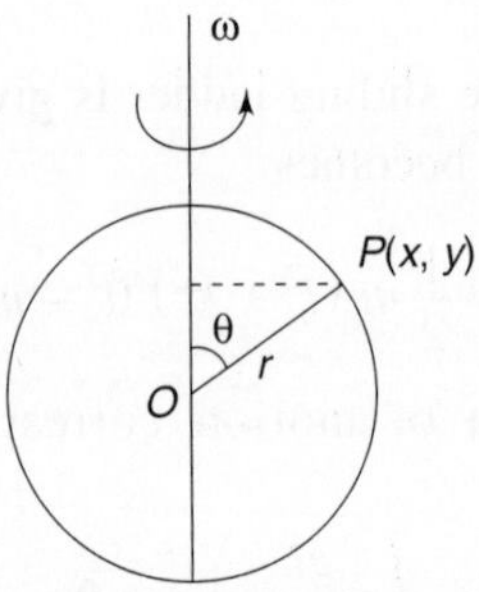

Solution: Let a particle of mass m be moving along a circular wire of radius r. Let $P(x, y)$ be the position of the particle at any instant and θ is the angle made by the radius with the vertical. Thus the constraint on the particle is scleronomic and is given by $x^2 + y^2 = r^2$. The corresponding transformation equations are given by

$$x = r \sin \theta, \quad y = r \cos \theta.$$

Clearly as r is the constant radius of the circle and θ is the only one generalized co-ordinate and hence the system has one degree of freedom. Therefore, the kinetic energy of the particle can be resolved in to the kinetic energy of motion of the particle of mass m and kinetic energy due to rotation of wire with angular velocity ω. Thus we have

$$T = \frac{1}{2} m \left(r\dot{\theta} \right)^2 + \frac{1}{2} I \omega^2, \qquad \ldots(365)$$

where the moment of inertia I of the particle is the product of the mass of the particle and the square of its perpendicular distance from the axis of rotation; and is given by

$$I = mx^2, \ i.e., \ I = mr^2 \sin^2 \theta.$$

Thus the kinetic energy of the system becomes

$$T = \frac{1}{2} mr^2 (\dot{\theta}^2 + \omega^2 \sin^2 \theta).$$

Similarly, the potential energy of the particle is given by

$$V = -mgr \, (1 - \cos \theta).$$

Hence the Lagrangian function becomes

$$L = \frac{1}{2} mr^2 (\dot{\theta}^2 + \omega^2 \sin^2 \theta) + mgr \, (1 - \cos \theta). \qquad \ldots(366)$$

The Lagrange's equation of motion corresponding to the generalized co-ordinate θ gives

$$\ddot{\theta} - \omega^2 \sin \theta \cos \theta - \frac{g}{r} \sin \theta = 0. \qquad \ldots(367)$$

Example 47:   A sphere of radius 'a' and mass m rests on the top of a fixed rough sphere of radius 'b'. The first sphere is slightly displaced so that it rolls without slipping. Obtain the equation of motion for rolling sphere.

Solution: Consider a sphere of radius 'a' rolling on another fixed sphere of radius 'b' without slipping. The system is shown in the figure. Let (x, y) be the position co-ordinates of the first sphere at any instant latter with respect to the co-ordinate system through the centre of the first sphere of radius 'b'. Thus we have

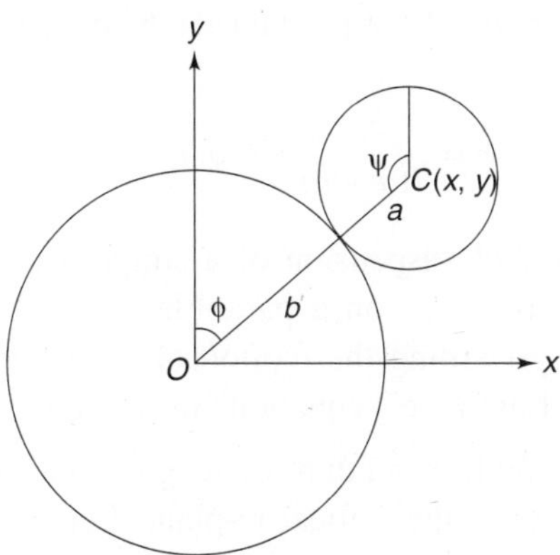

$$x = (a + b) \sin \phi, \; y = (a + b) \cos \phi.$$

The sphere of radius 'a' is constrained to move on another sphere of radius 'b'. The equation of constraint is $bd\phi = ad\psi$.

$$\Rightarrow \qquad b\dot{\phi} = a\dot{\psi}.$$

The rolling body has linear kinetic energy as well as rotational kinetic energy. Thus the kinetic energy of the rolling sphere is given by

$$T = \frac{1}{2} m \left(\dot{x}^2 + \dot{y}^2 \right) + \frac{1}{2} I\omega^2, \qquad\qquad ...(368)$$

where the moment of inertia I of the rolling sphere of radius 'a' is given by

$$I = \frac{2}{5} ma^2,$$

and ω is the angular velocity of the rolling sphere and is given by

$$\omega = \dot{\phi} + \dot{\psi} \text{ or } \omega = \frac{1}{a} \left(\dot{x}^2 + \dot{y}^2 \right)^{\frac{1}{2}}.$$

Thus the kinetic energy of the rolling sphere becomes

$$T = \frac{7}{10} m(a + b)^2 \, \dot{\phi}^2.$$

The potential energy of the system is given by

$$V = mg(a + b) \cos \phi.$$

Hence the Lagrangian of the rolling sphere becomes

$$L = \frac{7}{10} m \, (a + b)^2 \, \dot{\phi}^2 - mg(a + b) \cos \phi. \qquad\qquad ...(369)$$

The corresponding Lagrange's equation of motion gives

$$\frac{d}{dt} \left[\frac{7}{5} m \, (a + b)^2 \, \dot{\phi} \right] - mg \, (a + b) \sin \phi = 0.$$

$$\Rightarrow \qquad\qquad \ddot{\phi} = \frac{5g}{7(a + b)} \sin \phi. \qquad\qquad ...(370)$$

Example 48:   The point of suspension of a simple pendulum of length l and mass m is constrained to move on a parabola $z = ax^2$ in the vertical plane. Derive the Lagrangian governing the motion of the pendulum and its point of suspension. Obtain the Lagrange's equations of motion.

Solution: Consider a simple pendulum of length l and mass m constrained to move on a parabola $z = ax^2$ in the vertical xz-plane. Let (x_1, z_1) and (x_2, z_2) be the position co-ordinates of the point of suspension and the pendulum respectively. The system is shown in the figure. If θ is the angle made by the pendulum with the vertical, then we have

$$x_2 = x_1 + l \sin \theta,$$

$$z_2 = ax_1^2 - l \cos \theta.$$

The equations of the constraints are

$$z_1 = ax_1^2,$$

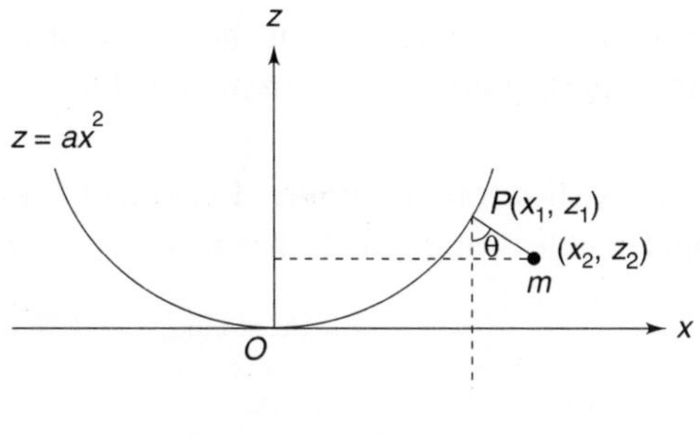

$$(z_2 - z_1)^2 + \left(x_2^2 - x_1^2\right)^2 = l^2.$$

We see that pendulum has two degrees of freedom and hence two generalized co-ordinates. The generalized co-ordinates are x_1 and θ. Hence the kinetic energy and potential energy of the pendulum are respectively given by

$$T = \frac{1}{2} m \left[\dot{x}_1^2 \left(1 + 4a^2 x_1^2\right) + l^2\dot{\theta}^2 + 2\dot{x}_1\dot{\theta}l \left(\cos\theta + 2ax_1 \sin\theta\right)\right],$$

$$V = mg \left(ax_1^2 - l \cos\theta\right).$$

Hence the Lagrangian of the motion of the system is given by

$$L = \frac{1}{2} m \left[\dot{x}_1^2 \left(1 + 4a^2 x_1^2\right) + l^2\dot{\theta}^2 + 2\dot{x}_1 \dot{\theta}l \left(\cos\theta + 2ax_1\sin\theta\right)\right] - $$
$$- mg (ax_1^2 - l \cos\theta). \qquad \ldots(371)$$

The x_1-Lagrange's equation of motion and θ-Lagrange's equation of motion give respectively

$$\frac{d}{dt}\left[m\dot{x}_1 \left(1 + 4a^2 x_1^2\right) + ml\dot{\theta} \left(\cos\theta + 2ax_1\sin\theta\right)\right] - $$
$$- m \left(4a^2x_1\dot{x}_1^2 + 2al\dot{x}_1\dot{\theta} \sin\theta - 2agx_1\right) = 0,$$

$$\ddot{x}_1 \left(1 + 4a^2x_1^2\right) + l\ddot{\theta} \left(\cos\theta + 2ax_1 \sin\theta\right) + 4a^2x_1 \dot{x}_1^2 + $$
$$+ \dot{\theta}^2l \left(2ax_1 \cos\theta - \sin\theta\right) + 2agx_1 = 0. \qquad \ldots(372)$$

and

$$\frac{d}{dt}\left[ml^2\dot{\theta} + ml\dot{x}_1(\cos\theta + 2ax_1 \sin\theta)\right] - $$
$$- m \left[l\dot{x}_1\dot{\theta} (2ax_1 \cos\theta - \sin\theta) - gl \sin\theta\right] = 0,$$

$$\ddot{x}_1l \left(\cos\theta + 2ax_1 \sin\theta\right) + l^2\ddot{\theta} + 2al \dot{x}_1^2 \sin\theta + lg \sin\theta = 0. \qquad \ldots(373)$$

Equations (372) and (373) describe the motion of the system completely.

Example 49: A mass m_2 hangs at one end of a string which passes over a fixed frictionless non-rotating pulley. At the other end of the string there is a non rotating pulley of mass m_1 over which there is a string carrying masses m_1' and m_2'. Set up the Lagrangian of the system and find the acceleration of the mass m_2.

Solution: The system is shown in the figure. Let l_1 and l_2 be the lengths of the upper and lower inextensible strings, such that $x_1 + x_2 = l_1$, $y_1 + y_2 = l_2$.

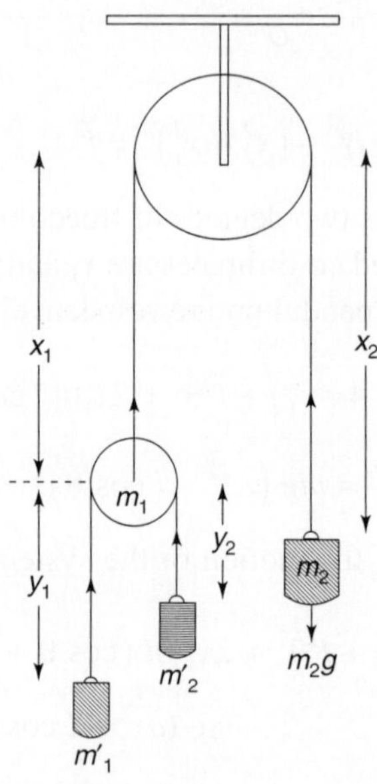

The total kinetic energy of the system is equal to the sum of the kinetic energies of masses m_1, m_2, m_1', m_2' and is given by

$$T = \frac{1}{2}\, m_1 \dot{x}_1^2 + \frac{1}{2}\, m_2 \dot{x}_2^2 + \frac{1}{2}\, m_1'(\dot{x}_1 + \dot{y}_1)^2 + \frac{1}{2}\, m_2'(\dot{x}_1 - \dot{y}_1)^2.$$

Similarly, the total potential energy of the system is equal to the sum of the potential energies of individual particles in the system and is given by

$$V = -m_1 g x_1 - m_2 g(l - x_1) - m_1' g(x_1 + y_1) - m_2'\, g(x_1 + l_2 - y_1).$$

Hence the Lagrangian of the motion becomes

$$L = \frac{1}{2}\, m_1 \dot{x}_1^2 + \frac{1}{2}\, m_2 \dot{x}_2^2 + \frac{1}{2}\, m_1'\, (\dot{x}_1 + \dot{y}_1)^2 + \frac{1}{2}\, m_2'\, (\dot{x}_1 - \dot{y}_1)^2 +$$

$$+ m_1 g x_1 + m_2 g(l - x_1) + m_1'\, g(x_1 + y_1) + m_2'\, g(x_1 + l_2 - y_1). \quad \ldots(374)$$

We see that x_1 and x_2 are the generalized co-ordinates. Hence the corresponding Lagrange's equations of motion reduce to the equations

$$(m_1 + m_2 + m_1' + m_2')\, \ddot{x}_1 + (m_1' - m_2')\, \ddot{y}_1 - g\, (m_1 - m_2 + m_1' + m_2') = 0, \quad \ldots(375)$$

and

$$(m_1' - m_2')\,\ddot{x}_1 + (m_1' + m_2')\,\ddot{y}_1 - g\,(m_1' - m_2') = 0. \qquad \ldots(376)$$

Solving these equations for $\ddot{x}_1,\ \ddot{y}_1$ we obtain

$$\ddot{x}_1 = g\,\frac{\left[(m_1 - m_2)\,(m_1' + m_2') + 4m_1'm_2'\right]}{(m_1 + m_2)\,(m_1' + m_2') + 4m_1'm_2'}. \qquad \ldots(377)$$

Thus the acceleration of mass m_2 is $\ddot{x}_2 = -\ddot{x}_1$, where $\ddot{x}_1$ is defined in equation (377). Similarly, the acceleration of mass m_1' is given by

$$\ddot{y}_1 = \frac{2m_2 g\,(m_1' - m_2')}{(m_1 + m_2)\,(m_1' + m_2') + 4m_1'm_2'}. \qquad \ldots(378)$$

Example 50: A particle is moving on a cycloid $s = 4a \sin\theta$ under the action of gravity. Obtain the Lagrangian and Lagrange's equation of motion.

Solution: A particle is moving on a cycloid under the action of gravity whose intrinsic equation is given by

$$s = 4a \sin\theta,$$

where θ is the angle made by the tangent to the curve with the horizontal. This gives

$$ds = 4a \cos\theta\,d\theta$$

$$\Rightarrow \qquad ds^2 = 16a^2 \cos^2\theta\,d\theta^2. \qquad \ldots(379)$$

Hence the kinetic energy of the particle is given by

$$T = \frac{1}{2}\,m\left(\frac{ds}{dt}\right)^2,$$

$$\Rightarrow \qquad T = 8a^2 m \cos^2\theta\,\dot{\theta}^2.$$

The infinitesimal distance between two neighbouring points on the cycloid is given by

$$ds^2 = dx^2 + dy^2, \text{ where } \frac{dy}{dx} = \tan\theta.$$

$$\Rightarrow \qquad \frac{dy}{ds} = \sin\theta \text{ and } \frac{dx}{ds} = \cos\theta,$$

$$\Rightarrow \qquad dy = \sin\theta\,ds,\ dx = \cos\theta\,ds,$$

$$\Rightarrow \qquad dy = 2a \sin 2\theta\,d\theta,\ dx = 4a \cos^2\theta\,d\theta.$$

Integrating these equations between the limits from the initial position $\theta = 0$ to the present position $\theta = \theta$, we get

$$x = a\,(2\theta + \sin 2\theta), \quad y = a\,(1 - \cos 2\theta).$$

These are the parametric equations of the cycloid. Hence the potential energy of the particle is therefore given by $V = mgy$. i.e.,

$$V = mga(1 - \cos 2\theta).$$

Thus the Lagrangian of motion of the particle is

$$L = 8a^2 m \cos^2 \theta\,\dot{\theta}^2 - mga(1 - \cos 2\theta). \qquad \ldots(380)$$

We see that the system has one degree of freedom and θ is the only generalized co-ordinate. Hence the Lagrange's equation of motion is obtained as

$$\ddot{\theta} - \tan\theta\,\dot{\theta}^2 + \left(\frac{g}{4a}\right)\tan\theta = 0. \qquad \ldots(381)$$

Example 51:   A bead of mass m sliding smoothly along a wire of shape $z = f(x)$, where z-axis and x-axis being respectively the vertical and horizontal. Show that the Lagrangian is given by $L = \dfrac{1}{2}\,m\dot{x}^2\left[1 + f'^2(x)\right] - mgf(x)$ and hence find the equation of motion.

Solution: A bead of mass m sliding smoothly along a plane curve $z = f(x)$ in the xz-plane, where x is horizontal and z vertical. Hence the system has only one degree of freedom and hence one generalized co-ordinate. x is the only generalized coordinate. The kinetic energy of the particle is given by

$$T = \frac{1}{2}\,m\,(\dot{x}^2 + \dot{z}^2),$$

where $$z = f(x) \Rightarrow \dot{z} = f'(x)\dot{x}.$$

Hence the kinetic energy becomes

$$T = \frac{1}{2}\,m\dot{x}^2\left[1 + f'^2(x)\right].$$

The potential energy is given by

$$V = mgf(x).$$

Hence the Lagrangian of the system is given by

$$L = \frac{1}{2}\,m\dot{x}^2\left[1 + f'^2(x)\right] - mgf(x). \qquad \ldots(382)$$

The corresponding Lagrange's equation of motion yield

$$m\ddot{x}\left[1 + f'^2(x)\right] + m\dot{x}^2 f'(x) f''(x) + mgf'(x) = 0. \qquad \ldots(383)$$

Example 52: A Lagrangian for a system with two degrees of freedom is

$$L = \dot{q}_1^2 + \dot{q}_2^2 - \dot{q}_1\dot{q}_2 - q_1^2 - q_2^2.$$

Obtain the Lagrange's equations of motion and solve them.

Solution: We see from the given Lagrangian that q_1, q_2 are the generalized co-ordinates. The corresponding Lagrange's equations of motion give two equations

$$2\ddot{q}_1 - \ddot{q}_2 = -2q_1, \qquad \ldots(384)$$

$$-\ddot{q}_1 + 2\ddot{q}_2 = -2q_2. \qquad \ldots(385)$$

Adding and subtracting equations (384) and (385) we get

$$\ddot{q}_1 + \ddot{q}_2 = -2(q_1 + q_2), \qquad \ldots(386)$$

$$3(\ddot{q}_1 - \ddot{q}_2) = -2(q_1 - q_2). \qquad \ldots(387)$$

Define $X = q_1 + q_2$ and $Y = q_1 - q_2$. Then the equations (386) and (387) reduce to

$$\ddot{X} + 2X = 0, \qquad \ldots(388)$$

and $$3\ddot{Y} + 2Y = 0. \qquad \ldots(389)$$

Equations (388) and (389) are the required Lagrange's equations of motion. These equations are ordinary second order differential equations, whose solutions are obtained as

$$X = q_1 + q_2 = c_1 \cos\left(\sqrt{2}t\right) + c_2 \sin\left(\sqrt{2}t\right), \qquad \ldots(390)$$

and $$Y = q_1 - q_2 = c_3 e^{\sqrt{\frac{2}{3}}t} + c_4 e^{-\sqrt{\frac{2}{3}}t} \qquad \ldots(391)$$

Now adding and subtracting equations (390) and (391), we get

$$q_1 = \frac{1}{2}\left[c_1 \cos\left(\sqrt{2}t\right) + c_2 \sin\left(\sqrt{2}t\right) + c_3 e^{\sqrt{\frac{2}{3}}t} + c_4 e^{-\sqrt{\frac{2}{3}}t}\right], \qquad \ldots(392)$$

$$q_2 = \frac{1}{2}\left[c_1 \cos\left(\sqrt{2}t\right) + c_2 \sin\left(\sqrt{2}t\right) - c_3 e^{\sqrt{\frac{2}{3}}t} - c_4 e^{-\sqrt{\frac{2}{3}}t}\right]. \qquad \ldots(393)$$

Example 53: Find the Lagrangian and the equations of motion for swinging Atwood machine.

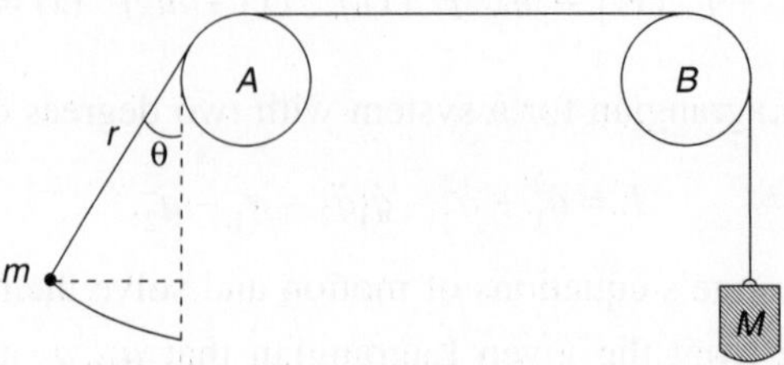

Solution: Consider two masses M and m tied at the ends of an inextensible string passes over a pair of horizontally placed frictionless and weightless pulleys. Let l be the length of the string. The heavier mass M can move only up or down, where as the lighter mass m can oscillate in a vertical plane as shown in the figure.

Let (x, y) be the position co-ordinates of the oscillating mass. Then from the figure, we have

$$x = r \sin \theta, \ y = r \cos \theta,$$

where as the position of the heavier mass M is determined by $l - (AB + r)$. We see that the system has two degrees of freedom and hence two generalized co-ordinates r and θ. Hence the kinetic energy of the system is given by

$$T = \frac{1}{2}(M + m)\dot{r}^2 + \frac{1}{2}mr^2\dot{\theta}^2.$$

The potential energy of the system is sum of the potential energies of two masses M and m and is given by

$$V = -mgr \cos \theta - Mg(l - r - AB).$$

We drop the constant term in the potential energy expression as it will not contribute in the equation of motion and write the Lagrangian of the system as

$$L = \frac{1}{2}(M + m)\dot{r}^2 + \frac{1}{2}mr^2\dot{\theta}^2 - gr(M - m \cos \theta) \qquad \ldots(394)$$

The Lagrange's equations of motion corresponding to two generalized co-ordinates r and θ are respectively given by

$$(M + m)\ddot{r} - mr\dot{\theta}^2 + g(M - m \cos \theta) = 0,$$

$$r\ddot{\theta} + 2\dot{r}\dot{\theta} - g \sin \theta = 0. \qquad \ldots(395)$$

Example 54: Find the equation of motion for the Lagrangian

$$L = \frac{1}{2}e^{\alpha t}\left(m\dot{q}^2 - kq^2\right), \ \alpha, \ m, \ k \text{ are constants.}$$

Suppose a point transformation $s = e^{\alpha t}q$ is made, what is the effective Lagrangian in terms of s. Find the equation of motion for s. Is there any change in the eigen values of motion?

Solution: We see from the Lagrangian given in the question that q is the only generalized co-ordinate. Hence the corresponding Lagrange's equation of motion gives

$$\ddot{q} + \alpha\dot{q} + \frac{k}{m}\,q = 0. \qquad \ldots(396)$$

Suppose now a point transformation from q to s is made by the transformation equation

$$s = e^{\alpha t}q. \qquad \ldots(397)$$

To find the equation of motion in s, differentiate equation (397) with respect to time t to get the equation

$$\dot{s} = e^{\alpha t}\dot{q} + \alpha e^{\alpha t}q. \qquad \ldots(398)$$

Eliminating q between the equations (397) and (398) we obtain

$$\dot{q} = (\dot{s} - \alpha s)e^{-\alpha t}. \qquad \ldots(399)$$

Now eliminating q and $\dot{q}$ from the given Lagrangian function L, we obtain the expression for the effective Lagrangian in the transformed co-ordinate s in the form

$$L = \frac{1}{2}\,e^{-\alpha t}\left[m(\dot{s} - \alpha s)^2 - ks^2\right]. \qquad \ldots(400)$$

We see that s is the generalized co-ordinate. Hence the s-Lagrange's equation of motion gives

$$\ddot{s} - \alpha\dot{s} + \left(\frac{k}{m}\right)s = 0. \qquad \ldots(401)$$

We notice from equations (396) and (401) that the equations of motion in q and s are not the same. Hence the eigen values of the two motion are not identical.

Example 55: Consider a Lagrangian $L(q_j, \dot{q}_j, t)$, where q_j, $j = 1, 2, 3, \ldots, n$ is a set of independent generalized co-ordinates for a system of n degrees of freedom. Suppose q_j are transformed to another set of independent co-ordinates $s_1, s_2, s_3, \ldots, s_n$, by means of a point transformation equation

$$q_i = q_i\,(s_1, s_2, \ldots, s_n, t), \; i = 1, 2, 3, \ldots, n.$$

Show that L satisfies Lagrange's equations with respect to the s co-ordinates

$$\frac{d}{dt}\left(\frac{\partial L}{\partial \dot{s}_j}\right) - \frac{\partial L}{\partial s_j} = 0.$$

Solution: A Lagrangian for a system of n degrees of freedom is given by

$$L = L(q_i, \dot{q}_i, t), \qquad \qquad \ldots(402)$$

where

$$q_i = q_i(s_1, s_2, \ldots, s_n, t), \, i = 1, 2, 3, \ldots, n. \qquad \ldots(403)$$

Differentiating equation (403) with respect to time t, we obtain

$$\dot{q}_i = \sum_k \frac{\partial q_i}{\partial s_k} \dot{s}_k + \frac{\partial q_i}{\partial t}. \qquad \qquad \ldots(404)$$

Now differentiating equation (404) with respect to s_j, we get

$$\frac{\partial \dot{q}_i}{\partial s_j} = \sum_k \frac{\partial^2 q_i}{\partial s_k \partial s_j} \dot{s}_k + \frac{\partial^2 q_i}{\partial t \partial s_j}. \qquad \qquad \ldots(405)$$

Also differentiating equation (402) with respect to s_j, we get

$$\frac{\partial L}{\partial s_j} = \sum_i \frac{\partial L}{\partial q_i} \frac{\partial q_i}{\partial s_j} + \sum_i \frac{\partial L}{\partial \dot{q}_i} \frac{\partial \dot{q}_i}{\partial s_j}. \qquad \ldots(406)$$

Using the equation (405) in to the equation (406), we get

$$\frac{\partial L}{\partial s_j} = \sum_i \frac{\partial L}{\partial q_i} \frac{\partial q_i}{\partial s_j} + \sum_i \frac{\partial L}{\partial \dot{q}_i}\left(\sum_k \frac{\partial^2 q_i}{\partial s_k \partial s_j} \dot{s}_k + \frac{\partial^2 q_i}{\partial t \partial s_j}\right),$$

$$\frac{\partial L}{\partial s_j} = \sum_i \frac{\partial L}{\partial q_i} \frac{\partial q_i}{\partial s_j} + \sum_{i,k} \frac{\partial L}{\partial \dot{q}_i} \frac{\partial^2 q_i}{\partial s_k \partial s_j} \dot{s}_k + \sum_i \frac{\partial L}{\partial \dot{q}_i} \frac{\partial^2 q_i}{\partial t \partial s_j}. \qquad \ldots(407)$$

Now we find from the equation (402)

$$\frac{\partial L}{\partial \dot{s}_j} = \sum_i \frac{\partial L}{\partial \dot{q}_i} \frac{\partial \dot{q}_i}{\partial \dot{s}_j},$$

where from the equation (404) we have

$$\frac{\partial \dot{q}_i}{\partial \dot{s}_j} = \frac{\partial q_i}{\partial s_j}.$$

Consequently, we have

$$\frac{\partial L}{\partial \dot{s}_j} = \sum_i \frac{\partial L}{\partial \dot{q}_i} \frac{\partial q_i}{\partial s_j}. \qquad \qquad \text{...(408)}$$

Differentiating the equation (408) with respect to time t, we obtain

$$\frac{d}{dt}\left(\frac{\partial L}{\partial \dot{s}_j}\right) = \sum_i \frac{d}{dt}\left(\frac{\partial L}{\partial \dot{q}_i}\right) \frac{\partial q_i}{\partial s_j} + \sum_{i,k} \frac{\partial L}{\partial \dot{q}_i} \frac{\partial^2 q_i}{\partial s_j \partial s_k} \dot{s}_k + \sum_i \frac{\partial L}{\partial \dot{q}_i} \frac{\partial^2 q_i}{\partial s_j \partial t}. \qquad \text{...(409)}$$

Subtracting equation (407) from (408), we get

$$\frac{d}{dt}\left(\frac{\partial L}{\partial \dot{s}_j}\right) - \frac{\partial L}{\partial s_j} = \sum_i \left[\frac{d}{dt}\left(\frac{\partial L}{\partial \dot{q}_i}\right) - \frac{\partial L}{\partial q_i}\right] \frac{\partial q_i}{\partial s_j} = 0,$$

$$\Rightarrow \qquad \frac{d}{dt}\left(\frac{\partial L}{\partial \dot{s}_j}\right) - \frac{\partial L}{\partial s_j} = 0.$$

This shows that the Lagrangian satisfies the Lagrange's equation of motion with respect to the co-ordinate s_j.

Example 56: Obtain the expression for kinetic energy of a particle constrained to move on a horizontal xy-plane which is rotating about the vertical z-axis with angular velocity ω. Show that

$$\dot{x}\frac{\partial T}{\partial \dot{x}} + \dot{y}\frac{\partial T}{\partial \dot{y}} = 2T_2 + T_1,$$

where

$$T_1 = m\omega \, (x\dot{y} - y\dot{x}), \quad T_2 = \frac{1}{2} m \, (\dot{x}^2 + \dot{y}^2).$$

Show also that the Lagrangian of the particle is given by

$$L = \frac{1}{2} m \left[(\dot{x} - \omega y)^2 + (\dot{y} + \omega x)^2 \right] - V \, (x, y),$$

hence find the equations of motion.

Solution: Consider a particle is moving on the xy-plane and the plane itself is rotating with respect to z-axis with angular velocity ω. Let (x_1, y_1, z_1) be the co-ordinates of the particle with respect to the fixed co-ordinate system and (x, y, z) the co-ordinates of the particle with respect to rotating axes. The co-ordinates with respect to the rotating axes are taken as the generalized co-ordinates. The transformation equations for rotation about z-axis are given by

$$x_1 = x \cos \omega t - y \sin \omega t,$$
$$y_1 = x \sin \omega t + y \cos \omega t, \qquad \qquad \text{...(410)}$$
$$z_1 = z. \qquad \qquad \text{...(411)}$$

The equation of the constraint is z = const., hence the system has only two degrees of freedom and hence only two generalized co-ordinates x and y. We note here that the transformation the equations (410) are time dependent, though the constraint equation (411) is not. Thus the kinetic energy of the particle is given by

$$T = \frac{1}{2} m \left(\dot{x}_1^2 + \dot{y}_1^2 \right). \qquad \qquad ...(412)$$

Differentiating the equations (410) with respect to time t and putting in the equation (412) we obtain

$$T = \frac{1}{2} m \left[(\dot{x}^2 + \dot{y}^2) + \omega^2 (x^2 + y^2) + 2\omega (x\dot{y} - y\dot{x}) \right],$$

$$T = \frac{1}{2} m \left[(\dot{x} - \omega y)^2 + (\dot{y} + \omega x)^2 \right]. \qquad \qquad ...(413)$$

We can also write this equation as

$$T = T_2 + T_1 + T_0,$$

where $\quad T_2 = \frac{1}{2} m \left(\dot{x}^2 + \dot{y}^2 \right), \ T_1 = m\omega \left(x\dot{y} - y\dot{x} \right), \ T_0 = \frac{1}{2} m\omega^2 \left(x^2 + y^2 \right).$

Differentiating the equation (413) with respect to $\dot{x}$, $\dot{y}$ we get

$$\frac{\partial T}{\partial \dot{x}} = m\dot{x} - \omega m y, \ \frac{\partial T}{\partial \dot{y}} = m\dot{y} + \omega m x.$$

This gives on solving

$$\dot{x} \frac{\partial T}{\partial \dot{x}} + \dot{y} \frac{\partial T}{\partial \dot{y}} = 2T_2 + T_1. \qquad \qquad ...(414)$$

Now if V is the potential energy of the particle which is function of the generalized co-ordinates x, y then the Lagrangian of the particle is given by

$$L = \frac{1}{2} m \left[(\dot{x} - \omega y)^2 + (\dot{y} + \omega x)^2 \right] - V(x, y). \qquad \qquad ...(415)$$

Direct calculations of the Lagrange's equations of motion (81) lead to the two equations of motion,

$$m\ddot{x} - m\omega\dot{y} - m\omega^2 x = -\frac{\partial V}{\partial x},$$

$$m\ddot{y} + m\omega\dot{x} - m\omega^2 y = -\frac{\partial V}{\partial y}.$$

We see from the equations of motion that there appear other forces with components $(-m\omega\dot{y}, m\omega\dot{x})$ and $(-m\omega^2 x, -m\omega^2 y)$ are respectively called the coriolis force and the centrifugal force.

Example 57: The Lagrangian of a system is

$$L = \frac{m}{2}\left(a\dot{x}^2 + 2b\dot{x}\dot{y} + c\dot{y}^2\right) - \frac{k}{2}\left(ax^2 + 2bxy + cy^2\right), \quad a, b, c$$

are arbitrary constants such that $b^2 - 4ac \neq 0$. Write down the equation of motion. Examine the two cases $a = 0$, $c = 0$ and $b = 0$, $c = -a$ and interpret physically.

Solution: We notice from the given Lagrangian that, x and y are the generalized co-ordinates. Hence the corresponding Lagrange's equations of motion are

$$m(a\ddot{x} + b\ddot{y}) + k(ax + by) = 0, \qquad\qquad \dots(416)$$

$$m(b\ddot{x} + c\ddot{y}) + k(bx + cy) = 0. \qquad\qquad \dots(417)$$

Case (i) Let $a = 0$, $c = 0$. The equations of motion (416) and (417) reduce to

$$\ddot{y} + \left(\frac{k}{m}\right)y = 0 \text{ and } \ddot{x} + \left(\frac{k}{m}\right)x = 0. \qquad\qquad \dots(418)$$

Case (ii) Let $b = 0$, $c = -a$. Putting this in equations (416) and (417), we get

$$\ddot{x} + \left(\frac{k}{m}\right)x = 0 \text{ and } \ddot{y} + \left(\frac{k}{m}\right)y = 0. \qquad\qquad \dots(419)$$

We see from the equations (418) and (419) that in both the cases we get the same set of equations of motion. These are the differential equations of particle performing a linear simple harmonic motion. The solution of these equations gives the displacement of the particle with the frequency of oscillation

$$\omega = \sqrt{\frac{k}{m}}.$$

Example 58: Find the equation of motion of a particle of mass m moving on the surface of a cone of semi-vertical angle θ under gravitational force.

Solution: A particle of mass m is moving under gravity on the surface of a cone given by

$$x^2 + y^2 = z^2 \tan^2 \theta, \qquad\qquad \dots(420)$$

where θ is a constant semi-vertical angle of the cone. If $P(x, y, z)$ is the position of the particle at any instant t then the parametric equations of the cone are given by

$$x = r \sin \theta \cos \phi, \quad y = r \sin \theta \sin \phi, \quad z = r \cos \theta, \qquad\qquad \dots(421)$$

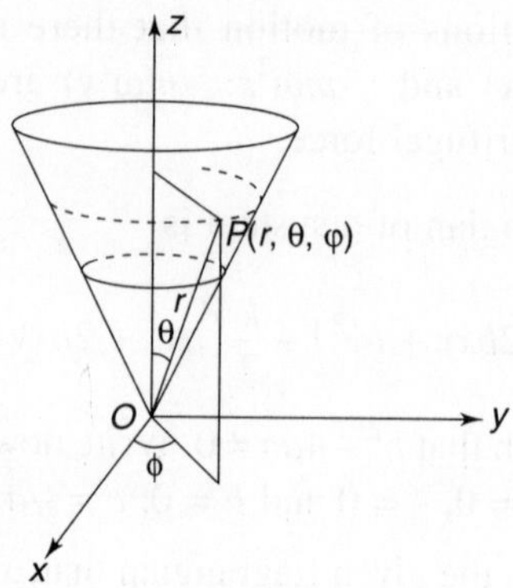

where θ is a constant. We see that r, ϕ are the generalized co-ordinates. The kinetic energy and the potential energy of the particle are given by

$$T = \frac{1}{2} m \left(\dot{r}^2 + r^2 \sin^2 \theta \dot{\phi}^2\right), \quad V = mgr \cos \theta.$$

Hence the Lagrangian of the particle is

$$L = \frac{1}{2} m \left(\dot{r}^2 + r^2 \sin^2 \theta \dot{\phi}^2\right) - mgr \cos \theta. \qquad \ldots(422)$$

We observe that ϕ is cyclic co-ordinate in the Lagrangian, hence the corresponding generalized momentum is conserved (refer the Theorem 11). *i.e.,*

$$p_\phi = \frac{\partial L}{\partial \dot{\phi}} = mr^2 \sin^2 \theta \dot{\phi} = \text{const.} \qquad \ldots(423)$$

The r - Lagrange's equation of motion gives

$$\frac{d}{dt} (m\dot{r}) - \left(m \sin^2 \theta \dot{\phi}^2 - mg \cos \theta\right) = 0$$

$$\Rightarrow \qquad m\ddot{r} - mr \sin^2 \theta \dot{\phi}^2 + mg \cos \theta = 0$$

$$\Rightarrow \qquad \ddot{r} - r \sin^2 \theta \dot{\phi}^2 + g \cos \theta = 0 \qquad \ldots(424)$$

Remark: Parametric representation of a surface is not unique. One can also represent cone (420) by the equations

$$x = r \cos \phi, \ y = r \sin \phi, \ z = r \cot \theta. \quad \ldots(425)$$

However, the Lagrangian function of the particle becomes

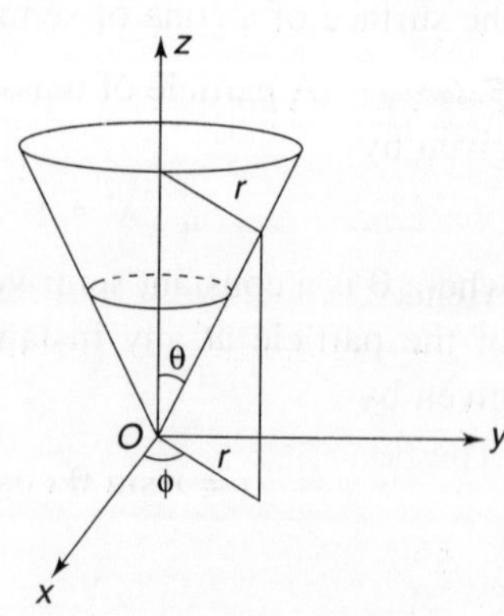

$$L = \frac{1}{2} m \, (\dot{r}^2 \, \text{cosec}^2 \, \theta + r^2 \, \dot{\phi}^2) - mgr \cot \theta, \qquad \qquad ...(426)$$

and the equation of motion is

$$\ddot{r} - r \sin^2 \theta \dot{\phi}^2 + g \cos \theta \sin \theta = 0. \qquad \qquad ...(427)$$

Example 59: Find the Lagrangian and the equations of motion of a particle of mass m constrained to move on the surface obtained by revolving the parabola $y^2 = 4ax$ about x-axis.

Solution: The surface obtained by revolving the parabola $y^2 = 4ax$ about x-axis is given by the equation

$$y^2 + z^2 = 4ax.$$

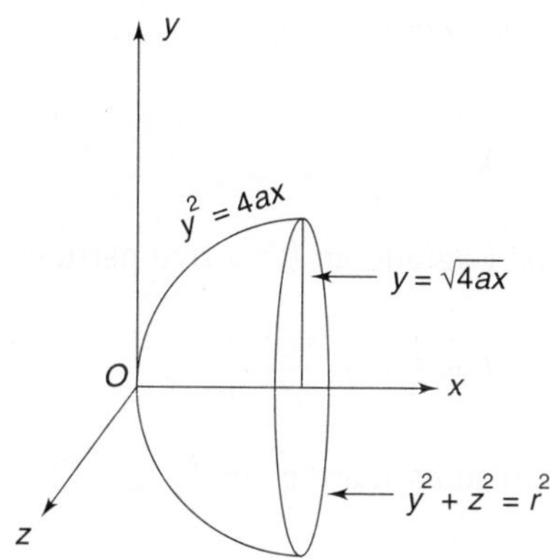

The parametric equations representing this surface are given by

$$x = au^2, \ y = 2au \sin v, \ z = 2au \cos v, \qquad \qquad ...(428)$$

where u, v are the parameters of the curve. The kinetic energy of the particle is given by

$$T = \frac{1}{2} m \, (\dot{x}^2 + \dot{y}^2 + \dot{z}^2). \qquad \qquad ...(429)$$

Where from equations (428) we obtain

$$\dot{x} = 2au\dot{u}, \ \dot{y} = 2a\dot{u} \sin v + 2au \cos v \dot{v}, \ \dot{z} = 2a\dot{u} \cos v - 2au \sin v \dot{v}.$$

Hence the kinetic energy of the particle becomes

$$T = 2a^2 m \left[u^2 \dot{v}^2 + (1 + u^2)\dot{u}^2 \right]. \qquad \qquad ...(430)$$

The potential energy of the particle is given by

$$V = mgy = mgau \sin v. \qquad \qquad ...(431)$$

Hence the Lagrange's function of the particle becomes

$$L = 2a^2m\left[u^2\dot{v}^2 + (1 + u^2)\dot{u}^2\right] - 2amgu\sin v. \qquad \text{...(432)}$$

We see from equation (432) that u and v are the generalized co-ordinates, hence the corresponding Lagrange's equations of motion are given by the equations

$$\frac{d}{du}\left[4a^2m(1 + u^2)\,\dot{u}\right] - \left[4a^2mu\,\dot{u}^2 + 4a^2mu\,\dot{v}^2 - 2amg\sin v\right] = 0,$$

$$\Rightarrow \quad (1 + u^2)\ddot{u} + u(\dot{u}^2 - \dot{v}^2) + \frac{g}{2a}\sin v = 0. \qquad \text{...(433)}$$

$$\frac{d}{du}\left(4a^2mu^2\dot{v}\right) + 2amgu\cos v = 0,$$

$$\Rightarrow \quad \ddot{v} + \frac{2\dot{v}}{u} + \frac{g\cos v}{2au} = 0. \qquad \text{...(434)}$$

Example 60:   Taking the Lagrangian for a free particle in the form

$$L = \frac{1}{2}\,mg_{ik}\,\frac{dx^i}{dt}\,\frac{dx^k}{dt},$$

show that the equations of motion are given by

$$\ddot{x}^i + \Gamma^i_{jk}\,\dot{x}^j\dot{x}^k = 0,$$

where $\Gamma^i_{jk} = g^{il}\,\Gamma_{jk,l} = \frac{1}{2}g^{il}\left(\dfrac{\partial g_{kl}}{\partial x^j} + \dfrac{\partial g_{jl}}{\partial x^k} - \dfrac{\partial g_{jk}}{\partial x^l}\right).$

Solution: A free particle is one upon which no external force is acting. Hence the potential energy of the particle is zero or at the most constant. Therefore the Lagrangian of the particle is essentially equal to its kinetic energy, and is given by

$$L = \frac{1}{2}\,m\left(\frac{ds}{dt}\right)^2. \qquad \text{...(435)}$$

However, the metric on the Riemannian space is given by

$$ds^2 = g_{ik}dx^idx^k,$$

where g_{ik} are called the components of the metric tensor and are functions of co-ordinates.

Hence the Lagrangian function becomes

$$L = \frac{1}{2} m g_{ik}\, \dot{x}^i \dot{x}^k, \ \dot{x}^i = \frac{dx^i}{dt} \qquad \ldots(436)$$

The Lagrange's equations of motion in this case are

$$\frac{d}{dt}\left(\frac{\partial L}{\partial \dot{x}^i}\right) - \frac{\partial L}{\partial x^i} = 0,\ i = 1, 2, 3, \ldots, n, \qquad \ldots(437)$$

where from equation (437) we find

$$\frac{\partial L}{\partial x^j} = \frac{1}{2}\, m\, \frac{\partial g_{ik}}{\partial x^j}\, \dot{x}^j\, \dot{x}^k \ \text{and} \ \frac{\partial L}{\partial \dot{x}^j} = m g_{ij}\dot{x}^j.$$

Consequently, equation of motion becomes

$$\frac{d}{dt}\left(g_{ij}\dot{x}^i\right) - \frac{1}{2}\frac{\partial g_{ik}}{\partial x^j}\, \dot{x}^i \dot{x}^k = 0,$$

$$\Rightarrow \qquad g_{ij}\ddot{x}^j + \frac{1}{2}\left(\frac{\partial g_{ij}}{\partial x^k} + \frac{\partial g_{kj}}{\partial x^i} - \frac{\partial g_{ik}}{\partial x^j}\right)\dot{x}^i\, \dot{x}^k = 0.$$

i.e. $\qquad g_{ij}\ddot{x}^i + \Gamma_{ik,\, j}\, \dot{x}^i\, \dot{x}^k = 0, \qquad \ldots(438)$

where $\qquad \Gamma_{ik,j} = \frac{1}{2}\left(\frac{\partial g_{kj}}{\partial x^i} + \frac{\partial g_{ij}}{\partial x^k} - \frac{\partial g_{ik}}{\partial x^j}\right)$

are the components of Christoffel symbol of first kind. Multiplying equation (438) by g^{jl}, and using the formula

$$g^{jl} g_{ij} = \delta_i^l,$$

where δ_i^l is the Kroneker delta symbol defined by

$$\delta_i^l = 1, \text{ when } i = l,$$

$$0 = \text{ when } i \neq l,$$

we get

$$\ddot{x}^l + \Gamma_{ik}^l\, \dot{x}^i \dot{x}^k = 0, \qquad \ldots(439)$$

where $\qquad \Gamma_{ik}^l = g^{jl}\, \Gamma_{ik,j}$

are the components of Christoffel symbol of second kind. Equations (439) are the required equations of motion of a free particle in a Riemannian space.

Generalized Momentum and Cyclic Co-ordinates

𝔇𝔢𝔣𝔦𝔫𝔦𝔱𝔦𝔬𝔫𝔰: In Newtonian mechanics the components of momentum (linear) are defined as the derivative of kinetic energy with respect to the corresponding components of velocity.

That is if

$$T = \frac{1}{2} m \left(\dot{x}^2 + \dot{y}^2 + \dot{z}^2 \right)$$

is the kinetic energy of a particle, then the components of momentum of the particle are defined as

$$p_x = \frac{\partial T}{\partial \dot{x}} = m\dot{x}, \; p_y = \frac{\partial T}{\partial \dot{y}} \, m\dot{y}, \; p_z = \frac{\partial T}{\partial \dot{z}} = m\dot{z} \qquad ...(440)$$

Generalized Momentum

The Newtonian momentum defined as the product of mass and the velocity is obtained by differentiating the kinetic energy with the components of velocity.

i.e.,
$$p_j = \frac{\partial T}{\partial \dot{q}_j}.$$

This momentum would corresponds to $\dfrac{\partial L}{\partial \dot{q}_j}$ provided the potential energy

V, which is a part of the Lagrangian, is independent of both the generalized velocities $\dot{q}_j$ and time t. Thus we have

$$p_j = \frac{\partial L}{\partial \dot{q}_j} = \frac{\partial T}{\partial \dot{q}_j}.$$

Thus for a conservative system in which the forces are derivable from a potential function V which is independent of generalized velocities, the quantity

$$p_j = \frac{\partial L}{\partial \dot{q}_j} \qquad\qquad ...(441)$$

is called the generalized momentum associated with the generalized co-ordinates q_j.

Note 1:   The definition of generalized momentum (441) is exactly analogous to the usual definition of momentum (440).

Note 2:   The word 'generalized' momentum subsumes linear momentum and angular momentum of the particle.

e.g. To illustrate the case, let a particle be moving in plane polar co-ordinates (r, θ). Then its kinetic energy is given by

$$T = \frac{1}{2} m \left(\dot{r}^2 + r^2 \dot{\theta}^2 \right).$$

Hence the generalized momentum corresponding to the generalized co-ordinate r and θ are respectively given by

$$p_r = \frac{\partial T}{\partial \dot{r}} = m\dot{r}, \; p_\theta = \frac{\partial T}{\partial \dot{\theta}} = mr\dot{\theta}. \qquad \text{...(442)}$$

We notice that p_r and p_θ respectively represent the linear momentum and angular momentum of the particle.

If on the other hand, the system is non-conservative, there arises a velocity dependent potential (generalized potential) U, then the Lagrangian function is given by $L = T - U$.

However, in a large number of realistic situations the generalized potential energy is given by

$$U = \sum_i V_i \dot{q}_i + V, \qquad \text{...(443)}$$

where V_i and the ordinary potential function V are all functions of generalized coordinates q_j only. In this case the generalized momentum becomes

$$p_j = \frac{\partial}{\partial \dot{q}_j} (T - U),$$

$$= \frac{\partial T}{\partial \dot{q}_j} - \frac{\partial}{\partial \dot{q}_j} \left(\sum_i V_i \dot{q}_i + V \right),$$

$$p_j = \frac{\partial T}{\partial \dot{q}_j} - V_j. \qquad \text{...(444)}$$

This shows that the generalized momentum does not arise totally from the kinetic energy term but also from generalized potential energy term.

The components of generalized force defined in the equation (66) in this case becomes

$$Q_j = \frac{d}{dt} \left[\frac{\partial}{\partial \dot{q}_j} \left(\sum_i V_i \dot{q}_i + V \right) \right] - \frac{\partial}{\partial q_j} \left(\sum_i V_i \dot{q}_i + V \right),$$

$$Q_j = \frac{dV_j}{dt} - \sum_i \frac{\partial V_i}{\partial q_j} \dot{q}_i - \frac{\partial V}{\partial q_j},$$

$$Q_j = \sum_i \left(\frac{\partial V_j}{\partial q_i} - \frac{\partial V_i}{\partial q_j} \right) \dot{q}_i - \frac{\partial V}{\partial q_j}. \qquad \ldots(445)$$

The first term in the equation (445) arises due the gyroscopic force while the second term is the usual potential force. Thus in this case the generalized force can be thought of as the sum of these two forces.

Cyclic or Ignorable Co-ordinates

Co-ordinates which are absent in the Lagrangian are called cyclic or ignorable co-ordinates, although the Lagrangian may contain the corresponding generalized velocity of the particle.

Conservation Theorem for Generalized Momentum

Theorem 11   Show that the generalized momentum corresponding to a cyclic co-ordinate is conserved.

Proof: If the generalized co-ordinate q_j is cyclic in L, then it must be absent in the Lagrangian. Obviously we have therefore, $\dfrac{\partial L}{\partial q_j} = 0$. Thus the Lagrange's equation of motion (81) becomes

$$\frac{d}{dt} \left(\frac{\partial L}{\partial \dot{q}_j} \right) = 0 \Rightarrow \dot{p}_j = 0.$$

$$\Rightarrow \qquad\qquad p_j = \text{const.}$$

This proves that the generalized momentum corresponding to the cyclic co-ordinate is conserved.

Conservation Theorem for Linear Momentum

We will show that the conservation Theorems are continued to be true for cyclic generalized co-ordinates.

Theorem 12   If the cyclic generalized co-ordinate q_j is such that dq_j represents the translation of the system, then prove that the total linear momentum is conserved.

Proof: Consider a conservative system so that the potential energy V is a function of generalized co-ordinates only.

i.e., $$V = V(q_j) \Rightarrow \frac{\partial V}{\partial \dot{q}_j} = 0.$$

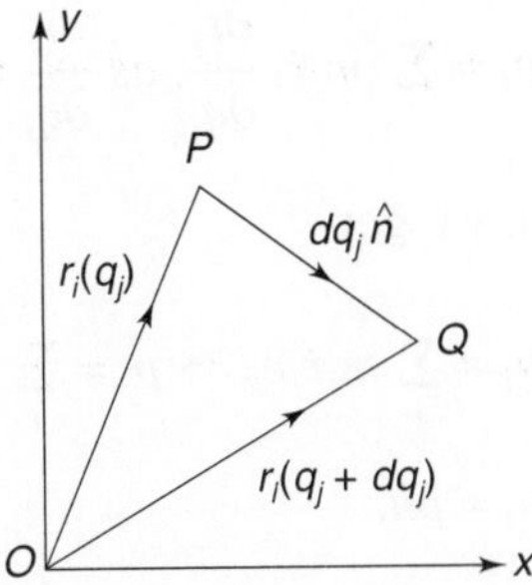

Let $P = \bar{r}_i(q_j)$ be the initial position of the system and let it be translated to a point $Q = \bar{r}_i(q_j + dq_j)$, so that $PQ = dq_j\,\hat{n}$, where $\hat{n}$ is the unit vector along the direction of translation and dq_j represents the translation of the system.

We know by the first principle that

$$\frac{\partial \bar{r}_i}{\partial q_j} = \lim_{dq_j \to 0} \frac{\bar{r}_i(q_j + dq_j) - \bar{r}_i(q_j)}{dq_j}$$

$$\frac{\partial \bar{r}_i}{\partial q_j} = \lim_{dq_j \to 0} \frac{PQ}{dq_j}$$

$$\frac{\partial \bar{r}_i}{\partial q_j} = \lim_{dq_j \to 0} \frac{dq_j}{dq_j}\,\hat{n}$$

$$\frac{\partial \bar{r}_i}{\partial q_j} = \hat{n}. \qquad\qquad ...(446)$$

Hence the generalized force given by $\overline{Q}_j = \sum_i \overline{F}_i \dfrac{\partial \bar{r}_i}{\partial q_j}$ becomes

$$\overline{Q}_j = \sum_i \overline{F}_i\hat{n} \Rightarrow \overline{Q}_j = \overline{F}\,\hat{n}, \qquad\qquad ...(447)$$

where $\overline{F}$ is the total force acting on the system. The equation (447) implies that $\overline{Q}_j$ are the components of the total force in the direction of translation $\hat{n}$. The generalized momentum p_j is defined in the equation (441) becomes

$$p_j = \frac{\partial}{\partial \dot{q}_j} \sum_i \frac{1}{2}\, m_i\, \dot{r}_i^2,$$

where $T = \sum_i \dfrac{1}{2}\, m_i\, \dot{r}_i^2$ is the kinetic energy of the system.

$$p_j = \sum_i m_i \dot{r}_i \frac{\partial \dot{r}_i}{\partial \dot{q}_j},$$

$$p_j = \sum_i m_i \dot{r}_i \frac{\partial r_i}{\partial q_j}, \ as \ \frac{\partial r_i}{\partial q_j} = \frac{\partial \dot{r}_i}{\partial \dot{q}_j}.$$

On using equation (446), we get

$$p_j = \sum_i m_i \dot{r}_i \hat{n}, \Rightarrow p_j = \sum_i p_i \hat{n},$$

$$p_j = p\hat{n},$$

where p is the total linear momentum of the system. This equation shows that the components of the generalized momentum p_j represent the components of total linear momentum of the system in the direction of displacement dq_j. Since in translation of the system, velocity is not affected and hence the kinetic energy of the system. This means that q_j will not appear in kinetic energy expression. That is, change in the kinetic energy due to change in q_j is zero. Consequently, we have $\dfrac{\partial T}{\partial q_i} = 0$. Thus from the Lagrange's equation of motion (81) we have,

$$\frac{d}{dt}\left(\frac{\partial T}{\partial \dot{q}_j}\right) + \frac{\partial V}{\partial q_j} = 0,$$

$$\Rightarrow \qquad \dot{p}_j = -\frac{\partial V}{\partial q_j} = \overline{Q}_j. \qquad \qquad \dots(448)$$

Now, if the co-ordinate q_j is cyclic in the Lagrangian, then

$$\frac{\partial L}{\partial q_j} = 0.$$

Consequently, we get

$$\frac{\partial V}{\partial q_j} = 0.$$

From equation (448) we have

$$\dot{p}_j = 0 \Rightarrow p_j = \text{const.}$$

This shows that corresponding to the cyclic co-ordinate q_j the total linear momentum is conserved.

Note: This can also be stated from equation (448) that if the components of total force Q_j are zero, then the total linear momentum is conserved. This is usual conservation theorem of linear momentum proved at the beginning.

Theorem 13   If the cyclic generalized co-ordinate q_j is such that dq_j represents the rotation of the system of particles around some axis $\hat{n}$, then prove that the total angular momentum is conserved along $\hat{n}$.

Proof: Consider a conservative system so that the potential energy V is a function of generalized co-ordinates only.

i.e.,
$$V = V(q_j) \Rightarrow \frac{\partial V}{\partial \dot{q}_j} = 0.$$

Let the system be rotated through an angle dq_j around a unit vector $\hat{n}$. This gives the rotation of the vector

$$OP = \bar{r}_i\,(q_j) \text{ to } OQ = \bar{r}_i\left(q_j + dq_j\right).$$

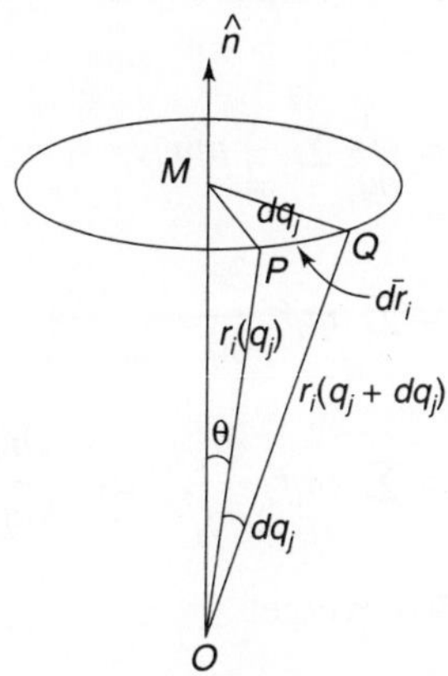

The magnitude of change in position vector $\bar{r}_i(q_j)$ due to rotation is given by

$$\left|d\bar{r}_i\right| = MP\,dq_j,$$

$$= OP \sin\theta \, dq_j,$$

$$\left|d\bar{r}_i\right| = r_i(q_j) \sin\theta \, dq_j,$$

$$\Rightarrow \qquad \left|\frac{d\bar{r}_i}{dq_j}\right| = r_i \sin\theta. \qquad\qquad \ldots(449)$$

Since $\bar{r}_i = \bar{r}_i(q_1, q_2, q_3, \ldots, q_n, t)$, we write therefore from the equation (449)

$$\frac{\partial \bar{r}_i}{\partial q_j} = \hat{n} \times \bar{r}_i. \qquad\qquad \ldots(450)$$

This shows that $\dfrac{\partial \bar{r}_i}{\partial q_j}$ is perpendicular to both the vectors $\hat{n}$ and $\bar{r}_i$. Hence the components of the generalized force (64) becomes

$$\overline{Q}_j = \sum_i \overline{F}_i \, (\hat{n} \times \bar{r}_i),$$

$$\Rightarrow \qquad \overline{Q}_j = \sum_i \hat{n} \, (\bar{r}_i \times \overline{F}_i),$$

$$\overline{Q}_j = \hat{n} \sum_i N_i \qquad\qquad \ldots(451)$$

where $N_i = \bar{r}_i \times \overline{F}_i$ is the torque on the ith particle. If $N = \sum_i N_i$ is the total torque acting on the system, then equation (451) shows that $\overline{Q}_j$ are the components of the total torque along the axis of rotation. Now the generalized momentum p_j is given by

$$p_j = \frac{\partial}{\partial \dot{q}_j} \sum_i \frac{1}{2} \, m_i \dot{r}_i^2,$$

$$p_j = \sum_i m_i \dot{r}_i \, \frac{\partial \dot{r}_i}{\partial \dot{q}_j},$$

$$p_j = \sum_i m_i \dot{r}_i \, \frac{\partial r_i}{\partial q_j}, \quad as \quad \frac{\partial r_i}{\partial q_j} = \frac{\partial \dot{r}_i}{\partial \dot{q}_j}.$$

On using the equation (450), we get

$$p_j = \sum_i m_i \dot{r}_i \, (\hat{n} \times \bar{r}_i),$$

$$p_j = \sum_i p_i \, (\hat{n} \times \bar{r}_i),$$

$$p_j = \sum_i \hat{n} \, (\bar{r}_i \times p_i),$$

$$p_j = \hat{n} \sum_i (\bar{r}_i \times p_i)$$

$$p_j = \hat{n} \sum_i L_i$$

where $L = \sum_i L_i$ is the total angular momentum of the system. Thus we have

$$p_j = \hat{n} L. \qquad\qquad \ldots(452)$$

This equation shows that the components of generalized momentum p_j are the components of total angular momentum of the system along the axis of rotation. Since the rotation of the system does not change the magnitude of the velocity and hence the kinetic energy of the system. This means that T does not depend on positions q_j. That is, change in the kinetic energy due to change in q_j is zero. Consequently, we have

$$\frac{\partial T}{\partial q_j} = 0.$$

Thus from the Lagrange's equation of motion (81) we have,

$$\frac{d}{dt}\left(\frac{\partial T}{\partial \dot{q}_j}\right) + \frac{\partial V}{\partial q_j} = 0,$$

$$\Rightarrow \qquad \dot{p}_j = -\frac{\partial V}{\partial q_j} = \overline{Q}_j. \qquad\qquad \ldots(453)$$

Now, if the co-ordinate q_j is cyclic in the Lagrangian, then

$$\frac{\partial L}{\partial q_j} = 0 \Rightarrow \frac{\partial V}{\partial q_j} = 0.$$

Consequently, from equation (453) we have

$$\dot{p}_j = 0 \Rightarrow p_j = \text{const.} \qquad\qquad \ldots(454)$$

This shows that corresponding to the cyclic co-ordinate q_j the total angular momentum is conserved.

Note: From equation (453) the Theorem can also be stated as, if the applied torque is zero then the total angular momentum is conserved.

Exercise

1. A particle has Lagrangian $L = \dfrac{1}{2}\left[f(\theta)\dot{\theta}^2 + 2g(\theta)\dot{\theta}\omega + \omega^2 h(\theta)\right] - V$,

   Show that $L = \dfrac{1}{2}\left[f(\theta)\dot{\theta}^2 - \omega^2 h(\theta)\right] + V = \text{constant}$. What does this constant represent?

2. Find the Lagrangian and the equation of motion for a particle of mass m constrained to move on the surface of a cone $y^2 + z^2 = x$.

   Ans: $L = \dfrac{1}{2}\left[u^2\dot{\theta}^2 + \left(1 + 4u^2\right)\dot{u}^2\right] - mgu \sin\theta$,

   Equation of Motion: $\ddot{\theta} + \dfrac{2}{u}\dot{u}\dot{\theta} + \dfrac{g}{u}\cos\theta = 0$.

3. A bead is sliding on a wire in the form of a cycloid

$$x = a(\theta - \sin\theta),\ y = a(1 + \cos\theta),\ 0 \le \theta \le 2\pi,$$

Find Lagrangian and the equation of motion.

Ans: $L = 2a^2 m\dot\theta^2 \sin^2\left(\dfrac{\theta}{2}\right) - 2mga\cos^2\left(\dfrac{\theta}{2}\right),$

Equation of Motion: $\ddot\theta + \dfrac{1}{2}\cot\left(\dfrac{\theta}{2}\right)\dot\theta^2 - \dfrac{g}{2a}\cot\left(\dfrac{\theta}{2}\right) = 0.$

4. Derive the Newton's equation of motion from the Lagrange's equation of motion for a particle moving under the action of the force F.

   Hint: The force is explicitly given, use Lagrange's equation of motion

$$\frac{d}{dt}\left(\frac{\partial T}{\partial \dot q_j}\right) - \frac{\partial T}{\partial q_j} = Q_j,$$

   where T is the kinetic energy of the particle.

5. Two mass m_1, m_2 are connected to the two ends of an inextensible light string passing over a smooth pulley at the edge of the inclined plane. The mass m_1 rests on the inclined plane and m_2 hangs vertically. If θ is the angle of inclination of the plane to the horizontal, then show by using D'Alembert's Principle that masses will be in equilibrium if $\sin\theta = \dfrac{m_2}{m_1}$. If μ is the coefficient of friction then show that

$$\text{Sin}\ \theta - \mu\cos\theta = \frac{m_2}{m_1}.$$

6. A particle is moving on a cycloid $s = 4a\sin\left(\dfrac{\theta}{2}\right)$ under the action of gravity. Obtain the Lagrangian and the equation of motion.

Ans: $L = 2a^2 m\cos^2\left(\dfrac{\theta}{2}\right)\dot\theta^2 - mga(1 - \cos\theta),$

Equation of Motion: $\ddot\theta - \dfrac{1}{2}\tan\left(\dfrac{\theta}{2}\right)\dot\theta^2 + \dfrac{g}{2a}\tan\left(\dfrac{\theta}{2}\right) = 0$

7. Show that the force F defined by

$$\overline{F} = \left(y^2 z^3 - 6xz^2\right)i + 2xyz^3 j + \left(3xy^2 z^2 - 6x^2 z\right)k$$

is conservative and hence find the total energy of the particle.

Ans: For conservative force

$$\nabla \times \overline{F} = 0,\ E = \frac{1}{2}m\left(\dot x^2 + \dot y^2 + \dot z^2\right) + 3x^2 z^2 - xy^2 z^3.$$

8. Show that Newton's equation of motion is the necessary condition for the action to have the stationary value.

9. Show that the two Lagrangians $L_1 = (q + \dot{q})^2$, $L_2 = (q^2 + \dot{q}^2)$ are equivalent.

    Ans: Both the Lagrangians produce the same equation of motion

    $$\ddot{q} - q = 0.$$

10. For a mechanical system the generalized co-ordinates appear separately in the kinetic energy and the potential energy such that

    $$T = \Sigma\, f_i(q_i)\, \dot{q}_i^2, \; V = \Sigma\, V_i(q_i).$$

    Show that the Lagrange's equations of motion reduce to the equations

    $$2f_i\ddot{q}_i + f_i'\dot{q}_i^2 + V_i' = 0, \; i = 1, 2, \ldots, n.$$

11. Find the Lagrangian and the equation of motion of a particle of mass m moving on the surface characterized by

    $$x = r \cos \phi, \; y = r \sin \phi, \; z = r \cot \theta.$$

    Ans: $L = \dfrac{1}{2} m \left(\dot{r}^2 \operatorname{cosec}^2\theta + r^2\dot{\phi}^2\right) - mgr \cot \theta.$

    Equation of motion: $\ddot{r} - r \sin^2 \theta\dot{\phi}^2 + g \cos \theta \sin \theta = 0.$

12. Find the Lagrangian and the equation of motion of a particle moving on the surface obtained by revolving the line $x = z$ about z-axis.

    Hint: Surface of revolution is a cone $x^2 + y^2 = z^2$.

    Ans: $L = \dfrac{1}{2} m \left(2\dot{r}^2 + r^2\dot{\phi}^2\right) - mgr.$

    Equation of motion: $\ddot{r} - \dfrac{1}{2} r\dot{\phi}^2 + \dfrac{1}{2} g = 0.$

13. Two equal masses m connected by a massless rigid rod of length l forming a dump-bell is rotated in the xy-plane. Set up the Lagrangian and the Lagrange's equation of motion.

    Hint: The system has three degrees of freedom and hence three generalized coordinates x, y, θ. The Lagrangian of the system is

    $$L = m \left(\dot{x}^2 + \dot{y}^2\right) + \dfrac{1}{2} m(l^2\dot{\theta}^2 - 2l\dot{x}\dot{\theta} \sin \theta + 2l\dot{y}\dot{\theta} \cos \theta) - mgy - mgl \sin \theta.$$

14. A mass $2m$ is suspended from a fixed support by a spring, of spring constant $2K$. From this mass another mass m is suspended by another spring of spring constant K. Find the equation of motion of the coupled system.

Ans: Lagrangian of the system is

$$L = m\dot{x}_1^2 + \frac{1}{2}\,m\dot{x}_2^2 - \frac{K}{2}\left(3x_1^2 + x_2^2 - 2x_1 x_2\right).$$

15. A particle is thrown horizontally from the top of a building of height h with initial velocity u. Neglecting all forces except gravity and resistance of the air assumed proportional to the velocity, find the Lagrangian and the equation of motion.

    Ans: Lagrangian of the system is $L = \dfrac{1}{2}\,m\left(\dot{x}^2 + \dot{y}^2\right) - mgy$.

    Equations of motion are $m\ddot{x} + k\dot{x} = 0$, $m\ddot{y} + k\dot{y} + mg = 0$.

16. A particle of mass m is projected vertically upward from point O with initial velocity u. Find Lagrangian, the equation of motion and hence the maximum height reached, assuming that the resistance of the air is proportional to the velocity.

    Ans: Equation of motion: $m\ddot{z} + k\dot{z} + mg = 0$.

    For maximum height put $\dot{z} = 0$ and $z_{\max} = \dfrac{m}{k}\left[u - \dfrac{mg}{k}\log\left(\dfrac{mg + ku}{mg}\right)\right]$.

17. Show that the oscillations of a particle along cycloid are isochronous.

    [Hint: refer remark on page 2.16]

2   Variational Principles

INTRODUCTION

We have seen in the Chapter 1 that co-ordinates are the tools in the hands of a mathematician. With the help of these co-ordinates the motion of a particle or a system of particles is discussed. The piece wise information of the traced path $y = f(x)$, whether it is minimum or maximum at a point can be obtained from differential calculus by putting the first derivative of the function equal to zero. The function is either maximum or minimum at the point depends upon the value of second derivative of the function at that point. However, if we want to know the information about the whole path, whether it is maximum or minimum, we use integral calculus. *i.e.*, the techniques of calculus of variation and are called variational principles. Thus the calculus of variation has its origin in the generalization of the elementary theory of maxima and minima of function of a single variable or more variables. The history of calculus of variations can be traced back to the year 1696, when John Bernoulli advanced the problem of the brachistochrone. In this problem one has to find the curve connecting two given points A and B that do not lie on a vertical line, such that a particle sliding down this curve under gravity from the height point A reaches the lowest point B in the shortest time.

Apart from the problem of brachistochrone, there are three other problems exerted great influence on the development of the subject and are:

1. the problem of geodesic,
2. the problem of minimum surface of revolution and
3. the isoperimetric problem.

Thus in calculus of variation we consider the motion of a particle or system of particles along a curve $y = f(x)$ joining two points $P(x_1, y_1)$ and $Q(x_2, y_2)$ of extremum length. To look for the curve of extremum length, out of infinitely many curves between the points, we consider two infinitesimally closed points on the curve and find distance between them. It is given by the Euclidean metric

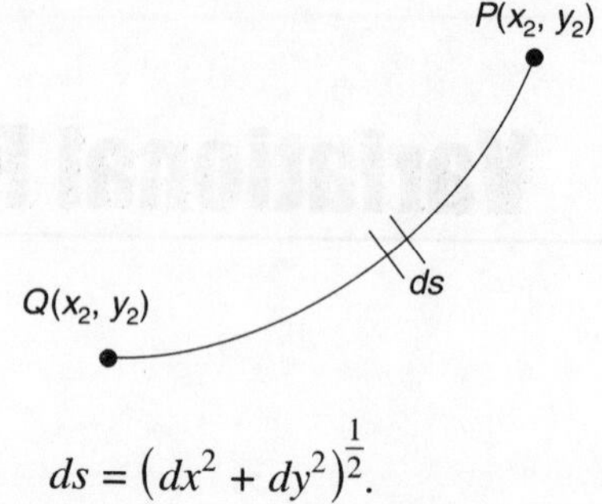

$$ds = \left(dx^2 + dy^2 \right)^{\frac{1}{2}}.$$

Hence the total distance between two points P and Q along the curve is the sum of such infinitesimal distances and it is given by the integral

$$I(y(x)) = \int_{x_1}^{x_2} \left(1 + y'^2 \right)^{\frac{1}{2}} dx, \; y' = \frac{dy}{dx} \, . \qquad \qquad \dots(1)$$

Our aim is to find the condition to be satisfied by the function $y(x)$ such that the functional $I(y(x))$ has extremum value. In general the integral (1) can be written as

$$I(y(x)) = \int_{x_1}^{x_2} f(x, y, y')dx \, . \qquad \qquad \dots(2)$$

Now the integral (2) can represent not only the path between two points along the curve, but also the time of decent form the highest point to the lowest point, the area of the surface of revolution of a curve about coordinate axis and so on and so forth depending upon the nature of the functional $f(x, y, y')$. The value of the integral is the functional I which depends on what function y is of x, the starting point (x_1, y_1) and the end point (x_2, y_2) of the curve. Thus we find the condition to be satisfied by $y(x)$ such that the value of the integral I must have extremum value for the actual path. The fascinating principle in calculus of variation paves the way to find the curve $y(x)$ which makes integral (2) extremum. A curve $y(x)$ which satisfies this condition is called an extremal of the integral I. Its object is to extremize the values of the functional. This is one of the most fundamental and beautiful principles in applied mathematics. Because from this principle one can determine the

    (a) Newton's equations of motion,

    (b) Lagrange's equations of motion,

    (c) Hamilton's equations of motion,

    (d) Schrödinger's equations of motion,

    (e) Einstein's field equations for gravitation,

    (f) Hoyle-Narlikar's equations for gravitation and so on and so forth by slightly modifying the integrand and many more such fundamental equations.

Note: A functional means a quantity whose values are determined by one or several functions. *i.e.*, domain of a functional is a set of all admissible functions. *e.g.* The length of the path I between two points is a function of curves $y(x)$, which it self is a function of x. Such functions are called functional.

Basic Lemma

If x_1 and x_2 $(> x_1)$ are fixed constants and $G(x)$ is a particular continuous function for $x_1 \le x \le x_2$ and if $\int_{x_1}^{x_2} G(x)\eta(x)dx = 0$ for every choice of continuous differentiable function $\eta(x)$ such that $\eta(x_1) = 0 = \eta(x_2)$, then $G(x) = 0$ identically in $x_1 \le x \le x_2$.

Proof: Let the lemma be not true. Let us assume that there is a particular value x' of x in the interval such that $G(x') \ne 0$. Let us assume that $G(x') > 0$.

Since $G(x)$ is continuous function in $x_1 \le x \le x_2$ and in particular it is continuous at $x = x'$. Hence there must exist an interval surrounding x' say $x_1' \le x \le x_2'$ in which $G(x) > 0$ everywhere. Let us now see whether the integral $\int_{x_1}^{x_2} G(x)\eta(x)dx = 0$ for every permissible choice of $\eta(x)$. We choose $\eta(x)$ such that

$$
\begin{aligned}
\eta(x) &= 0 && \text{for} && x_1 \le x \le x_1', \\
&= (x - x_1')^2 (x - x_2')^2 && \text{for} && x_1' \le x \le x_2', \\
&= 0 && \text{for} && x_2' \le x \le x_2. && \ldots(3)
\end{aligned}
$$

For this choice of $\eta(x)$ which also satisfies $\eta(x_1') = \eta(x_2') = 0$, the integral

$$\int_{x_1}^{x_2} G(x)\,\eta(x)\,dx$$

becomes

$$\int_{x_1}^{x_2} G(x)\eta(x)dx = \int_{x_1}^{x_1'} G(x)\eta(x)dx + \int_{x_1'}^{x_2'} G(x)\eta(x)dx + \int_{x_2'}^{x_2} G(x)\eta(x)dx$$

$$\Rightarrow \quad \int_{x_1}^{x_2} G(x)\eta(x)dx = \int_{x_1'}^{x_2'} (x - x_1')^2 (x - x_2')^2 \, G(x)\, dx. \qquad \ldots(4)$$

Since

$$G(x) > 0 \text{ in } x_1' \le x \le x_2',$$

$\Rightarrow$ Right hand side of equation (4) is definitely positive.

$$\Rightarrow \qquad \int_{x_1}^{x_2} \eta(x)G(x)\,dx > 0.$$

This is contradiction to the hypothesis

$$\int_{x_1}^{x_2} G(x)\eta(x)dx = 0.$$

If $G(x') < 0$, we obtain the similar contradiction. This contradiction arises because of our assumption that $G(x) \neq 0$ for x in $x_1 \leq x \leq x_2$. This implies that $G(x) = 0$ identically in $x_1 \leq x \leq x_2$. This completes the proof.

Euler-Lagrange's Equations: Euler-Lagrange's differential equations are very important tools to study the extremization of a given definite integral. We derive these equations in the following theorem.

Theorem 1   Find the Euler-Lagrange differential equation satisfied by twice differentiable function $y(x)$ which extremizes the functional

$$I(y(x)) = \int_{x_1}^{x_2} f(x,\ y,\ y')dx$$

where y is prescribed at the end points.

Proof: Let $P(x_1,\ y_1)$ and $Q(x_2,\ y_2)$ be two fixed points in xy-plane. The points P and Q can be joined by infinitely many curves. Accordingly the value of the integral I will be different for different paths. We shall look for a curve along which the functional I has an extremum value. Let c be a curve between P and Q given by the equation $y = y\ (x,\ 0)$ along which the functional I has extremum value.

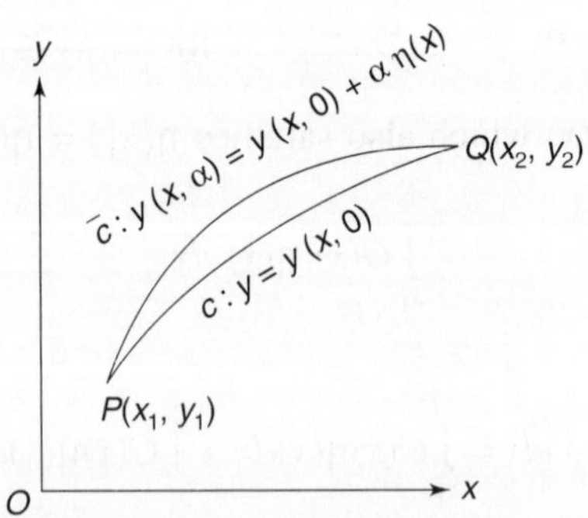

Let the extremum value of the functional along the curve c be given by the integral

$$I(y(x)) = \int_{x_1}^{x_2} f(x,\ y,\ y')dx \qquad\qquad \dots(5)$$

We can label all possible paths starting from P and ending at Q by the family of equations $y = y(x, \alpha)$, where

$$y(x, \alpha) = y(x, 0) + \alpha\eta(x), \qquad \qquad ...(6)$$

α is a parameter and $\eta(x)$ is any differentiable function of x. For different values of α we get different curves. Accordingly, the value of the integral I will be different for different paths. Since y is prescribed at the end points, this implies that there is no variation in y at the end points. *i.e.*, all the curves of the family must be identical at fixed points P and Q.

$$\Rightarrow \qquad \qquad \eta(x_1) = 0 = \eta(x_2). \qquad \qquad ...(7)$$

Conversely, the condition (7) ensures us that the curves of the family that all pass through the points P and Q. Let the value of the functional along the neighboring curve be given by the integral

$$I(y(x, \alpha)) = \int_{x_1}^{x_2} f(x, y(x, \alpha), y'(x, \alpha))dx. \qquad \qquad ...(8)$$

From differential calculus, we know the integral I is extremum if $\left(\dfrac{\partial I}{\partial \alpha}\right)_{\alpha=0} = 0$, since for $\alpha = 0$ the neighboring curve coincides with the curve c which gives extremum values of I.

Thus
$$\left(\frac{\partial I}{\partial \alpha}\right)_{\alpha=0} = 0, \Rightarrow \int_{x_1}^{x_2} \left(\frac{\partial f}{\partial y}\eta(x) + \frac{\partial f}{\partial y'}\eta'(x)\right)dx = 0.$$

Integrating the second integration by parts, we get

$$\int_{x_1}^{x_2} \frac{\partial f}{\partial y}\eta(x)dx + \left(\frac{\partial f}{\partial y'}\eta(x)\right)\Big|_{x_1}^{x_2} - \int_{x_1}^{x_2} \frac{d}{dx}\left(\frac{\partial f}{\partial y'}\right)\eta(x)dx = 0 \qquad \qquad ...(9)$$

As y is prescribed at the end points, hence on using equations (7), we get

$$\int_{x_1}^{x_2} \left(\frac{\partial f}{\partial y} - \frac{d}{dx}\left(\frac{\partial f}{\partial y'}\right)\right)\eta(x)dx = 0.$$

By using the basic lemma of calculus of variation we get

$$\frac{\partial f}{\partial y} - \frac{d}{dx}\left(\frac{\partial f}{\partial y'}\right) = 0. \qquad \qquad ...(10)$$

This is required Euler-Lagrange differential equation to be satisfied by $y(x)$ for which the functional I has extremum value. The solution of the equation (10) gives us the function $y(x)$ which makes the integral I stationary. However, whether the function $y(x)$ maximizes I, minimizes I or do neither has to be shown separately.

Important Note

If however, y is not prescribed at the end points then there is a difference in y even at the end points and hence $(\eta(x))_{x_1}^{x_2} \neq 0$. As the value of the functional I is taken only on the extremal between two points and hence we must have the Euler-Lagrange equation is true. Consequently, in this case we must have from equation (9) that

$$\left(\frac{\partial f}{\partial y'}\right)_{x_1}^{x_2} = 0 \Rightarrow \left(\frac{\partial f}{\partial y'}\right)_{x_1} = 0 \text{ and } \left(\frac{\partial f}{\partial y'}\right)_{x_2} = 0. \qquad \dots(11)$$

These conditions can be used to find the value of the arbitrary constants from the extremal of the problem. We will prove this result a little latter in Theorem 2.

Alternate Proof of the Theorem 1 Let the curve c be slightly deformed from the original position such that any point y on the curve c is displaced to $y + \delta y$, where δy is the small variation in the path for an arbitrary choice of α, except at the end points, as y is prescribed there. Mathematically this means that $y(x_1) = y_1$, $y(x_2) = y_2$,

$$\Rightarrow \qquad (\delta y)_{x_1}^{x_2} = 0. \qquad \dots(12)$$

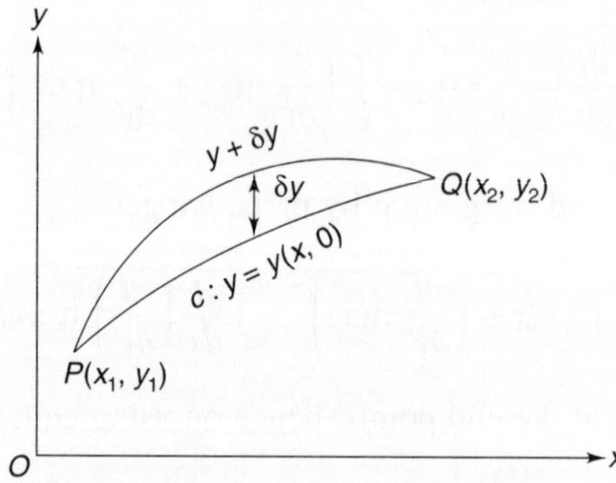

Thus the value of the functional along the varied path is given by

$$I' = \int_{x_1}^{x_2} f(x, y + \delta y, y' + \delta y')dx. \qquad \dots(13)$$

Hence the change in the value of the functional due to change in the path is given by

$$I' - I = \int_{x_1}^{x_2} [f(x, y + \delta y, y' + \delta y') - f(x, y, y')]dx,$$

Let
$$I' - I = \delta I = \int_{x_1}^{x_2} [f(x, y + \delta y, y' + \delta y') - f(x, y, y')]dx. \quad \ldots(14)$$

We recall the Taylor's series expansion for the function of two variables'

$$f(x, y + \delta y, y' + \delta y') = f(x, y, y') + \left(\frac{\partial f}{\partial y} \delta y + \frac{\partial f}{\partial y'} \delta y'\right) + \ldots$$

Since δy is very small, therefore by neglecting the higher order terms than the first in δy and $\delta y'$, we have

$$f(x, y + \delta y, y' + \partial y') - f(x, y, y') = \left(\frac{\partial f}{\partial y} \delta y + \frac{\partial f}{\partial y'} \delta y'\right).$$

Substituting this value in the equation (14), we get

$$\delta I = \int_{x_1}^{x_2} \left(\frac{\partial f}{\partial y} \delta y + \frac{\partial f}{\partial y'} \delta y'\right)dx,$$

where we know
$$\delta \frac{dy}{dx} = \frac{d}{dx} \delta y.$$

Consequently, we have

$$\delta I = \int_{x_1}^{x_2} \left(\frac{\partial f}{\partial y} \delta y + \frac{\partial f}{\partial y'} \frac{d}{dx} (\delta y)\right)dx.$$

Integration of the second integral by parts gives

$$\delta I = \int_{x_1}^{x_2} \frac{\partial f}{\partial y} \delta y \, dx + \left(\frac{\partial f}{\partial y'} \delta y\right)_{x_1}^{x_2} - \int_{x_1}^{x_2} \frac{d}{dx} \left(\frac{\partial f}{\partial y'}\right) \delta y \, dx.$$

On using equation (12), we get

$$\delta I = \int_{x_1}^{x_2} \left[\frac{\partial f}{\partial y} - \frac{d}{dx} \left(\frac{\partial f}{\partial y'}\right)\right] \delta y \, dx. \qquad \ldots(15)$$

If I is extremum along the curve $y = y(x)$ then change in I is zero. (*i.e.*, $\delta I = 0$). This resembles very closely with a similar condition of extremum of a function in differential calculus.

$$\Rightarrow \qquad \int_{x_1}^{x_2} \left[\frac{\partial f}{\partial y} - \frac{d}{dx} \left(\frac{\partial f}{\partial y'}\right)\right] \delta y \, dx = 0.$$

Since δy is arbitrary, we have

$$\frac{\partial f}{\partial y} - \frac{d}{dx}\left(\frac{\partial f}{\partial y'}\right) = 0.$$

This is the Euler-Lagrange's differential equation (10) to be satisfied by $y(x)$ for the extremum of the functional between two points.

Generalization of Theorem 1: Euler-Lagrange's Equations for several Dependent Variables

Theorem 1a   Derive the Euler-Lagrange's equations that are to be satisfied by twice differential functions $y_1, y_2, \ldots y_n$ that extremize the integral

$$I = \int_{x_1}^{x_2} f(x, y_1, \ldots y_2, \ldots, y_n, y_1', y_2', \ldots, y_n')dx$$

with respect to those functions y_1, y_2, ..., y_n which achieve prescribed values at the fixed points x_1, x_2.

Proof: The functional which is to be extremized can be written as

$$I = \int_{x_1}^{x_2} f(x, y_i, y_i') \, dx, \; i = 1, 2, \ldots, n.$$

Choose the family of neighboring curves defined by

$$y_i(x, \alpha) = y_i(x, 0) + \alpha \eta_i (x)$$

and repeating the procedure delineated either in the Theorem 1 or in the alternate proof we arrive at the following set of Euler-Lagrange's equations

$$\frac{\partial f}{\partial y_i} - \frac{d}{dx}\left(\frac{\partial f}{\partial y_i'}\right) = 0, \; i = 1, 2, \ldots, n. \qquad \ldots(16)$$

Geodesic: Geodesic is defined as a curve of stationary (extremum) length between two points. It is also defined to be a curve of unchanging direction. But we will use the former definition to characterize the geodesic in a given space. *A* curve can be represented either in Cartesian form, polar form or even in parametric form. We use the principle of calculus of variation to find the geodesic in a given space. Since, the parametric representation of a curve in a space is not unique, therefore, the corresponding functional which is to be extremized for the geodesic will be different. Consequently, the Euler-Lagrange's differential equations used to evaluate the geodesic will be different. Each one of these cases is illustrated in the examples (3), (25) and (26). These examples are repeatedly solved in the text (refer examples (44), (45) and (46)) by totally different approach illustrated in the Theorem 8. However, it is proved in the Theorem 12 that the solution of the Euler-Lagrange's equations derived on the assumption of the single valudness of y as a function x also satisfies the system of Euler-Lagrange's equations derived on the basis of the parametric relationship between x and y.

WORKED EXAMPLES

Example 1: Show that the geodesic (shortest distance between two points) in a Euclidian plane is a straight line.

Solution: Take $p(x_1, y_1)$ and $Q(x_2, y_2)$ be two fixed points in a Euclidean plane. Let $y = f(x)$ be the curve between P and Q. Then the element of distance between two neighboring points on the curve $y = f(x)$ joining P and Q is given by the Euclidean metric

$$ds^2 = dx^2 + dy^2 \qquad \ldots(17)$$

Hence the total distance between the point P and Q along the curve is given by

$$I = \int_P^Q ds$$

$$\Rightarrow \qquad I = \int_{x_1}^{x_2} \left(1 + y'^2\right)^{\frac{1}{2}} dx, \ y' = \frac{dy}{dx} \qquad \ldots(18)$$

Here the functional I is extremum if the integrand

$$f = \left(1 + y'^2\right)^{\frac{1}{2}} \qquad \ldots(19)$$

must satisfy the Euler-Lagrange's differential equation (10). Now from equation (19) we find that

$$\frac{\partial f}{\partial y} = 0 \text{ and } \frac{\partial f}{\partial y'} = \frac{y'}{\sqrt{1 + y'^2}},$$

$$\Rightarrow \qquad \frac{d}{dx}\left(\frac{y'}{\sqrt{1 + y'^2}}\right) = 0.$$

Integrating we get

$$y' = c \sqrt{1 + y'^2}.$$

Squaring we get

$$y' = c_1, \text{ where } c_1 = \frac{c}{\sqrt{1 + c^2}}.$$

Integrating we get

$$y = c_1 x + c_2 \qquad \ldots(20)$$

This is proves that the shortest distance between two points in a Euclidean plane is a straight line.

Example 2: Show that the geodesic $\theta = \theta(r)$ in polar co-ordinates is a straight line.

Solution: Define a curve in 2-dimensional polar plane as a functional relation between polar coordinates r and θ. If $A(x, y)$ and $B(x + dx, y + dy)$ are Cartesian co-ordinates of infinitesimal points on the curve, then an element of distance between A and B is given by equation (17). Let $\theta = \theta(r)$ be the polar equation of the curve and $P(r_1, \theta_1)$ and $Q(r_2, \theta_2)$ be two polar points on it. The relations $x = r \cos \theta$, $y = r \sin \theta$ between the Cartesian and polar coordinates reduce the metric (17) to the polar form

$$ds^2 = dr^2 + r^2 d\theta^2. \qquad \qquad \text{...(21)}$$

Thus the total distance between the points P and Q becomes

$$I = \int_{r_1}^{r_2} \left(1 + r^2\theta'^2\right)^{\frac{1}{2}} dr, \ \theta' = \frac{d\theta}{dr}. \qquad \text{...(22)}$$

The functional I is shortest if the integrand

$$f = \left(1 + r^2\theta'^2\right)^{\frac{1}{2}} \qquad \qquad \text{...(23)}$$

must satisfy the Euler-Lagrange's differential equation

$$\frac{\partial f}{\partial \theta} - \frac{d}{dr}\left(\frac{\partial f}{\partial \theta'}\right) = 0, \qquad \qquad \text{...(24)}$$

$$\Rightarrow \qquad \frac{d}{dr}\left(\frac{r^2\theta'}{\sqrt{1 + r^2\theta'^2}}\right) = 0,$$

$$\Rightarrow \qquad r^2\theta' = h \sqrt{1 + r^2\theta'^2}.$$

Squaring and solving for θ' we get

$$\frac{d\theta}{dr} = \pm \frac{h}{r\left(r^2 - h^2\right)^{\frac{1}{2}}}.$$

On integrating we get

$$\theta = \pm \cos^{-1}\left(\frac{h}{r}\right) + \theta_0,$$

where θ_0 is a constant of integration. We write this as

$$h = r \cos(\theta - \theta_0). \qquad \qquad \text{...(25)}$$

This is the polar form of the equation of straight line. Hence the shortest distance between two polar points in a plane is a straight line.

Note: One can choose the polar equation of the curve as $r = r(\theta)$, then the length of the curve in this case becomes

$$I = \int_{\theta_0}^{\theta_1} \sqrt{r^2 + \left(\frac{dr}{d\theta}\right)^2}\ d\theta.$$

Since the integrand $f = \sqrt{r^2 + r'^2}$ does not contain θ, we therefore have (refer Example 14 for proof)

$$f - r'\ \frac{\partial f}{\partial r'} = h.$$

Solving this equation we readily obtain the same polar equation of straight line as the geodesic.

Example 3: Show that the geodesic $\phi = \phi(\theta)$ on the surface of a sphere of radius r is an arc of the great circle.

Solution: Consider a sphere of radius r described by the parametric equations
$x = r \sin \theta \cos \phi$, $y = r \sin \theta$, $\sin \phi$, $z = r \cos \theta$, where r is constant. $\qquad$ ...(26)

Define the curve $\phi = \phi(\theta)$ on the sphere. Let $P(\theta_1, \phi_1)$ and $Q(\theta_2, \phi_2)$ be two points on it. If $A(x, y, z)$ and $B(x + dx, y + dy, z + dz)$ are the Cartesian coordinates of the two neighboring points on the curve joining the points P and Q, then the infinitesimal distance between A and B along the curve is given by the 3-dimensional Euclidean metric

$$ds^2 = dx^2 + dy^2 + dz^2, \qquad\qquad ...(27)$$

where from equation (26) we find

$$dx = r \cos \theta \cos \phi\, d\theta - r \sin \theta \sin \phi\, d\phi,$$

$$dy = r \cos \theta \sin \phi\, d\theta + r \sin \theta \cos \phi\, d\phi,$$

$$dz = -r \sin \theta\, d\theta. \qquad\qquad ...(28)$$

Squaring and adding these equations we readily obtain 2-dimensional metric on the surface of a sphere of radius r as

$$ds^2 = r^2 d\theta^2 + r^2 \sin^2 \theta\, d\phi^2. \qquad\qquad ...(29)$$

Hence the total distance between the points P and Q along the curve $\phi = \phi(\theta)$ becomes

$$I = \int_{\theta_1}^{\theta_2} r \left(1 + \sin^2 \theta \phi'^2\right)^{\frac{1}{2}}\ d\theta, \quad \phi' = \frac{d\phi}{d\theta} \qquad ...(30)$$

where

$$f = r \left(1 + \sin^2 \theta \phi'^2\right)^{\frac{1}{2}}. \qquad\qquad ...(31)$$

The curve is geodesic if the functional I is stationary. This is true if the function f must satisfy the Euler-Lagrange's equation.

$$\frac{\partial f}{\partial \phi} - \frac{d}{d\theta}\left(\frac{\partial f}{\partial \phi'}\right) = 0 \qquad \qquad ...(32)$$

$$\Rightarrow \qquad \frac{d}{d\theta}\left(\frac{r \sin^2 \theta \phi'}{\sqrt{1 + \sin^2 \theta \phi'^2}}\right) = 0.$$

Integrating we get

$$\Rightarrow \qquad \frac{\sin^2 \theta \phi'^2}{\sqrt{1 + \sin^2 \theta \phi'^2}} = c$$

Solving for ϕ' we get

$$\phi' = \frac{c \, \mathrm{cosec}^2 \theta}{\left(1 - c^2 \, \mathrm{cosec}^2\theta\right)^{\frac{1}{2}}}.$$

On simplifying we get

$$\frac{d\phi}{d\theta} = \frac{k \, \mathrm{cosec}^2 \theta}{\left(1 - k^2 \cot^2 \theta\right)^{\frac{1}{2}}}, \qquad \qquad ...(33)$$

where $k = \dfrac{c}{\sqrt{1 - c^2}}$ is a constant. Put $k \cot \theta = t \Rightarrow -\mathrm{cosec}^2 \theta \, d\theta = dt$. Therefore, we have

$$d\phi = -\frac{dt}{\sqrt{1 - t^2}}.$$

Integrating we get

$$\phi = \alpha - \sin^{-1} t$$

or $$\phi = \alpha - \sin^{-1}(k \cot \theta)$$

$$\Rightarrow \qquad k \cot \theta = \sin(\alpha - \phi)$$

$$\Rightarrow \qquad k \cos \theta = \sin \alpha \sin \theta \cos \phi - \cos \alpha \sin \theta \sin \phi$$

$$\Rightarrow \qquad kz = x \sin \alpha - y \cos \alpha. \qquad \qquad ...(34)$$

This is the first-degree equation in x, y, z which represents a plane. This plane passes through the origin, hence cutting the sphere in a great circle. Hence the geodesic on the surface of a sphere is an arc of a great circle.

Minimum Surface of Revolution

Example 4: Show that the curve is a catenary for which the area of surface of revolution is minimum when revolved about y-axis.

Solution: Consider a curve between two points (x_1, y_1) and (x_2, y_2) in the xy-plane whose equation is $y = y(x)$. We form a surface by revolving the curve about y-axis. Our claim is to find the nature of the curve for which the surface area is minimum. Consider a small strip at a point A formed by revolving the arc length ds of the curve about y-axis. If the distance of the point A on the curve from y-axis is x, then the surface area of the strip is equal to $2\pi x ds$.

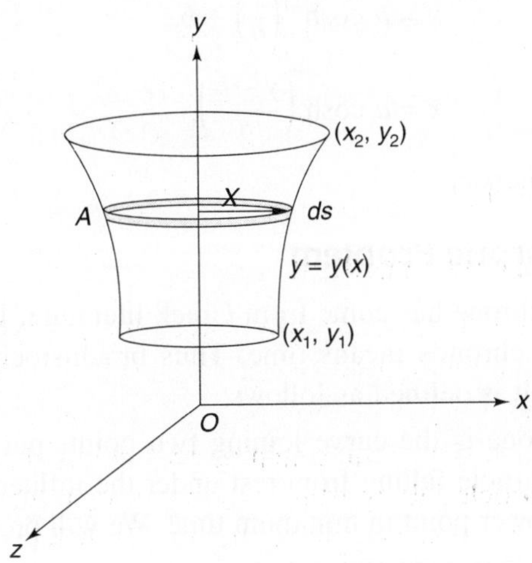

But we know the element of arc ds is given by

$$ds = \sqrt{1 + y'^2}\ dx.$$

Thus the surface area of the strip ds is equal to $2\pi x \sqrt{1 + y'^2}\ dx$.

Hence the total area of the surface of revolution of the curve $y = y(x)$ about y-axis is given by the integral

$$I = \int_{x_1}^{x_2} 2\pi x \sqrt{1 + y'^2}\ dx. \qquad \qquad ...(35)$$

This surface area will be minimum if the integrand $f = 2\pi x \sqrt{1 + y'^2}$ must satisfy Euler-Lagrange's differential equation (10). Hence the equation (10) reduces to

$$\Rightarrow \qquad \frac{d}{dx}\left(\frac{2\pi x y'}{\sqrt{1 + y'^2}} \right) = 0,$$

$$\Rightarrow \qquad \frac{d}{dx}\left(\frac{x y'}{\sqrt{1 + y'^2}} \right) = 0.$$

Integrating we get

$$xy' = a\sqrt{1 + y'^2}.$$

Solving for y' we get

$$\frac{dy}{dx} = \frac{a}{\sqrt{x^2 - a^2}}.$$

Integrating we get

$$y = a \cosh^{-1}\left(\frac{x}{a}\right) + b.$$

or
$$x = a \cosh\left(\frac{y - b}{a}\right). \qquad \qquad \ldots(36)$$

This curve is a catenary.

The Brachistochrone Problem

The word brachistochrone has come from Greek literature. In Greek brachistos means shortest and chronos means time. Thus brachistochrone curve means shortest time curve. It is defined as follows:

The Brachistochrone is the curve joining two points not lie on the vertical line, such that the particle falling from rest under the influence of gravity from higher point to the lower point in minimum time. We will prove in the following example that the curve is a cycloid.

Example 5: A particle slides down a curve in the vertical plane under gravity. Find the curve such that it reaches the lowest point in shortest time.

Solution: Let A and B be two points on the curve not lie on the vertical line. Let $v = \dfrac{ds}{dt}$ be the speed of the particle along the curve. Then the time required to fall an arc length ds is given by

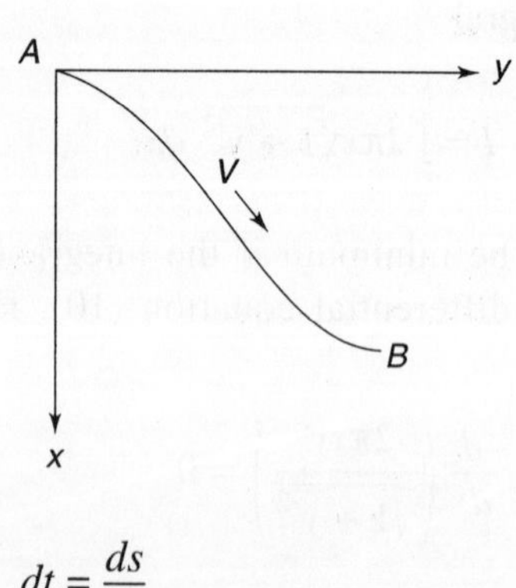

$$dt = \frac{ds}{v}$$

$$\Rightarrow \qquad dt = \frac{\sqrt{1 + y'^2}}{v}\, dx.$$

Therefore the total time of decent from A to B becomes

$$t_{AB} = \int_A^B \frac{\sqrt{1 + y'^2}}{v}\, dx. \qquad \ldots(37)$$

Since the particle falls freely under gravity, therefore its potential energy goes on decreasing and at distance x below the point A it is given by $V = -mgx$, and the kinetic energy is $T = \frac{1}{2} mv^2$. Now from the principle of conservation of energy we have $T + V = $ constant. Initially at point A, we have $x = 0$ and $v = 0$. Hence the constant is zero.

$$\Rightarrow \qquad \frac{1}{2} mv^2 = mgx,$$

$$\Rightarrow \qquad v = \sqrt{2gx}. \qquad \ldots(38)$$

Hence the time integral (37) becomes

$$t_{AB} = \int_{x_1}^{x_2} \frac{\sqrt{1 + y'^2}}{\sqrt{2gx}}\, dx. \qquad \ldots(39)$$

Thus t_{AB} is minimum if the integrand $f = \dfrac{\sqrt{1+y'^2}}{\sqrt{2gx}}$ must satisfy Euler-Lagrange's equation (10). This gives

$$\Rightarrow \qquad \frac{d}{dx}\left(\frac{y'}{\sqrt{x(1 + y'^2)}} \right) = 0.$$

Integrating we get

$$y' = c\sqrt{x(1 + y'^2)}.$$

Solving the equation for y' we get

$$\frac{dy}{dx} = \frac{\sqrt{x}}{\sqrt{a - x}}.$$

Integrating we get

$$y = \int \frac{\sqrt{x}}{\sqrt{a - x}}\, dx + b. \qquad \ldots(40)$$

Put $\qquad\qquad x = a\, \sin^2(\theta/2). \qquad \ldots(41)$

$$\Rightarrow \qquad dx = 2a\, \sin(\theta/2)\, \cos(\theta/2)d\theta$$

$$y = a \int \sin^2(\theta/2)d\theta + b.$$

$$\Rightarrow \qquad y = \frac{a}{2}\, (\theta - \sin\theta) + b.$$

If $y = 0$, $\theta = 0 \Rightarrow b = 0$, hence $y = \dfrac{a}{2}\, (\theta - \sin\theta)$. $\qquad \ldots(42)$

Thus from equations (41) and (42), we have

$$x = b(1 - \cos\theta),$$

$$y = b(\theta - \sin\theta), \quad \text{for } b = \frac{a}{2}.$$

This is a cycloid. Thus the curve is a cycloid for which the time of decent is minimum.

Remark: We have shown in the Example 50 of Chapter 1 that the equation of motion of a particle along the cycloid is governed by the second order differential equation given by

$$\ddot{\theta} - \tan\theta\,\dot{\theta}^2 + \left(\frac{g}{4a}\right)\tan\theta = 0.$$

For the small angle θ, we have $\tan\theta = \theta$. Hence by neglecting the higher order terms in θ than the first, the equation of motion becomes

$$\ddot{\theta} + \left(\frac{g}{4a}\right)\theta = 0.$$

This is the equation for the angular simple harmonic motion with period $T = 4\pi\sqrt{\dfrac{a}{g}}$. We see that the period is independent of the amplitude and hence the oscillations of the particle along Cycloid are isochronous.

Example 6:   Find the extremal of the functional

$$I = \int_{0}^{\frac{\pi}{2}} \left(y'^2 - y^2 + 2xy\right)dx$$

subject to the conditions that $y(0) = 0$, $y\left(\dfrac{\pi}{2}\right) = 0$.

Solution: For the given functional the Euler-Lagrange's equation reduces to the equation

$$2(x - y) - \frac{d}{dx}(2y') = 0$$

$$\Rightarrow \qquad\qquad y'' + y = x. \qquad\qquad\qquad \ldots(44)$$

This is second order differential equation, whose complementary function is given by

$$y = c_1 \cos x + c_2 \sin x, \qquad\qquad\qquad \ldots(45)$$

where c_1 and c_2 are arbitrary constants. The particular integral of the equation (44) is

$$y = \left(1 + D^2\right)^{-1}x \Rightarrow y = x.$$

Hence the general solution is given by

$$y = c_1 \cos x + c_2 \sin x + x. \qquad \ldots(46)$$

This shows that the extremals of the functional are the two-parameter family of curves. On using the boundary conditions we obtain

$$y(0) = 0 \Rightarrow c_1 = 0, \text{ and } y\left(\frac{\pi}{2}\right) = 0 \Rightarrow c_2 = -\frac{\pi}{2}.$$

Hence the required extremal is

$$y = x - \frac{\pi}{2} \sin x. \qquad \ldots(47)$$

Example 7: Find the extremal of the functional

$$I = \int_1^2 \left(\frac{x^3}{y'^2}\right) dx$$

subject to the conditions that $y(1) = 0$, $y(2) = 3$.

Solution: For the extremal of the functional I, the Euler-Lagrange's equation (10) becomes

$$\frac{d}{dx}\left(\frac{x^3}{y'^3}\right) = 0.$$

Integrating we get $x^3 = cy'^3 \Rightarrow y' = ax$. Integrating we get

$$y = \frac{a}{2} x^2 + b. \qquad \ldots(48)$$

Now using the boundary conditions we get

$$y(1) = 0 \Rightarrow \frac{a}{2} + b = 0, \text{ and } y(2) = 3 \Rightarrow 2a + b = 3.$$

Solving these two equations we obtain $a = 2$, $b = -1$. Hence the required functional becomes

$$y = x^2 - 1. \qquad \ldots(49)$$

Example 8: Show that the time taken by a particle moving along a curve $y = y(x)$ with velocity $\dfrac{ds}{dt} = x$, from the point $(0, 0)$ to the point $(1, 1)$ is minimum if the curve is a circle having its center on the x-axis.

Solution: Let a particle be moving along a curve $y = y(x)$ from the point $(0, 0)$ to the point $(1, 1)$ with velocity $x = \dfrac{ds}{dt} \Rightarrow dt = \dfrac{ds}{x}$. Therefore, the total time required for the particle to move from the point $(0, 0)$ to the point $(1, 1)$ is given by

$$t = \int_0^1 \frac{ds}{x}, \qquad \ldots(50)$$

where the infinitesimal distance between two points on the path is given by

$$ds = \sqrt{1 + y'^2}\, dx, \quad y' = \frac{dy}{dx}$$

Hence the equation (50) becomes

$$t = \int_0^1 \frac{\sqrt{1 + y'^2}}{x}\, dx \qquad \qquad \ldots(51)$$

This time is to be minimum the integrand must satisfy the Euler-Lagrange's equation (10). This gives

$$\Rightarrow \qquad \frac{d}{dx}\left(\frac{y'}{x\sqrt{1 + y'^2}}\right) = 0,$$

$$\Rightarrow \qquad \qquad y' = cx\sqrt{1 + y'^2}$$

Solving for y' we get

$$y' = \frac{x}{\sqrt{a^2 - x^2}}, \quad \text{for } a = \frac{1}{c}.$$

Integrating we get

$$y = \int \frac{x}{\sqrt{a^2 - x^2}}\, dx + b$$

Put $\qquad\qquad\qquad x = a \sin\theta \Rightarrow dx = a \cos\theta d\theta \qquad\qquad \ldots(52)$

Therefore, $\qquad\qquad y = \int a \sin\theta d\theta + b,$

$$y = -a \cos\theta + b. \qquad\qquad \ldots(53)$$

Squaring and adding equations (52) and (53), we get $x^2 + (y - b)^2 = a^2$, which is the circle having center on y-axis.

Example 9: Show that the geodesic on a right circular cylinder is a helix.

Solution: We know the right circular cylinder is characterized by the equations

$$x^2 + y^2 = a^2, \, z = z \qquad\qquad \ldots(54)$$

The parametric equations of the right circular cylinder are

$$x = a \cos\theta, \, y = a \sin\theta, \, z = z,$$

where a is a constant. The element of the distance (metric) $ds^2 = dx^2 + dy^2 + dz^2$ on the surface of the cylinder becomes

$$ds^2 = a^2 d\theta^2 + dz^2. \qquad\qquad \ldots(55)$$

Hence the total length of the curve on the surface of the cylinder is given by

$$s = \int_{\theta_0}^{\theta} \sqrt{a^2 + z'^2} \; dz, \text{ for } z' = \frac{dz}{d\theta}. \qquad ...(56)$$

For extremum of s the Euler-Lagrange's equation gives

$$\Rightarrow \qquad \frac{d}{d\theta}\left(\frac{z'}{f}\right) = 0.$$

Integrating the equation and solving for z' we get $z' = a$, where a is constant. Integrating we get

$$z = a\theta + b, \; a \neq 0, \qquad ...(57)$$

where a, b are constants. Equation (57) gives the required equation of helix on the right circular cylinder. Thus the geodesic on the surface of a cylinder is a helix.

Example 10: Find the differential equation of the geodesic on the surface of an inverted cone with semi-vertical angle θ.

Solution: The surface of the cone is characterized by the equation

$$x^2 + y^2 = z^2 \tan^2 \theta, \; \theta = \text{const.}$$

The parametric equations of the cone are given by

$$x = ar \cos \phi, \; y = ar \sin \phi, \; z = br, \qquad ...(58)$$

where for $a = \sin \theta$, $b = \cos \theta$ are constant. On the surface of the cone the metric becomes

$$ds^2 = dr^2 + a^2 r^2 d\phi^2. \qquad ...(59)$$

Hence the total length of the curve $\phi = \phi(r)$ on the surface of the cone is given by

$$s = \int \sqrt{1 + a^2 r^2 \phi'^2} \; dr, \; \phi' = \frac{d\phi}{dr} \qquad ...(60)$$

The Euler-Lagrange's equation gives

$$\Rightarrow \qquad \frac{d}{dr}\left(\frac{a^2 r^2 \phi'}{f}\right) = 0.$$

Solving for ϕ' we get

$$\frac{d\phi}{dr} = \frac{c_1}{ar\sqrt{a^2 r^2 - c_1^2}}, \qquad ...(61)$$

where $c_1 = \text{constant}$.

This is the required differential equation of geodesic, and the geodesic on the surface of the cone is obtained by integrating equation (61). This gives

$$\phi = \int \frac{c_1}{ar\sqrt{a^2r^2 - c_1^2}}\, dr + \alpha.$$

$$\phi = \frac{1}{a}\, \sec^{-1}\!\left(\frac{ar}{c_1}\right) + \alpha$$

$$\Rightarrow \qquad r = \frac{c_1}{a}\, \sec\left[a(\phi - \alpha)\right].$$

Example 11:   Find the curve for which the functional

$$I\,[y(x)] = \int\limits_{0}^{\frac{\pi}{4}} \left(y^2 - y'^2\right)dx$$

can have extrema, given that $y(0) = 0$, while the right-hand end point can vary along the line $x = \dfrac{\pi}{4}$.

Solution: The Euler-Lagrange's equation for the given functional becomes

$$y'' + y = 0. \qquad\qquad\qquad ...(62)$$

This is the second order differential equation, whose solution is given by

$$y = a \cos x + b \sin x. \qquad\qquad ...(63)$$

The boundary condition $y(0) = 0$ gives $a = 0$.

$$\Rightarrow \qquad\qquad y = b \sin x. \qquad\qquad\qquad ...(64)$$

The second boundary point is not fixed but moves along the line $x = \dfrac{\pi}{4}$. In this case we have

$$\Rightarrow \qquad\qquad \left(\frac{\partial f}{\partial y'}\right)_{x=\frac{\pi}{4}} = 0,$$

$$\Rightarrow \qquad\qquad (y')_{x=\frac{\pi}{4}} = 0,$$

where from equation (64) we have $y' = b \cos x$. Thus y' at $x = \dfrac{\pi}{4}$ gives $b = 0$. This implies the extremal is attained on the x-axis.

Example 12:   Find the extremals of the functional

$$I[y(x)] = \int\limits_{0}^{\log 2} \left(e^{-x}\, y'^2 - e^{x}y^2\right)dx.$$

Solution: For the extremization of the functional in question, the Euler-Lagrange's equation yields

$$y'' - y' + e^{2x}y = 0. \qquad\qquad ...(65)$$

However, the solution of (65) cannot be easily obtained. Therefore we convert the original problem into a simple problem by substituting

$$x = \log u \text{ and } y = v.$$

When $x = 0$ we have $u = 1$, and when $x = \log 2 \Rightarrow u = 2$.

Thus for
$$u = e^x \Rightarrow \frac{du}{dx} = e^x$$

$$\Rightarrow \qquad dx = \frac{du}{u}.$$

Also
$$y = v \Rightarrow \frac{dy}{dx} = \frac{dv}{dx} = \frac{dv}{du}\frac{du}{dx}$$

$$\Rightarrow \qquad \frac{dy}{dx} = v'u, \; v' = \frac{dv}{du}.$$

Hence the functional in the question becomes

$$I(v(u)) = \int_1^2 \left(v'^2 - v^2\right)du. \qquad \qquad ...(66)$$

For extremization of (66), the corresponding Euler-Lagrange's equation gives

$$v'' + v = 0. \qquad \qquad ...(67)$$

The solution of this equation is given by

$$v = a \cos u + b \sin u. \qquad \qquad ...(68)$$

Putting it in to original co-ordinates we get the required curve

$$y = a \cos(e^x) + b \sin(e^x). \qquad \qquad ...(69)$$

Example 13:   If f satisfies Euler-Lagrange's equation (10), then show that f is the total derivative $\dfrac{dg}{dx}$ of some function of x and y and conversely.

Solution: Given that f satisfies Euler-Lagrange's equation (10). We claim that $f = \dfrac{dg}{dx}$, where $g = g(x, y)$. We write the Euler-Lagrange's equation (10) explictly as

$$\frac{\partial f}{\partial y} - \frac{\partial^2 f}{\partial x \partial y'} - \frac{\partial^2 f}{\partial y \partial y'}\, y' - \frac{\partial^2 f}{\partial^2 y'^2}\, y'' = 0. \qquad \qquad ...(70)$$

We see from equation (70) that the first three terms on the left hand side of (70) contain at the highest the first derivative of y. Therefore equation (70) is satisfied identically if the coefficient of y'' vanishes identically.

$$\Rightarrow \qquad \frac{\partial^2 f}{\partial y'^2} = 0.$$

Integrating with respect to y' we get

$$\frac{\partial f}{\partial y'} = q(x, y).$$

Integrating once again we get

$$f = q(x, y)y' + p(x, y), \qquad \qquad ...(71)$$

where $p(x, y)$ and $q(x, y)$ are constants of integration and may be function of x and y only. Then the function f so determined must satisfy the Euler-Lagrange's equation (10). From equation (71), we find

$$\frac{\partial f}{\partial y} = y'\frac{\partial q}{\partial y} + \frac{\partial p}{\partial y}, \text{ and } \frac{\partial f}{\partial y'} = q(x, y).$$

Therefore equation (10) becomes

$$y'\frac{\partial q}{\partial y} + \frac{\partial p}{\partial y} - \frac{d}{dx}(q(x, y)) = 0,$$

$$\Rightarrow \qquad y'\frac{\partial q}{\partial y} + \frac{\partial p}{\partial y} - \frac{\partial q}{\partial x} - \frac{\partial q}{\partial y}y' = 0.$$

$$\Rightarrow \qquad \frac{\partial p}{\partial y} = \frac{\partial q}{\partial x}. \qquad \qquad ...(72)$$

Equation (72) is the condition that the equation $pdx + qdy$ is an exact differential equation dg.

$$\Rightarrow \qquad dg = pdx + qdy,$$

$$\Rightarrow \qquad \frac{dg}{dx} = p + qy' = f \quad \text{by (71)}$$

Therefore,

$$f = \frac{dg}{dx}. \qquad \qquad ...(73)$$

This proves the necessary part.

Conversely, we assume that $f = \dfrac{dg}{dx}$, and prove f satisfies the Euler-Lagrange's equation (10). From equation (73), we have

$$f = \frac{dg}{dx} = \frac{\partial g}{\partial x} + \frac{\partial g}{\partial y}y'.$$

Therefore, from this we find

$$\frac{\partial f}{\partial y} = \frac{\partial^2 g}{\partial x\,\partial y} + \frac{\partial^2 g}{\partial y^2}y',$$

$$\frac{\partial f}{\partial y'} = \frac{\partial g}{\partial y}.$$

Consider now

$$\frac{\partial f}{\partial y} - \frac{d}{dx}\left(\frac{\partial f}{\partial y'}\right) = \frac{\partial^2 g}{\partial x \partial y} + \frac{\partial^2 g}{\partial y^2}\, y' - \frac{d}{dx}\left(\frac{\partial g}{\partial y}\right).$$

$$\frac{\partial f}{\partial y} - \frac{d}{dx}\left(\frac{\partial f}{\partial y'}\right) = \frac{\partial^2 g}{\partial x \partial y} + \frac{\partial^2 g}{\partial y^2}\, y' - \frac{\partial^2 g}{\partial x \partial y} - \frac{\partial^2 g}{\partial y^2}\, y'.$$

$$\frac{\partial f}{\partial y} - \frac{d}{dx}\left(\frac{\partial f}{\partial y'}\right) = 0.$$

This proves that the function $f = \dfrac{dg}{dx}$ satisfies Euler-Lagrange's equation.

Example 14:   Show that the Euler-Lagrange's equation of the functional

$$I(y(x)) = \int_{x_1}^{x_2} f(x, y, y')\,dx$$

has the first integral $f - y'\,\dfrac{\partial f}{\partial y'} = $ const, if the integrand does not depend on x.

Solution: Let the functional which satisfies the Euler-Lagrange's equation (10) be independent of x, then $\dfrac{\partial f}{\partial x} = 0$. Hence the Euler-Lagrange's equation (10) which is represented in the Equation (70) becomes

$$\frac{\partial f}{\partial y} - \frac{\partial^2 f}{\partial y \partial y'}\, y' - \frac{\partial^2 f}{\partial y'^2}\, y'' = 0. \qquad\qquad \text{...(74)}$$

Multiplying equation (74) by y' we get

$$\frac{\partial f}{\partial y}\, y' - \frac{\partial^2 f}{\partial y \partial y'}\, y'^2 - \frac{\partial^2 f}{\partial y'^2}\, y''y' = 0. \qquad\qquad \text{...(75)}$$

Now consider

$$\frac{d}{dx}\left(f - y'\,\frac{\partial f}{\partial y'}\right) = y'\,\frac{\partial f}{\partial y} + y''\,\frac{\partial f}{\partial y'} - y''\,\frac{\partial f}{\partial y'} - y'^2\,\frac{\partial^2 f}{\partial y' \partial y} - y'y''\,\frac{\partial^2 f}{\partial y'^2},$$

$$\frac{d}{dx}\left(f - y'\,\frac{\partial f}{\partial y'}\right) = y'\,\frac{\partial f}{\partial y} - y'^2\,\frac{\partial^2 f}{\partial y' \partial y} - y'y''\,\frac{\partial^2 f}{\partial y'^2}, \qquad\qquad \text{...(76)}$$

From equations (75) and (76) we see that

$$\frac{d}{dx}\left(f - y'\,\frac{\partial f}{\partial y'}\right) = 0,$$

$$\Rightarrow \qquad f - y'\,\frac{\partial f}{\partial y'} = \text{const.} \qquad \qquad \text{...(77)}$$

This is the first integral of Euler-Lagrange's equation, when the functional $f = f(y, y')$.

Example 15: Find the extremals of the functional

$$I(y(x)) = \int_{x_0}^{x_1} \frac{(1 + y^2)}{y'^2}\, dx.$$

Solution: We notice that the integrand f is independent of x. Hence the first integral of Euler's equation is given by the equation (77).

$$\Rightarrow \qquad \frac{1 + y^2}{y'^2} + 2y'\left(\frac{1 + y^2}{y'^3}\right) = \text{const.}$$

$$\Rightarrow \qquad \frac{1 + y^2}{y'^2} = a,$$

$$\Rightarrow \qquad \frac{dy}{\sqrt{1 + y^2}} = a\,dx.$$

Integrating we get

$$\sinh^{-1}(y) = ax + b.$$

$$y = \sinh(ax + b). \qquad \qquad \text{...(78)}$$

This is the required extremals for the given functional.

A Case of Variable end Points along Vertical Lines $x = a$ and $x = b$

Theorem 2   Among all curves whose end points lie on two given vertical lines $x = a$ and $x = b$, find the curve for which the functional

$$I(y(x)) = \int_{a}^{b} f(x, y, y')\,dx$$

has an extremum.

Proof: Let among all curves whose end points lie on two given vertical lines $x = a$ and $x = b$ be given by $y = y(x)$. Let the value of the integral along this curve be

$$I(y(x)) = \int_{a}^{b} f(x, y, y')\,dx. \qquad \qquad \text{...(79)}$$

Let
$$y = y(x) + \eta(x), \qquad \text{...(80)}$$

be the neighboring path, where $\eta(x)$ is a differentiable function of x such that

$$\eta(a) = \eta(b) = 0. \qquad \text{...(81)}$$

Therefore the value of the integral along the varied path is

$$I(y(x) + \eta(x)) = \int_a^b f(x, y + \eta, y' + \eta')dx \qquad \text{...(82)}$$

Hence the change in the value of the integral along the varied path is

$$\delta I = I[y + \eta] - I[y] = \int_a^b \left[f(x, y + \eta, y' + \eta') - f(x, y, y') \right]dx \qquad \text{...(83)}$$

Using the Taylor series expansion of the integrand we have

$$f(x, y + \eta, y' + \eta') = f(x, y, y') + \left(\frac{\partial f}{\partial y} \eta + \frac{\partial f}{\partial y'} \eta' \right) + \cdots$$

As η and η' are very small, by neglecting the higher order terms in η and η' we have therefore the value of the integral (83) as

$$\delta I = \int_a^b \left[\eta \frac{\partial f}{\partial y} + \eta' \frac{\partial f}{\partial y'} \right]dx.$$

Integrating the second integral by parts we obtain

$$\delta I = \int_a^b \eta \frac{\partial f}{\partial y} dx + \left[\eta(x) \frac{\partial f}{\partial y'} \right]_{x=a}^{x=b} - \int_a^b \frac{d}{dx}\left(\frac{\partial f}{\partial y'} \right)\eta(x)dx$$

$$\delta I = \int_a^b \left[\frac{\partial f}{\partial y} - \frac{d}{dx}\left(\frac{\partial f}{\partial y'} \right) \right]\eta(x)dx + \left[\eta(x) \frac{\partial f}{\partial y'} \right]_{x=b} - \left[\eta(x)\frac{\partial f}{\partial y'} \right]_{x=a} \qquad \text{...(84)}$$

Note here that unlike the fixed end points problem, $\eta(x)$ need no longer vanish at the points a and b. In order that the curve $y = y(x)$ to be a solution of the variable end point problem, y must be an extremal. *i.e.,* y must be a solution of Euler's equation. Thus for the extremal we have $\delta I = 0$

$$\Rightarrow \qquad \left[\frac{\partial f}{\partial y'} \right]_{x=a} = 0, \quad \left[\frac{\partial f}{\partial y'} \right]_{x=b} = 0. \qquad \text{...(85)}$$

Thus to solve the variable end point problem, we must first solve a general integral of Euler's equation and then use the condition (85) to determine the values of the arbitrary constants.

WORKED EXAMPLES

Example 16: Find the minimum of the functional

$$I(y(x)) = \int_0^1 \left(\frac{1}{2} y'^2 + yy' + y' + y \right) dx$$

if the values at the end points are not given.

Solution: The Euler-Lagrange's equation gives

$$y' + 1 - \frac{d}{dx}(y' + y + 1) = 0,$$

$$\Rightarrow \qquad\qquad y'' = 1. \qquad\qquad\qquad ...(86)$$

Integrating we get

$$y' = x + c_1. \qquad\qquad\qquad ...(87)$$

Further integrating gives

$$y = \frac{x^2}{2} + c_1 x + c_2, \qquad\qquad\qquad ...(88)$$

where c_1, c_2 are constants of integration and are to be determined. However, note that the values of y at the end points are not prescribed. In this case the constants are determined from the conditions.

$$\left(\frac{\partial f}{\partial y'} \right)_{x=0} = 0, \text{ and } \left(\frac{\partial f}{\partial y'} \right)_{x=1} = 0, \qquad\qquad ...(89)$$

$$\Rightarrow \qquad (y' + y + 1)_{x=0} = 0, \text{ and } (y' + y + 1)_{x=1} = 0, \qquad ...(90)$$

where from equations (87) and (88), we have

$$y'(0) = c_1 \text{ and } y(0) = c_2.$$

Similarly, $\qquad\qquad y'(1) = 1 + c_1 \text{ and } y(1) = \frac{1}{2} + c_1 + c_2.$

Thus the conditions (90) give

$$c_1 + c_2 + 1 = 0, \quad 2c_1 + c_2 + \frac{5}{2} = 0.$$

Solving these equations for c_1 and c_2 we obtain $c_1 = -\frac{3}{2}$ and $c_2 = \frac{1}{2}$.

Hence the required curve for which the functional given in (79) becomes minimum is

$$y = \frac{1}{2} \left(x^2 - 3x + 1 \right). \qquad\qquad\qquad ...(91)$$

Theorem 3 Find the Euler-Lagrange differential equation satisfied by four times differentiable function $y(x)$ which extremizes the functional

$$I(y(x)) = \int_{x_1}^{x_2} f(x, y, y', y'')dx$$

under the conditions that both y and y' are prescribed at the end points.

Proof: Let $P(x_1, y_1)$ and $Q(x_2, y_2)$ be two fixed points in xy-plane. The points P and Q can be joined by infinitely many curves. Accordingly the value of the integral I will be different for different paths. We shall look for a curve along which the functional I has an extremum value. Let c be a curve between P and Q whose equation is given by $y = y(x, 0)$. Let also the value of the functional along the curve c be extremum and is given by

$$I(y(x)) = \int_{x_1}^{x_2} f(x, y, y', y'')dx \qquad \qquad ...(92)$$

We can label all possible paths starting from P and ending at Q by the family of equations

$$y(x, \alpha) = y(x, 0) + \alpha\eta(x), \qquad \qquad ...(93)$$

where α is a parameter and $\eta(x)$ is any differentiable function of x. For different values of α we get different curves. Accordingly the value of the integral I will be different for different paths. Since y and y' are prescribed at the end points, this implies that there is no variation in y and y' at the end points. *i.e.*, all the curves of the family and their derivative must be identical at fixed points P and Q. This implies that

$$y(x_1, \alpha) = y(x_1, 0) = y_1, \qquad y'(x_1, \alpha) = y'(x_1, 0) = y_1',$$

$$y(x_2, \alpha) = y(x_2, 0) = y_2, \text{ and } \quad y'(x_2, \alpha) = y'(x_2, 0) = y_2',$$

$$\Rightarrow \qquad \eta(x_1) = 0 = \eta(x_2) \quad \text{ and } \eta'(x_1) = 0 = \eta'(x_2). \qquad \qquad ...(94)$$

Conversely, the conditions (94) ensure us that the curves of the family that all pass through the points P and Q. Let the value of the functional along the neighboring curve be given by

$$I(y(x, \alpha)) = \int_{x_1}^{x_2} f(x, y(x, \alpha), y'(x, \alpha), y''(x, \alpha))dx. \qquad \qquad ...(95)$$

From differential calculus, we know the integral I is extremum if $\left(\dfrac{\partial I}{\partial \alpha}\right)_{\alpha=0} = 0$, because for $\alpha = 0$ the neighboring curve coincides with the curve which gives extremum values of I. Thus

$$\left(\frac{\partial I}{\partial \alpha}\right)_{\alpha=0} = 0, \Rightarrow \int_{x_1}^{x_2}\left(\frac{\partial f}{\partial y}\,\eta(x) + \frac{\partial f}{\partial y'}\,\eta'(x) + \frac{\partial f}{\partial y''}\,\eta''(x)\right)dx = 0.$$

Integrating the second and the third integrations by parts, we get

$$\int_{x_1}^{x_2} \frac{\partial f}{\partial y}\, \eta(x)\,dx + \left(\frac{\partial f}{\partial y'}\, \eta(x)\right)_{x_1}^{x_2} - \int_{x_1}^{x_2} \frac{d}{dx}\left(\frac{\partial f}{\partial y'}\right)\eta(x)\,dx + \left(\frac{\partial f}{\partial y''}\, \eta'(x)\right)_{x_1}^{x_2} -$$

$$-\left(\frac{d}{dx}\left(\frac{\partial f}{\partial y''}\right)\eta(x)\right)_{x_1}^{x_2} + \int_{x_1}^{x_2} \frac{d^2}{dx^2}\left(\frac{\partial f}{\partial y''}\right)\eta(x)\,dx = 0 \qquad \text{...(96)}$$

$$\int_{x_1}^{x_2} \frac{\partial f}{\partial y}\, \eta(x)\,dx + \left(\left[\frac{\partial f}{\partial y'} - \frac{d}{dx}\left(\frac{\partial f}{\partial y''}\right)\right]\eta(x)\right)_{x_1}^{x_2} - \int_{x_1}^{x_2} \frac{d}{dx}\left(\frac{\partial f}{\partial y'}\right)\eta(x)\,dx +$$

$$+ \left(\frac{\partial f}{\partial y''}\, \eta'(x)\right)_{x_1}^{x_2} + \int_{x_1}^{x_2} \frac{d^2}{dx^2}\left(\frac{\partial f}{\partial y''}\right)\eta(x)\,dx = 0 \qquad \text{...(97)}$$

As y and y' are both prescribed at the end points, hence on using equations (94) we obtain

$$\int_{x_1}^{x_2} \left(\frac{\partial f}{\partial y} - \frac{d}{dx}\left(\frac{\partial f}{\partial y'}\right) + \frac{d^2}{dx^2}\left(\frac{\partial f}{\partial y''}\right)\right)\eta(x)\,dx = 0. \qquad \text{...(98)}$$

By using the basic lemma of calculus of variation we get

$$\frac{\partial f}{\partial y} - \frac{d}{dx}\left(\frac{\partial f}{\partial y'}\right) + \frac{d^2}{dx^2}\left(\frac{\partial f}{\partial y''}\right) = 0. \qquad \text{...(99)}$$

This is required Euler-Lagrange differential equation to be satisfied by $y(x)$ for which the functional I has extremum value.

Note: If the functions y and y' are not prescribed at the end points then we must have unlike the fixed end point problem, $\eta(x)$ and $\eta'(x)$ need no longer vanish at the points x_1 and x_2. In order that the curve $y = y(x)$ to be a solution of the variable end point problem, y must be an extremal. *i.e.*, y must be a solution of Euler's equation (99). Thus for the extremal of $y(x)$ we have from equation (97)

$$\Rightarrow \qquad \left[\frac{\partial f}{\partial y''}\right]_{x=a}^{x=b} = 0, \quad \left[\frac{\partial f}{\partial y'} - \frac{d}{dx}\left(\frac{\partial f}{\partial y''}\right)\right]_{x=a}^{x=b} = 0. \qquad \text{...(100)}$$

Thus to solve the variable end point problem, we must first solve Euler's equation (99) and then use the conditions (100) to determine the values of the arbitrary constants.

The above result can be summarized in the following theorem (3a).

Theorem (3a)   Derive the differential equation satisfied by four times differentiable function $y(x)$, which extremizes the integral

$$I = \int_{x_0}^{x_1} f(x, y, y', y'')dx$$

under the condition that both y, y' are prescribed at both the ends. Show that if neither y nor y' is prescribed at either end points then

$$\left(\frac{\partial f}{\partial y''}\right)_{x=x_0} = \left(\frac{\partial f}{\partial y''}\right)_{x=x_1} = 0, \quad \left[\frac{\partial f}{\partial y'} - \frac{d}{dx}\left(\frac{\partial f}{\partial y''}\right)\right]_{x=x_0}^{x=x_1} = 0. \qquad ...(101)$$

General case

If the integrand of the definite integral which is to be extremized is of the form $f = f(x, y, y', y'', \ldots , y^n)$ with the boundary conditions

$$y(x_1) = y_1, \, y'(x_1) = y'_1, \, \ldots\ldots, \, y^{n-1}(x_1) = y_1^{n-1},$$

$$y(x_2) = y_2, \, y'(x_2) = y'2, \, \ldots\ldots, \, y^{n-1}(x_2) = y_2^{n-1},$$

then the Euler-Lagrange's differential equation is

$$\frac{\partial f}{\partial y} - \frac{d}{dx}\left(\frac{\partial f}{\partial y'}\right) + \frac{d^2}{dx^2}\left(\frac{\partial f}{\partial y''}\right) + \ldots\ldots + (-1)^n \frac{d^n}{dx^n}\left(\frac{\partial f}{\partial y^n}\right) = 0. \qquad ...(102)$$

WORKED EXAMPLES

Example 17:   Find the curve, which extremizes the functional

$$I(y(x)) = \int_0^{\frac{\pi}{4}} \left(y''^2 - y^2 + x^2\right)dx$$

under the conditions that $y(0) = 0$, $y'(0) = 1$, $y\left(\frac{\pi}{4}\right) = y'\left(\frac{\pi}{4}\right) = \frac{1}{\sqrt{2}}$.

Solution: To find the curve which extremizes the given definite integral, the integrand $f = y''^2 - y^2 + x^2$ must satisfy the Euler–Lagrange's equation (99). This gives

$$-2y + \frac{d^2}{dx^2}(2y'') = 0$$

$$\Rightarrow \qquad\qquad \frac{d^4 y}{dx^4} - y = 0. \qquad\qquad ...(103)$$

The solution of equation (103) is given by

$$y = ae^x + be^{-x} + c \cos x + d \sin x \qquad \qquad ...(104)$$

where a, b, c, d are constants of integration and are to be determined. The conditions

$$y(0) = 0 \qquad \Rightarrow a + b + c = 0,$$

$$y\left(\frac{\pi}{4}\right) = \frac{1}{\sqrt{2}} \qquad \Rightarrow ae^{\frac{\pi}{4}} + be^{-\frac{\pi}{4}} + \frac{1}{\sqrt{2}} c + \frac{1}{\sqrt{2}} d = \frac{1}{\sqrt{2}},$$

$$y'(0) = 1 \qquad \Rightarrow a - b + d = 1,$$

$$y'\left(\frac{\pi}{4}\right) = \frac{1}{\sqrt{2}} \qquad \Rightarrow ae^{\frac{\pi}{4}} - be^{-\frac{\pi}{4}} - \frac{1}{\sqrt{2}} c + \frac{1}{\sqrt{2}} d = \frac{1}{\sqrt{2}}.$$

Solving these equations we get $a = b = c = 0$ and $d = 1$. Hence the required curve is $y = \sin x$.

Example 18: Minimize the functional $I = \dfrac{1}{2} \displaystyle\int_0^2 (\ddot{x})^2 dt$ satisfying the conditions

$$x(0) = 1, \; \dot{x}(0) = 1, \; x(2) = 1, \; \dot{x}(2) = 0.$$

Solution: The Euler-Lagrange's equation (99) in this case reduces to the fourth order equation

$$\frac{d^4 x}{dt^4} = 0. \qquad \qquad ...(105)$$

Integrating we get

$$x = c_1 \frac{t^3}{6} + c_2 \frac{t^2}{2} + c_3 t + c_4. \qquad \qquad ...(106)$$

where x given in (106) must satisfy the conditions

$$x(0) = 1 \Rightarrow c_4 = 1, \; \dot{x}(0) = 1 \Rightarrow c_3 = 1,$$

$$x(2) = 1 \Rightarrow 4c_1 + 6c_2 = -3, \; \dot{x}(2) = 0 \Rightarrow 2c_1 + 2c_2 = -1.$$

Solving for c_1 and c_2 we get the required functional in the form

$$x = -\frac{t^2}{4} + t + 1.$$

Example 19: Find the function on which the functional

$$I(y(x)) = \int_0^1 \left(y''^2 - 2xy\right)dx,$$

can be extremized such that $y(0) = y'(0)$, $y(1) = \dfrac{1}{120}$ and $y'(1)$ is not prescribed.

Solution: The Euler-Lagrange's differential equation (99) yields

$$-2x + \frac{d^2}{dx^2}(2y'') = 0,$$

$$\Rightarrow \qquad \frac{d^4y}{dx^4} = x.$$

Integrating we obtain

$$y''' = \frac{x^2}{2} + a,$$

$$\Rightarrow \qquad y'' = \frac{x^3}{6} + ax + b,$$

$$\Rightarrow \qquad y' = \frac{x^4}{24} + \frac{a}{2}x^2 + bx + c,$$

$$\Rightarrow \qquad y = \frac{x^5}{120} + \frac{a}{6}x^3 + \frac{b}{2}x^2 + cx + d, \qquad \qquad ...(107)$$

where a, b, c, d are constants to be determined. Using the given conditions, we obtain

$$y(0) = 0 \Rightarrow d = 0,$$

$$y'(0) = 0 \Rightarrow c = 0,$$

$$y(1) = \frac{1}{120} \Rightarrow a + 3b = 0.$$

This is the only one equation and two unknowns to be determined. As $y'(1)$ is not prescribed *i.e.*, y' at $x = 1$ is not given, then in this case we use the condition given by

$$\left(\frac{\partial f}{\partial y''}\right)_{x=1} = 0 \Rightarrow y''(1) = 0.$$

This gives another equation $6a + 6b = -1$. Solving these two equations in a and b, we get

$$a = -\frac{1}{4} \text{ and } b = \frac{1}{12}.$$

Substituting these values in equation (107), we get the required curve

$$y = \frac{x^5}{120} + \frac{1}{24}\left(x^2 - x^3\right).$$

Example 20: Find the function on which the functional

$$I(y(x)) = \int_0^1 (y''^2 - 2xy)\,dx,$$

can be extremized such that $y(0) = 0$, $y(1) = \dfrac{1}{120}$ and y' is not prescribed at both the ends.

Solution: The Euler-Lagrange's differential equation (99) gives

$$\frac{d^4 y}{dx^4} = x. \qquad\qquad ...(108)$$

The solution of this differential equation is given by

$$\Rightarrow \qquad\qquad y = \frac{x^5}{120} + \frac{a}{6} x^3 + \frac{b}{2} x^2 + cx + d, \qquad\qquad ...(109)$$

where a, b, c, d are constants to be determined. The conditions

$$y(0) = 0 \Rightarrow d = 0,$$

$$y(1) = \frac{1}{120} \Rightarrow a + 3b + 6c = 0.$$

To determine these three constants we need two more relations. However, y' is not prescribed at both the ends. In this case we have

$$\left(\frac{\partial f}{\partial y''} \right)_{x=0} = 0 \Rightarrow y''(0) = (0) \Rightarrow b = 0,$$

$$\left(\frac{\partial f}{\partial y''} \right)_{x=1} = 0 \Rightarrow y''(1) = 0 \Rightarrow a = -\frac{1}{6}.$$

Hence $c = \dfrac{1}{36}$. Thus the required curve is

$$y = \frac{x^5}{120} + \frac{1}{36}\left(x - x^3 \right).$$

Example 21: Establish the Euler equation corresponding to the variational problem

$$I(y(x)) = \int_{x_1}^{x_2} f(y,\, y'')\,dx,$$

where y and y' are prescribed and if $f = y^2 - y''^2$, determine the function satisfying the conditions $y(0) = 1$, $y\left(\dfrac{\pi}{2}\right) = 0$, $y'\left(\dfrac{\pi}{2}\right) = -1$.

Solution: We see that the integrand does not involve y'. Hence the corresponding Euler-Lagrange's differential equation will be

$$\frac{\partial f}{\partial y} + \frac{d^2}{dx^2}\left(\frac{\partial f}{\partial y''}\right) = 0.$$ ...(110)

This equation yield

$$\frac{d^4 y}{dx^4} - y = 0.$$ ...(111)

The solution of this equation is given by

$$y = ae^x + be^{-x} + c \cos x + d \sin x.$$ ...(112)

The boundary conditions give

$$y(0) = 1 \Rightarrow a + b + c = 1,$$

$$y\left(\frac{\pi}{2}\right) = 0 \Rightarrow ae^{\frac{\pi}{2}} + be^{\frac{-\pi}{2}} + d = 0,$$

$$y'(0) = 0 \Rightarrow a - b + d = 0,$$

$$y'\left(\frac{\pi}{2}\right) = -1 \Rightarrow ae^{\frac{\pi}{2}} - be^{\frac{-\pi}{2}} - c = -1.$$

Solving these equations we readily obtain the values of the constants as
$$a = b = d = 0 \text{ and } c = 1.$$

Hence the required curve becomes $y = \cos x$.

Example 22: Show that the Euler-Lagrange's equation of the functional

$$I(y(x)) = \int_{x_1}^{x_2} f(x, y, y', y'')dx$$

has the first integral $\dfrac{\partial f}{\partial y'} - \dfrac{d}{dx}\left(\dfrac{\partial f}{\partial y''}\right) = $ const., if the integrand does not depend on

y, and the first integral $f - y'\left[\dfrac{\partial f}{\partial y'} - \dfrac{d}{dx}\left(\dfrac{\partial f}{\partial y''}\right)\right] - y''\dfrac{\partial f}{\partial y''} = $ const., if the integrand

does not depend on x.

Solution: The Euler-Lagrange's differential equation of the given functional is given in equation (99)

Case 1: Let the integrand does not depend on y. This implies

$$f = f(x, y', y''). \Rightarrow \frac{\partial f}{\partial y} = 0.$$

Hence equation (99) becomes $\dfrac{d}{dx}\left(\dfrac{\partial f}{\partial y'}\right) - \dfrac{d^2}{dx^2}\left(\dfrac{\partial f}{\partial y''}\right) = 0,$

$$\frac{d}{dx}\left(\frac{\partial f}{\partial y'} - \frac{d}{dx}\left(\frac{\partial f}{\partial y''}\right)\right) = 0.$$

Integrating we get $\dfrac{\partial f}{\partial y'} - \dfrac{d}{dx}\left(\dfrac{\partial f}{\partial y''}\right) = \text{const.}$ 　　　　　...(113)

Case 2: Let the integrand f does not contain x explicitly. This means $f = f(y, y', y'')$. $\Rightarrow \dfrac{\partial f}{\partial x} = 0$. Consider

$$\frac{d}{dx}\left[f - y'\left[\frac{\partial f}{\partial y'} - \frac{d}{dx}\left(\frac{\partial f}{\partial y''}\right)\right] - y''\frac{\partial f}{\partial y''}\right] = \frac{df}{dx} - y''\left[\frac{\partial f}{\partial y'} - \frac{d}{dx}\left(\frac{\partial f}{\partial y''}\right)\right] -$$

$$-y'\frac{d}{dx}\left[\frac{\partial f}{\partial y'} - \frac{d}{dx}\left(\frac{\partial f}{\partial y''}\right)\right] - y'''\frac{\partial f}{\partial y''} - y''\frac{d}{dx}\left(\frac{\partial f}{\partial y''}\right).$$

$$\frac{d}{dx}\left[f - y'\left[\frac{\partial f}{\partial y'} - \frac{d}{dx}\left(\frac{\partial f}{\partial y''}\right)\right] - y''\frac{\partial f}{\partial y''}\right] = y'\frac{\partial f}{\partial y} + y''\frac{\partial f}{\partial y'} + y'''\frac{\partial f}{\partial y''} -$$

$$- y''\left[\frac{\partial f}{\partial y'} - \frac{d}{dx}\left(\frac{\partial f}{\partial y''}\right)\right] - y'\frac{d}{dx}\left[\frac{\partial f}{\partial y'} - \frac{d}{dx}\left(\frac{\partial f}{\partial y''}\right)\right] - y'''\frac{\partial f}{\partial y''} - y''\frac{d}{dx}\left(\frac{\partial f}{\partial y''}\right),$$

$$\frac{d}{dx}\left[f - y'\left[\frac{\partial f}{\partial y'} - \frac{d}{dx}\left(\frac{\partial f}{\partial y''}\right)\right] - y''\frac{\partial f}{\partial y''}\right] = y'\left[\frac{\partial f}{\partial y} - \frac{d}{dx}\left(\frac{\partial f}{\partial y'}\right) + \frac{d^2}{dx^2}\left(\frac{\partial f}{\partial y''}\right)\right].$$

On using equation (99) we obtain

$$\frac{d}{dx}\left[f - y'\left[\frac{\partial f}{\partial y'} - \frac{d}{dx}\left(\frac{\partial f}{\partial y''}\right)\right] - y''\frac{\partial f}{\partial y''}\right] = 0.$$ 　　　　　...(114)

Integrating we get

$$f - y'\left[\frac{\partial f}{\partial y'} - \frac{d}{dx}\left(\frac{\partial f}{\partial y''}\right)\right] - y''\frac{\partial f}{\partial y''} = \text{const.}$$

This is the first integral of the Euler's equation when the integrand does not contain x explicitly.

Example 23: Find the extremal of the functional

$$I(y(x)) = \int_a^b f(x, y) \sqrt{1 + y'^2}\, dx.$$

Solution: For the extremization of the functional in question the Euler-Lagrange's equation (10) gives

$$\frac{\partial f}{\partial y} \sqrt{1 + y'^2} - \frac{d}{dx}\left(\frac{fy'}{\sqrt{1 + y'^2}}\right) = 0.$$

$$\frac{\partial f}{\partial y} \sqrt{1 + y'^2} - \frac{\partial f}{\partial x}\frac{y'}{\sqrt{1 + y'^2}} - \frac{\partial f}{\partial y}\frac{y'^2}{\sqrt{1 + y'^2}} - f\frac{y''}{\sqrt{1 + y'^2}} + f\frac{y'y''}{\left(1 + y'^2\right)^{\frac{3}{2}}} = 0.$$

$$\frac{\partial f}{\partial y} - \frac{\partial f}{\partial x}y' - f\frac{y''}{\left(1 + y'^2\right)} = 0. \qquad\qquad ...(115)$$

This is the required equation.

When Integrand is a Function of More than Two Dependent Variables

Example 24: Prove that the shortest distance between two points in a Euclidean 3-space is a straight line.

Solution: In the 3-dimentional Euclidean space define a curve by the equations $y = y(x)$, $z = z(x)$. Let $P(x, y, z)$ and $Q(x + dx, y + dy, z + dz)$ be two neighboring points on the curve joining the points $A(x_1, y_1, z_1)$ and $B(x_2, y_2, z_2)$. The total distance between the points A and B along the curve is given by

$$I = \int_{x_1}^{x_2} \left(1 + \dot{y}^2 + \dot{z}^2\right)^{\frac{1}{2}} dx, \quad \dot{y} = \frac{dy}{dx}. \qquad\qquad ...(116)$$

The functional I is shortest if the integrand must satisfy the Euler-Lagrange's differential equations (10) corresponding to two dependent variables y and z. These equations give

$$\frac{d}{dx}\left(\frac{\dot{y}}{f}\right) = o \Rightarrow \dot{y} = af, \quad a = \text{const.}$$

$$\Rightarrow \qquad \dot{y}^2\left(1 - a^2\right) - a^2\dot{z}^2 = a^2, \qquad\qquad ...(117)$$

and $\qquad \dot{z}^2\left(1 - b^2\right) - b^2\dot{y}^2 = b^2. \qquad\qquad ...(118)$

Solving equations (117) and (118) for $\dot{y}$ and $\dot{z}$ we obtain

$$\dot{y} = \pm \frac{a}{\sqrt{1 - a^2 - b^2}}, \text{ and } \dot{z} = \pm \frac{b}{\sqrt{1 - a^2 - b^2}},$$

i.e.,
$$\dot{y} = \pm c_1, \quad \text{for } c_1 = a\left(1 - a^2 - b^2\right)^{-\frac{1}{2}} \qquad \text{...(119)}$$

and
$$\dot{z} = \pm c_2 \quad \text{for } c_2 = b\left(1 - a^2 - b^2\right)^{-\frac{1}{2}}. \qquad \text{...(120)}$$

Integrating equations (119) and (120), we get

$$y = \pm c_1 x + \phi(z), \; z = \pm c_2 x + \psi(y), \qquad \text{...(121)}$$

where $\phi(z)$ and $\psi(y)$ are constants of integration and may be functions of z and y respectively. Thus the required curve is given by equation (121). But these equations represent a pair of planes. The common point of intersection of these planes is the straight line. Hence the shortest distance between two points in Euclidean 3-space is a straight line.

Example 25: Show that the geodesic defined in the 3-dimentional Euclidean space by the equations $x = x(t)$, $y = y(t)$, $z = z(t)$ is a straight line.

Solution: The representation of a curve in the parametric form is not unique. A curve in 3-dimensional Euclidean space can also be represented by the equations

$$x = x(t), \; y = y(t), \; z = z(t),$$

where t is a parameter of the curve. The infinitesimal distance between two neighboring points on the curve becomes

$$ds^2 = \left(\dot{x}^2 + \dot{y}^2 + \dot{z}^2\right)dt^2.$$

Hence the total length of the curve between the points $P(t_0)$ and $P(t_1)$ is given by the integral

$$I = \int_{t_0}^{t_1} \left(\dot{x}^2 + \dot{y}^2 + \dot{z}^2\right)^{\frac{1}{2}} dt. \qquad \text{...(122)}$$

The curve is geodesic if the length of the curve I is extremum. This is true if the integrand must satisfy the Euler-Lagrange's equations (16) given by the equations

$$\frac{\partial f}{\partial x_i} - \frac{d}{dt}\left(\frac{\partial f}{\partial \dot{x}_i}\right) = 0, \; \forall \; i = 1, 2, 3 \text{ with } x_i = (x, y, z), \qquad \text{...(123)}$$

where $\dfrac{\partial f}{\partial x_i} = 0$ and $\dfrac{\partial f}{\partial \dot{x}_i} = \dfrac{\dot{x}_i}{f}$, $\forall \; i = 1, 2, 3. \Rightarrow \dot{x} = af, \dot{y} = bf, \dot{z} = cf.$

Simplifying these equations, we obtain

$$\left(a^2 - 1\right)\dot{x}^2 + a^2\dot{y}^2 + a^2\dot{z}^2 = 0,$$

$$b^2\dot{x}^2 + \left(b^2 - 1\right)\dot{y}^2 + b^2\dot{z}^2 = 0,$$

$$c^2\dot{x}^2 + c^2\dot{y}^2 + \left(c^2 - 1\right)\dot{z}^2 = 0. \qquad \qquad ...(124)$$

These equations are consistent provided

$$\begin{vmatrix} a^2 - 1 & a^2 & a^2 \\ b^2 & b^2 - 1 & b^2 \\ c^2 & c^2 & c^2 - 1 \end{vmatrix} = 0.$$

$$\Rightarrow \qquad \qquad a^2 + b^2 + c^2 = 1.$$

Solving equations (124) we obtain

$$\dot{x} = \frac{a}{\sqrt{1 - a^2 - b^2}}\, \dot{z} \text{ and } \dot{y} = \frac{b}{\sqrt{1 - a^2 - b^2}}\, \dot{z}, \dot{z} \neq 0. \qquad ...(125)$$

Integrating equations (125) we obtain

$$x = c_1 z + \phi(y), \quad c_1 = \frac{a}{\sqrt{1 - a^2 - b^2}}, \qquad \qquad ...(126)$$

$$y = c_2 z + \psi(x), \quad c_2 = \frac{b}{\sqrt{1 - a^2 - b^2}}, \qquad \qquad ...(127)$$

where $\psi(x)$ and $\phi(y)$ are constants of integration and may be functions of x and y respectively. Equations (126) and (127) represent planes. The locus of the common points of these planes is the straight line. Hence the geodesic in 3-dimentional Euclidean space is the straight line.

Example 26: Show that the functional

$$I = \int_0^{\frac{\pi}{2}} \left(2xy + x'^2 + y'^2\right)dt, \ x' = \frac{dx}{dt}$$

such that $x(0) = 0,\ x\!\left(\frac{\pi}{2}\right) = -1,\ y(0) = 0,\ y\!\left(\frac{\pi}{2}\right) = 1$ is stationary for

$$x = -\sin t, \ y = \sin t.$$

Solution: The Euler-Lagrange's differential equations for the functional I to be stationary are given by

$$x'' - y = 0, \ y'' - x = 0. \qquad \qquad ...(128)$$

A single equation obtained from equations (128) is

$$\frac{d^4x}{dt^4} - x = 0. \qquad \qquad \dots(129)$$

The solution of this equation is given by

$$x = ae^t + be^{-t} + c \cos t + d \sin t. \qquad \dots(130)$$

$$\Rightarrow \qquad y = ae^t + be^{-t} - c \cos t - d \sin t. \qquad \dots(131)$$

The boundary conditions can be used to determine these constants.

$$x(0) = 0 \Rightarrow a + b + c = 0,$$

$$x\left(\frac{\pi}{2}\right) = -1 \Rightarrow ae^{\frac{\pi}{2}} + be^{-\frac{\pi}{2}} + d = -1,$$

$$y(0) = 0 \Rightarrow a + b - c = 0,$$

$$y\left(\frac{\pi}{2}\right) = 1 \Rightarrow ae^{\frac{\pi}{2}} + be^{-\frac{\pi}{2}} - d = 1.$$

Solving we get $a = b = c = 0$ and $d = -1$. Thus the required solutions are

$$x = -\sin t, \; y = \sin t.$$

Isoperimetric Problems

The problems in which the functional which is eligible for the extremization of a given definite integral is required to confirm with certain restrictions that are given as the boundary conditions. Such problems are called isoperimetric problems. The method of extremization of an isoperimetric problem is exactly analogous to the method of finding stationary value of a function under certain conditions by Lagrange's multipliers method in differential calculus. We prove the isoperimetric problem in the following theorem.

Theorem 4 Obtain the differential equation, which is satisfied by the functional $f(x, y, y')$ which extremizes the integral $I(y(x)) = \int_{x_1}^{x_2} f(x, y, y')dx$

subject to the conditions $y(x_1) = y_1$, $y(x_2) = y_2$, and the integral

$$J = \int_{x_1}^{x_2} g(x, y, y')dx = \text{constant}.$$

Proof: Consider the functional between two points $P(x_1, x_1)$ and $Q(x_2, y_2)$ given by

$$I(y(x)) = \int_{x_1}^{x_2} f(x, y, y')dx \qquad \dots(132)$$

subject to the conditions

$$y(x_1) = y_1, \ y(x_2) = y_2, \qquad\qquad \text{...(133)}$$

and
$$J = \int_{x_1}^{x_2} g(x, y, y')dx = \text{constant.} \qquad\qquad \text{...(134)}$$

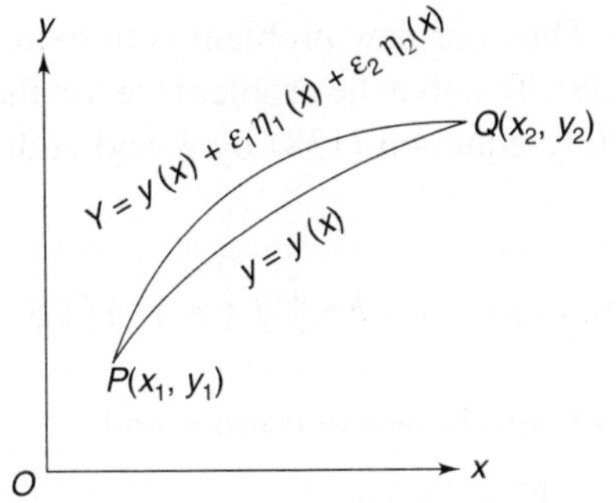

The points P and Q can be joined by infinitely many curves. Accordingly the value of the integral I will be different for different paths. Let all possible paths starting from P and ending at Q be given by two parameters family of curves

$$Y(x) = y(x) + \varepsilon_1\eta_1(x) + \varepsilon_2\eta_2(x) \qquad\qquad \text{...(135)}$$

where ε_1, ε_2 are parameters and $\eta_1(x)$, $\eta_2(x)$ are arbitrary differentiable functions of x such that

$$\eta_1(x_1) = 0 = \eta_1(x_2), \ \eta_2(x_1) = 0 = \eta_2(x_2). \qquad\qquad \text{...(136)}$$

These conditions ensure us that the curves of the family that all pass through the points P and Q. Note that, we cannot however, express $Y(x)$ as merely a one parameter family of curves, because any change in the value of the single parameter would in general alter the value of J, whose constancy must be maintained as prescribed. For this reason we introduce two parameter families of curves. We shall look for a curve along which the functional I has an extremum value under the condition (134). Let c be such a curve between P and Q whose equation is given by $y = y(x)$ such that the functional (132) along the curve c has extremum value. The values of the integrals (132) and (134) along the neighboring curve (135) are obtained by replacing y by Y in both the equations (132) and (134). Thus we have

$$I(\varepsilon_1, \varepsilon_2) = \int_{x_1}^{x_2} f(x, Y, Y')dx \qquad\qquad \text{...(137)}$$

and
$$J(\varepsilon_1, \varepsilon_2) = \int_{x_1}^{x_2} g(x, Y, Y')dx = \text{const.} \qquad\qquad \text{...(138)}$$

Equation (138) shows that ε_1 and ε_2 are not independent, but are related by equation

$$J(\varepsilon_1, \varepsilon_2) = \text{const.} \qquad \qquad ...(139)$$

Thus the changes in the value of the parameters are such that the constancy of (138) is maintained. Thus our new problem is to extremize the integral (137) under the restriction (138). To solve the problem we use the method of Lagrange's multipliers. Thus multiply equation (138) by λ and adding it to equation (137) we get

$$I^*(\varepsilon_1, \varepsilon_2) = I + \lambda J = \int_{x_1}^{x_2} f^*(x, Y, Y')dx \qquad \qquad ...(140)$$

where λ is Lagrange's undetermined multiplier and

$$f^* = f + \lambda g. \qquad \qquad ...(141)$$

Thus extremization of (132) subject to the condition (134) is equivalent to the extremization of (140). Thus from differential calculus, the integral I^* is extremum if $\left(\dfrac{\partial I^*}{\partial \varepsilon_j}\right)_{\varepsilon_1=0,\varepsilon_2=0} = 0$. Thus from equation (140) we have

$$\left(\frac{\partial I^*}{\partial \varepsilon_j}\right)_{\varepsilon_1=0,\varepsilon_2=0} = 0 \Rightarrow \int_{x_1}^{x_2}\left(\frac{\partial f^*}{\partial y}\, \eta_j(x) + \frac{\partial f^*}{\partial y'}\, \eta_j'(x) \right)dx = 0. \, J = 1,\, 2 \quad ...(142)$$

Note here that by setting $\varepsilon_1 = \varepsilon_2 = 0$, we replace Y and Y' to y, y'. Integrating the second integration by parts, we get

$$\int_{x_1}^{x_2} \frac{\partial f^*}{\partial y}\, \eta_j(x)dx + \left(\frac{\partial f^*}{\partial y'}\, \eta_j(x) \right)_{x_1}^{x_2} - \int_{x_1}^{x_2} \frac{d}{dx}\left(\frac{\partial f^*}{\partial y'} \right)\eta_j(x)dx = 0 \qquad ...(143)$$

As any curve is prescribed at the end points, hence on using conditions (136) we obtain

$$\int_{x_1}^{x_2}\left(\frac{\partial f^*}{\partial y} - \frac{d}{dx}\left(\frac{\partial f^*}{\partial y'} \right) \right)\eta_j(x)\, dx = 0, \, j = 1,\, 2.$$

By using the basic lemma of calculus of variation we get

$$\frac{\partial f^*}{\partial y} - \frac{d}{dx}\left(\frac{\partial f^*}{\partial y'} \right) = 0. \qquad \qquad ...(144)$$

This is required Euler-Lagrange's differential equation to be satisfied by $y = y(x)$ for which the functional I has extremum value under the conditions (133) and (134).

Remark: If y is not prescribed at either of the end points then from equation (143) we have

$$\frac{\partial f^*}{\partial y'} = 0 \text{ at the end points.}$$

Generalization of Theorem 4: Euler-Lagrange's Equations for Several Dependent Variable

Theorem 4a Obtain the differential equations which must be satisfied by the function which extremize the integral $I = \int_{x_1}^{x_2} f(x, y_1, y_2, \ldots, y_n, y'_1, y'_2, \ldots, y'_n)dx$ with respect to the twice differentiable functions $y_1, y_2, \ldots, y_n$ for which

$$J = \int_{x_1}^{x_2} g(x, y_1, y_2, \ldots, y_n, y'_1, y'_2, \ldots, y'_n)dx = \text{const.}$$

and y_i, y'_i are all prescribed at points x_1, x_2.

Proof: Repeating the procedure described in the Theorem 4, we arrive at the following set of Euler-Lagrange's equations

$$\frac{\partial f^*}{\partial y_i} - \frac{d}{dx}\left(\frac{\partial f^*}{\partial y'_t}\right) = 0, \ i = 1, 2, \ldots, n, \qquad \ldots(145)$$

where $\qquad\qquad\qquad f^* = f + \lambda g.$

Theorem 5 Obtain the differential equation, which is satisfied by four times differential function $y(x)$ which extremizes the functional

$$I(y(x)) = \int_{x_1}^{x_2} f(x, y, y', y'')dx$$

subject to the conditions that the integral

$$J = \int_{x_1}^{x_2} g(x, y, y', y'')dx = \text{constant}$$

and both y and y' are prescribed at the end points.

Proof: Proof of the Theorem 5 runs exactly in the same manner as that of the proof of Theorem 3 and Theorem 4. The required Euler-Lagrange's differential equation in this case is obtained in the form

$$\frac{\partial f^*}{\partial y} - \frac{d}{dx}\left(\frac{\partial f^*}{\partial y'}\right) + \frac{d^2}{dx^2}\left(\frac{\partial f^*}{\partial y''}\right) = 0, \qquad \ldots(146)$$

where $\qquad\qquad\qquad f^* = f + \lambda g.$

Remarks:

1. If y is not prescribed at either end point, then we have the condition
$$\frac{\partial f^*}{\partial y'} - \frac{d}{dx}\left(\frac{\partial f^*}{\partial y''}\right) = 0 \text{ at that end point.}$$

2. If y' is not prescribed at either end point then we have $\dfrac{\partial f^*}{\partial y''} = 0$ at that point.

3. In general if $f = f(x, y, y', y'', ..., y^n)$, $g = g(x, y, y', y'', ..., y^n)$ with the boundary conditions that $y, y', y'', ..., y^{n-1}$ are prescribed at both the ends, then in this case the Euler-Lagrange's equation is

$$\frac{\partial f^*}{\partial y} - \frac{d}{dx}\left(\frac{\partial f^*}{\partial y'}\right) + \frac{d^2}{dx^2}\left(\frac{\partial f^*}{\partial y''}\right) + ... + (-1)^n \frac{d^n}{dx^n}\left(\frac{\partial f^*}{\partial y''}\right) = 0. \qquad ...(147)$$

WORKED EXAMPLES

The following problem is supposed to have arisen from the gift of a king who was happy with a person and promised to give him all the land that he could enclose by running round in a day. The perimeter of his path was fixed.

Example 27: Find the plane curve of fixed perimeter that encloses maximum area.

Solution: Let a curve c be $y = y(x)$ with fixed perimeter l. *i.e.,*

$$l = \int_{x_1}^{x_2} ds,$$

where the infinitesimal distance between two points on the curve is given by

$$ds = \sqrt{1 + y'^2}\, dx, \quad y' = \frac{dy}{dx}.$$

Hence the total length of the curve between two points P and Q becomes

$$\int_{x_1}^{x_2} \sqrt{1 + y'^2}\, dx = l. \qquad ...(148)$$

The area bounded by the curve c and the x-axis is given by

$$A(y(x)) = \int_{x_1}^{x_2} y\, dx. \qquad ...(149)$$

Thus we maximize the area defined in (149) subject to the condition (148). Hence the respective Euler-Lagrange's equation (144) becomes

$$1 - \frac{d}{dx}\left(\frac{\lambda y'}{\sqrt{1 + y'^2}}\right) = 0,$$

$$\Rightarrow \qquad x - \frac{\lambda y'}{\sqrt{1 + y'^2}} = a.$$

Solving the equation for y' we get

$$y' = \frac{x - a}{\sqrt{\lambda^2 - (x - a)^2}}. \qquad \qquad ...(150)$$

Integrating we get

$$y = \int \frac{(x - a)}{\left[\lambda^2 - (x - a)^2\right]^{\frac{1}{2}}}\, dx + b. \qquad \qquad ...(151)$$

Put $x - a = \lambda \sin t$, $\Rightarrow dx = \lambda \cos t\, dt$. Hence equation (151) becomes

$$y = \int \lambda \sin t\, dt + b,$$

$$y = -\lambda \cos t + b.$$

From these equations we obtain a curve as a circle centered at (a, b) and of radius λ

$$(x - a)^2 + (y - b)^2 = \lambda^2. \qquad \qquad ...(152)$$

The perimeter of the circle is $2\pi\lambda = l$. From which the value of λ is obtained $\lambda = \dfrac{l}{2\pi}$.

Example 28: Find the shape of the plane curve of fixed length l whose end points lie on the x-axis and area enclosed by it and the x-axis is maximum.

Solution: Let $y = y(x)$ be a required plane curve of fixed length l whose end points lie on the x-axis and the curve lies in the upper half plane. The area bounded by the curve and the x-axis is given by the integral (149). The length of the curve is fixed and is given by the integral (148).

The area is maximum under the given condition, provided

$$1 - \frac{d}{dx}\left(\frac{\lambda y'}{\sqrt{1 + y'^2}}\right) = 0,$$

where λ is Lagrange's multiplier. This gives on integrating

$$x - \frac{\lambda y'}{\sqrt{1 + y'^2}} = a. \qquad \qquad ...(153)$$

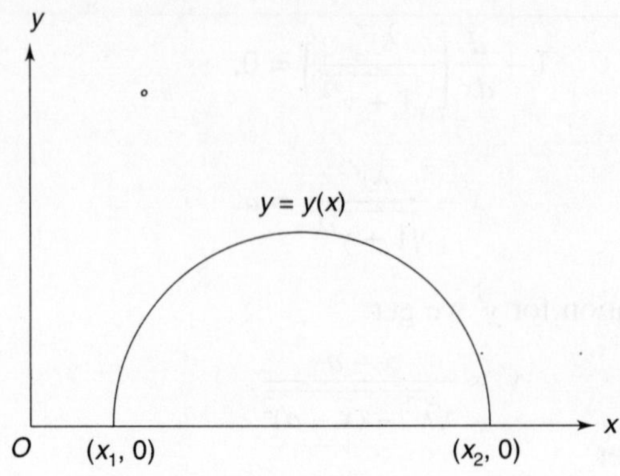

Solving equation (153) for y' we get

or
$$y' = x - \frac{a}{\sqrt{\lambda^2 - (x - a)^2}}. \qquad \qquad ...(154)$$

Integrating we get

$$y = \int (x - a) \left[\lambda^2 - (x - a)^2\right]^{-1/2} dx + b. \qquad ...(155)$$

Put
$$x - a = \lambda \sin t, \qquad \qquad ...(156)$$

$\Rightarrow$
$$dx = \lambda \cos t \, dt,$$

Hence we have

$$y = \int \lambda \sin t \, dt + b,$$

$$y = -\lambda \cos t + b. \qquad \qquad ...(157)$$

Squaring and adding equations (156) and (157), we get a circle centered at (a, b) and of radius $\left(\lambda = \dfrac{l}{\pi}\right)$ in the form

$$(x - a)^2 + (y - b)^2 = \lambda^2. \qquad \qquad ...(158)$$

Example 29: Find the extremals for an isoperimetric problem

$$I(y(x)) = \int_0^1 \left(y'^2 + x^2\right) dx$$

subject to the conditions that

$$\int_0^1 y^2 dx = 2, \ y(0) = 0, \ y(1) = 0.$$

Solution: The Euler-Lagrange's differential equation for the extremization of an isoperimetric problem yields the equation

$$y'' - \lambda y = 0. \qquad \qquad \text{...(159)}$$

This equation has roots $\pm \sqrt{\lambda}$ for $\lambda > 0$ or $\pm i\sqrt{\lambda}$ for $\lambda < 0$.

Case 1: Let $\lambda > 0$. Take $\lambda = \alpha^2$, $\alpha > 0$.

The solution of equation (159) is

$$y = ae^{\alpha x} + be^{-\alpha x}. \qquad \qquad \text{..(160)}$$

The given conditions give

$$y(0) = 0 \Rightarrow a + b = 0,$$

$$y(1) = 0 \Rightarrow ae^{\alpha} + be^{-\alpha} = 0.$$

Solving for a and b we have for $b \neq 0$,

$$e^{2\alpha} = 1, \Rightarrow e^{2\alpha + 2in\pi} = 1,$$

$$\Rightarrow \qquad 2\alpha + 2in\pi = 0, \Rightarrow \alpha = -in\pi.$$

This contradicts that $\alpha > 0$. This leads to the conclusion that both a and b are zero, and hence the solution in this case is trivial. Hence we assume $\lambda < 0$.

Case 2: Let $\lambda < 0$. We take $\lambda = -\alpha^2$, $\alpha > 0$.

The solution of the equation (159) in this case becomes

$$y = a \cos \alpha x + b \sin \alpha x. \qquad \qquad \text{...(161)}$$

The boundary conditions give

$$y(0) = 0 \Rightarrow a = 0,$$

$$y(1) = 0 \Rightarrow b \sin \alpha x = 0,$$

$$\Rightarrow \qquad \alpha = n\pi, \, n = 1 \; 2, \, ...$$

Hence the required solution becomes

$$y = b \sin n\pi x. \qquad \qquad \text{...(162)}$$

The integral condition $\int_{0}^{1} y^2 dx = 2$ gives $b = \pm 2$.

Therefore, the required solution is $y = \pm 2 \sin n\pi x$. The values $\alpha = n\pi, n = 1, 2, ...$ are called the eigen values of the differential equation, while the solutions for each n are called the corresponding eigen functions.

Example 30: Find the extremals for an isoperimetric problem

$$I(y(x)) = \int_{0}^{1} (y'^2 - y^2) \, dx$$

subject to the conditions that

$$\int_0^\pi ydx = 1, \ y(0) = 0, \ y(\pi) = 1.$$

Solution: The Euler-Lagrange's equation (144) gives the differential equation

$$y'' + y = \frac{\lambda}{2}. \qquad \qquad ...(163)$$

The general solution becomes

$$y = a \cos x + b \sin x + \frac{\lambda}{2}. \qquad \qquad ...(164)$$

To determine the arbitrary constants of integration, we use the boundary conditions.

$$y(0) = 0 \Rightarrow a = -\frac{\lambda}{2},$$

$$y(\pi) = 1 \Rightarrow \lambda = 1, \ a = -\frac{1}{2}.$$

To determine other constant of integration, we use the integral condition

$$\int_0^\pi ydx = 1.$$

$$\Rightarrow \qquad -\frac{1}{2}\int_0^\pi \cos xdx + b \int_0^\pi \sin xdx + \frac{1}{2}\int_0^\pi dx = 1,$$

This gives the value of b as $b = \left(\dfrac{2-\pi}{4}\right)$. Hence the required curve is

$$y = \frac{1}{2}(1 - \cos x) + \frac{1}{4}(2 - \pi)\sin x.$$

Example 31: Prove that the extremal of the isoperimetric problem $I = \int_1^4 y'^2 \, dx$, subject to the condition $\int_1^4 ydx = 36$, and $y(1) = 3$, $y(4) = 24$ is a parabola.

Solution: The corresponding Euler-Lagrange's differential equation yields

$$2y'' - \lambda = 0. \qquad \qquad ...(165)$$

Integrating two times we obtain a solution in the form

$$y = \frac{\lambda}{4}x^2 + ax + b, \qquad \qquad ...(166)$$

where the constants of integration a and b are to be determined. Now the boundary conditions yield

$$y(1) = 3 \Rightarrow \lambda + 4a + 4b = 12,$$

$$y(4) = 24 \Rightarrow 4\lambda + 4a + b = 24. \qquad \text{...(167)}$$

Also the condition $\int_1^4 y\,dx = 36$ gives

$$\int_1^4 \left(\frac{\lambda}{4} x^2 + ax + b\right) dx = 36,$$

$$\Rightarrow \qquad 21\lambda + 30a + 12b = 144. \qquad \text{...(168)}$$

Solving equations (167) and (168) we obtain $a = 2$, $b = 0$, $\lambda = 4$. Thus the required curve is obtained by putting these values in equation (166) in the form $y = x^2 + 2x$. We write this solution in the form

$$(x + 1)^2 = y + 1.$$

Or $\qquad\qquad X^2 = Y^2$, for $X = (x + 1)$, $Y = (y + 1)$.

Hence the curve is a parabola.

Example 32: Find the extremals for the isoperimetric problem

$$I = \frac{1}{2} \int_{t_1}^{t_2} (x\dot{y} - y\dot{x})\, dt \text{ with the conditions that}$$

$$J = \int_{t_1}^{t_2} \sqrt{\dot{x}^2 + \dot{y}^2}\, dt = l,\ x(t_1) = x(t_2) = x_0,\ y(t_1) = y(t_2) = y_0.$$

Solution: In this case the integrand involves two dependent variables; hence we have two Euler-Lagrange's differential equations. Hence the corresponding Euler-Lagrange's equations yield

$$\frac{\dot{y}}{2} - \frac{d}{dt}\left(-\frac{y}{2} + \frac{\lambda \dot{x}}{\sqrt{\dot{x}^2 + \dot{y}^2}}\right) = 0,$$

$$-\frac{\dot{x}}{2} - \frac{d}{dt}\left(\frac{x}{2} + \frac{\lambda \dot{y}}{\sqrt{\dot{x}^2 + \dot{y}^2}}\right) = 0.$$

Integrating these equations with respect to time t, we get

$$\frac{y}{2} - \left(-\frac{y}{2} + \frac{\lambda \dot{x}}{\sqrt{\dot{x}^2 + \dot{y}^2}}\right) = b, \Rightarrow y - b = \frac{\lambda \dot{x}}{\sqrt{\dot{x}^2 + \dot{y}^2}},$$

and $\qquad\qquad \dfrac{x}{2} + \left(\dfrac{x}{2} + \dfrac{\lambda \dot{y}}{\sqrt{\dot{x}^2 + \dot{y}^2}}\right) = a, \Rightarrow x - a = \dfrac{\lambda \dot{y}}{\sqrt{\dot{x}^2 + \dot{y}^2}}.$

Squaring and adding these equations we get

$$(x - a)^2 + (y - b)^2 = \lambda^2.$$

The closed curve, for which the enclosed area is maximum, is a circle of radius λ and centered at (a, b). The length of the circle is $2\pi\lambda = l \Rightarrow \lambda = \dfrac{l}{2\pi}$. This gives the radius of the curve.

Example 33:   Find the extremals for an isoperimetric problem

$$I(y(x), z(x)) = \int_0^1 \left(y'^2 + z'^2 - 4xz' - 4z\right) dx$$

subject to the conditions

$$\int_0^1 \left(y'^2 - xy' - z'^2\right) dx = 2, \ y(0) = 0, \ z(0) = 0, \ y(1) = 1, \ z(1) = 1.$$

Solution: The two Euler-Lagrange's equations corresponding to two dependent variable y and z yield the differential equations

$$y'' = \frac{\lambda}{2(1 + \lambda)}, \qquad\qquad\qquad ...(169)$$

and
$$z'' = 0. \qquad\qquad\qquad ...(170)$$

Integrating equations (169) and (170) we obtain

$$y = \frac{\lambda}{4(1 + \lambda)} x^2 + ax + b, \qquad\qquad ...(171)$$

$$z = cx + d. \qquad\qquad\qquad ...(172)$$

The boundary conditions give $y(0) = 0 \Rightarrow b = 0$, $y(1) = 1 \Rightarrow a = \dfrac{3\lambda + 4}{4(\lambda + 1)}$

and

$$z(0) = 0 \Rightarrow d = 0. \ z(1) = 1 \Rightarrow c = 1.$$

Hence the solutions become $y = \dfrac{\lambda}{4(1 + \lambda)} x^2 + \dfrac{3\lambda + 4}{4(1 + \lambda)} x, \ \lambda \neq -1$ and

$z = x.$

The value of λ is to be determined. To determine λ we find the values of y'

and z' from solutions and then substituting in integral $\int_0^1 (y'^2 - xy' - z'^2) dx = 2$, we obtain

$$121\lambda^2 + 242\lambda + 120 = 0.$$

The roots of the equation are given by $\lambda = -\dfrac{10}{11}$ and $\lambda = -\dfrac{12}{11}$. Substituting these values in the solution we get $y = -\dfrac{5}{2}x^2 + \dfrac{7}{2}x$, $z = x$.

Example 34: Find the curve $r = r(\theta)$ of given length which encloses maximum area.

Solution: The polar equation of the curve is given by $r = r(\theta)$, $0 \le \theta \le 2\pi$. The area bounded by the curve is given by the integral

$$A = \frac{1}{2}\int_0^{2\pi} r^2 d\theta. \qquad \ldots(173)$$

The length of the curve is prescribed by the equation

$$\int_0^{2\pi} ds = l,$$

where the metric in polar coordinates is given by $ds^2 = dr^2 + r^2 d\theta^2$. Hence above integral becomes

$$\int_0^{2\pi} (r^2 + r'^2)^{\frac{1}{2}} d\theta = l. \qquad \ldots(174)$$

Thus our problem is to maximize the area given in equation (173) subject to the condition (174) that the length of the curve is constant. The area bounded by the curve is maximum if the functional $f^* = \dfrac{r^2}{2} + \lambda\sqrt{r^2 + r'^2}$, where λ is the Lagrange's multiplier, must satisfy the Euler-Lagrange's equation

$$\frac{\partial f^*}{\partial r} - \frac{d}{d\theta}\left(\frac{\partial f^*}{\partial r'}\right) = 0. \qquad \ldots(175)$$

This gives

$$\frac{r}{\sqrt{r^2 + r'^2}}\left(\lambda + \sqrt{r^2 + r'^2}\right) - \frac{d}{d\theta}\left(\frac{\lambda r'}{\sqrt{r^2 + r'^2}}\right) = 0.$$

Simplifying we obtain

$$\left(r^2 + r'^2\right)^{\frac{3}{2}} = -\lambda(r^2 + 2r'^2 - rr''),$$

or $\qquad \dfrac{\left(r^2 + r'^2\right)^{\frac{3}{2}}}{(r^2 + 2r'^2 - rr'')} = \lambda_1, \ \lambda_1 = -\lambda \qquad \ldots(176)$

The left hand side of equation (176) represents the curvature of the curve and is constant. Hence the required curve is a curve of constant curvature and it is a circle.

Example 35: Find the solid of revolution with a given surface area and maximum volume.

Solution: The surface area of revolution obtain by revolving a curve $y = y(x)$ about x-axis is given by

$$S = 2\pi \int y\, ds,$$

$$S = 2\pi \int y\sqrt{1 + y'^2}\, dx. \qquad \qquad ...(177)$$

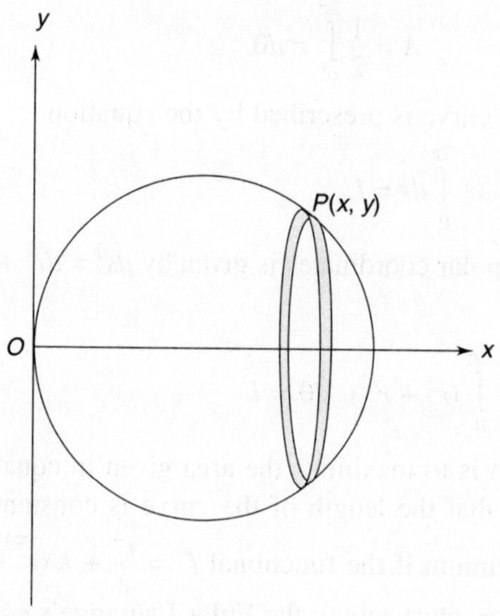

The volume of the solid so formed is given by

$$V = \pi \int y^2 dx. \qquad \qquad ...(178)$$

Now we maximize V subject to the condition that the surface area of revolution S is constant. Now V is maximum if the functional

$f^* = \pi y^2 + 2\pi\lambda y\sqrt{1 + y'^2}$, where λ is the Lagrange's multiplier, must satisfy the Euler-Lagrange's equation. However, we see that f^* does not involve the independent variable x, hence the Euler-Lagrange's equation gives

$$f^* - y'\frac{\partial f^*}{\partial y'} = a.$$

This gives after simplifying

$$y^2 + \frac{2\lambda y}{\sqrt{1 + y'^2}} = a. \qquad \qquad ...(179)$$

Integration of equation (179) involves elliptical functions. Hence as a special case we choose the constant $a = 0$. In this case we have from equation (179) that

$$y\sqrt{1 + y'^2} = -2\lambda.$$

Squaring and solving for y' we get

$$y' = \frac{dy}{dx} = \frac{\sqrt{4\lambda^2 - y^2}}{y},$$

$$dx = \frac{ydy}{\sqrt{4\lambda^2 - y^2}}.$$

Integrating the equation we obtain

$$x = \int \frac{ydy}{\sqrt{4\lambda^2 - y^2}} + c \qquad \qquad ...(180)$$

Put $\qquad\qquad 4\lambda^2 - y^2 = t \Rightarrow ydy = -\dfrac{1}{2}\,dt.$

Hence we get from equation (180)

$$(x - c)^2 + y^2 = 4\lambda^2.$$

This is a circle centered at $(c, 0)$ and of radius 2λ. Thus the surface obtained by revolving a circle is a sphere.

Example 36: Find the shape of a freely hanging uniform heavy string under gravity when two ends of it are fixed.

Solution: Consider a heavy uniform string of length l and weight w per unit length, hangs freely under gravity between two points A and B. Let ds be an infinitesimal portion of the string at a distance y from the origin. Then the potential energy of the portion ds is given by $mgyds$. Hence the total potential energy of the string becomes

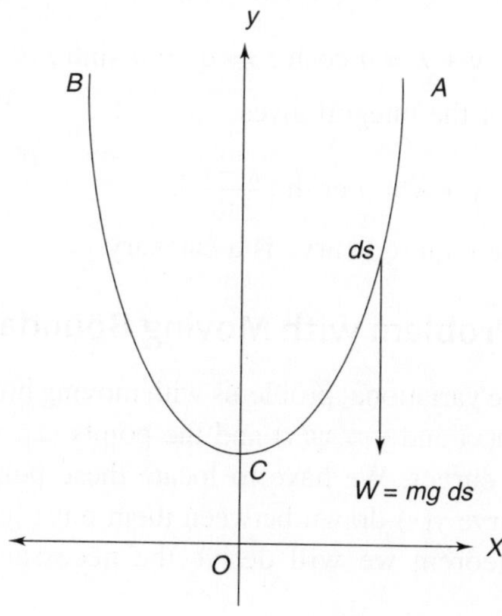

$$V = mg \int y\sqrt{1 + y'^2}\ dx. \qquad\qquad ...(181)$$

The length of the string is given by

$$l = \int \sqrt{1 + y'^2}\ dx. \qquad\qquad ...(182)$$

From the principle of minimum potential energy we have, in stable equilibrium, the potential energy is least. Hence to find the shape of a freely hanging uniform heavy string under gravity, we minimize V subject to the condition that the length of the string is constant. Thus V is minimum if the functional

$$f^* = mgy\sqrt{1 + y'^2} + \lambda\sqrt{1 + y'^2}, \qquad\qquad ...(183)$$

where λ is the Lagrange's multiplier, must satisfy the Euler-Lagrange's equation. However, we see from equation (183) that the functional f^* does not involve the independent variable x, hence the Euler-Lagrange's equation gives

$$f^* - y' \frac{\partial f^*}{\partial y'} = a.$$

This gives after simplifying

$$\frac{dy}{dx} = \frac{\left[(y - \lambda)^2 - a^2\right]^{\frac{1}{2}}}{a},$$

$$\Rightarrow \qquad dx = \frac{a\,dy}{\sqrt{(y + \lambda)^2 - a^2}}.$$

Integrating we obtain

$$x = \int \frac{a\,dy}{\sqrt{(y + \lambda)^2 - a^2}} + c$$

Put $\qquad\qquad y + \lambda = a \cosh t \Rightarrow dy = a \sinh t\ dt$

Thus the solution of the integral gives

$$y + \lambda = a \cosh\left(\frac{x - c}{a}\right).$$

This shows that the required curve is a catenary.

3. Variational Problem with Moving Boundaries

Introduction: In the variational problems with moving boundaries we are given two curves say $y = \phi(x)$ and $y = \psi(x)$ and the points (x_1, y_1) and (x_2, y_2) move along the respective curves. We have to locate these points on the respective curves such that a curve $y(x)$ drawn between them must have extremum length. In the following Theorem we will derive the necessary conditions for the existence of the extremal between points (x_1, y_1) and (x_2, y_2) move along the curves $y = \phi(x)$ and $y = \psi(x)$.

Theorem 6   Find the extremal of the functional $I(y(x)) = \int\limits_{x_1}^{x_2} f(x, y, y')\, dx$ with movable boundaries.

Proof: For simplicity sake, let us assume that one of the boundary points $A(x_1, y_1)$ be fixed, while the other boundary point $B(x_2, y_2)$ moves along the curve $y = \phi(x)$ to a neighboring point $C(x_2 + \delta x_2, y_2 + \delta y_2)$, where δx_2 and δy_2 are small increments in x and y respectively. Then the extremals passing through $A(x_1, x_2)$ form a pencil of extremals

$$y = y(x, \alpha), \hspace{3cm} ...(184)$$

where α is a parameter, for different value of α we get different curves passing through $A(x_1, x_1)$. Out of the curves of the family (184), let $c: y = y(x)$ be a curve between the fixed point $A(x_1, x_1)$ and the movable point $B(x_2, y_2)$ on the curve $y = \phi(x)$, along which the functional

$$I = \int\limits_{x_1}^{x_2} f(x, y, y')\, dx, \hspace{3cm} ...(185)$$

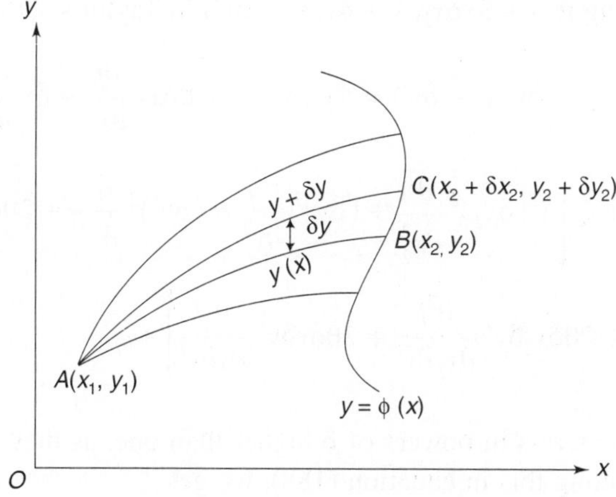

has extremum value. Thus to find the condition to be satisfied by $y(x)$ for the extremum of (185), consider a varied path from $A(x_1, y_1)$ to $C(x_2 + \delta x_2, y_2 + \delta y_2)$. Then the value of the functional (185) along the varied path when the boundary point moves from $B(x_2, y_2)$ to $C(x_2 + \delta x_2, y_2 + \delta y_2)$ is given by

$$I' = \int\limits_{x_1}^{x_2 + \delta x_2} f(x, y + \delta y, y' + \delta y')\, dx, \hspace{2cm} ...(186)$$

where δy is the variation in the path y with respect to the arbitrary parameter α. Then the increment in the value of the functional is given by

$$\delta I = I' - I = \int_{x_1}^{x_2+\delta x_2} f(x, y + \delta y, y' + \delta y')\, dx - \int_{x_1}^{x_2} f(x, y, y')\, dx.$$

We write this integral in the form

$$\delta I = \int_{x_1}^{x_2} [f(x, y + \delta y, y' + \delta y') - f(x, y, y')]\, dx +$$

$$+ \int_{x_2}^{x_2+\delta x_2} f(x, y + \delta y, y' + \delta y')\, dx. \qquad \ldots(187)$$

From the Mean Value Theorem of integral calculus, we have

$$\int_{x_2}^{x_2+\delta x_2} f(x, y + \delta y, y' + \delta y')\, dx = f(x_2 + \theta \cdot \delta x_2, y + \delta y, y' + \delta y')\, \delta x_2, \quad 0 < \theta < 1.$$

$$\int_{x_2}^{x_2+\delta x_2} f(x, y + \delta y, y' + \delta y')\, dx = (f)_{x=x_2+\theta\delta x_2}\, \delta x_2, \quad 0 < \theta < 1. \qquad \ldots(188)$$

Now expanding $f(x_2 + \theta \cdot \delta x_2, y + \delta y, y' + \delta y')$ in Taylor's series, we have

$$f(x_2 + \theta \cdot \delta x_2, y + \delta y, y' + \delta y') = f(x_2, y, y') + \theta\delta x_2 \frac{\partial f}{\partial x_2} + \delta y \frac{\partial f}{\partial y} + \delta y' \frac{\partial f}{\partial y'} +$$

$$+ \frac{1}{2}\left[\theta^2(\delta x_2)^2 \frac{\partial^2 f}{\partial x_2^2} + (\delta y)^2 \frac{\partial^2 f}{\partial y^2} + (\delta y')^2 \frac{\partial^2 f}{\partial y'^2} + 2\theta\delta x_2\, \delta y \frac{\partial^2 f}{\partial x_2 \partial y} + \right.$$

$$\left. + 2\theta\delta x_2 \delta y' \frac{\partial^2 f}{\partial x_2 \partial y'} + 2\theta\delta y \delta y' \frac{\partial^2 f}{\partial y \partial y'} \right] + \ldots.$$

Neglecting the terms in powers of δ higher than one, as they are very small, and then substituting this in equation (188), we get

$$\int_{x_2}^{x_2+\delta x_2} f(x, y + \delta y, y' + \delta y')\, dx = [f(x, y, y')]_{x=x_2}\, \delta x_2. \qquad \ldots(189)$$

Now consider the Taylor series expansion of

$$f(x, y + \delta y, y' + \delta y') = f(x, y, y') + \left(\delta y \frac{\partial f}{\partial y} + \delta y' \frac{\partial f}{\partial y'} \right) + \ldots$$

Neglecting the terms in δy and $\delta y'$ of order higher than first, we obtain

$$f(x, y + \delta y, y' + \delta y') - f(x, y, y') = \left(\delta y \, \frac{\partial f}{\partial y} + \delta y' \, \frac{\partial f}{\partial y'}\right).$$

Hence the value of the integral becomes

$$\int_{x_1}^{x_2} [f(x, y + \delta y, y' + \delta y') - f(x, y, y')] \, dx = \int_{x_1}^{x_2} \left(\frac{\partial f}{\partial y} \, \delta y + \frac{\partial f}{\partial y'} \, \delta y'\right) dx .$$

Since it is always true that $\delta y' = \dfrac{d}{dx} \, (\delta y)$. Hence, the above integration becomes

$$\int_{x_1}^{x_2} [f(x, y + \delta y, y' + \delta y') - f(x, y, y')] \, dx = \int_{x_1}^{x_2} \left(\frac{\partial f}{\partial y} \, \delta y + \frac{\partial f}{\partial y'} \, \frac{d}{dx}(\delta y)\right) dx.$$

Integrating the second term on the right hand side by parts we get

$$\int_{x_1}^{x_2} [f(x, y + \delta y, y' + \delta y') - f(x, y, y')] \, dx = \int_{x_1}^{x_2} \left(\frac{\partial f}{\partial y} - \frac{d}{dx}\left(\frac{\partial f}{\partial y'}\right)\right) \delta y \, dx +$$

$$+ \left(\frac{\partial f}{\partial y'}(\delta y)\right)\Bigg|_{x_1}^{x_2}. \qquad\qquad ...(190)$$

Now between points $A(x_1, y_1)$ and $B(x_2, y_2)$ the functional is extremal, hence Euler-Lagrange's equation (10) must be true. Hence equation (190) reduces to

$$\int_{x_1}^{x_2} [f(x, y + \delta y, y' + \delta y') - f(x, y, y')] \, dx = \left(\frac{\partial f}{\partial y'}(\delta y)\right)\Bigg|_{x_1}^{x_2}. \qquad ...(191)$$

Using equations (189) and (191) in equation (187), we get

$$\delta I = \left(\frac{\partial f}{\partial y'}\delta y\right)\Bigg|_{x_1}^{x_2} + [f(x, y, y')]_{x=x_2} \, \delta x_2. \qquad ...(192)$$

Further, y is prescribed at point A, hence at $x = x_1 \Rightarrow (\delta y)_{x=x_1} = 0$. Therefore, we have

$$\delta I = \left(\frac{\partial f}{\partial y'}\right)_{x=x_2} (\delta y)_{x=x_2} + [f(x, y, y')]_{x=x_2} \, \delta x_2. \qquad ...(193)$$

We note here that $(\delta y)_{x=x_2} \neq \delta y_2$, where δy_2 is increment of y_2 when boundary point is displaced to $C(x_2 + \delta x_2, y_2 + \delta y_2)$, while $(\delta y)_{x_2}$ is the increment in the ordinate y at the point x_2. Now by the first principle of derivative of y with respect to x at point D, we have

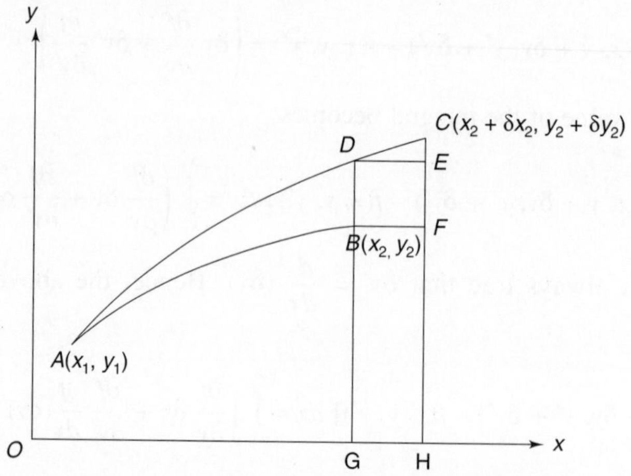

$$(y')_{x=x_2} = \text{change in } y \text{ due to change in } x,$$

$$(y')_{x=x_2} = \frac{CE}{DE},$$

$$(y')_{x=x_2} = \frac{(CH - EH)}{\delta x_2},$$

$$(y')_{x=x_2} = \frac{[CH - (BD + BG)]}{\delta x_2},$$

where from figure we have $CH = y_2 + \delta y_2$, $EF = DB =$ variation in y at B $= (\delta y)_{x_2}$ and $BG = y_2$. Hence, the value of (y') at $x = x_2$ becomes

$$(y')_{x=x_2} = \frac{\delta y_2 - (\delta y)_{x_2}}{\delta x_2}, \Rightarrow (\delta y)_{x=x_2} = \delta y_2 - y'(x_2) \cdot \delta x_2.$$

Hence equation (193) becomes

$$\delta I = \left(\frac{\partial f}{\partial y'}\right)_{x=x_2} [\delta y_2 - y'(x_2) \cdot \delta x_2] + [f(x, y, y')]_{x=x_2} \delta x_2. \quad ...(194)$$

Since δx_2 and δy_2 are not independent because the boundary point $B(x_2, y_2)$ varies along the curve $y = \phi(x)$. $\Rightarrow y_2 = \phi(x_2)$,

$$\Rightarrow \qquad \delta y_2 = \phi'(x_2)\, \delta x_2. \qquad ...(195)$$

Hence equation (194) becomes

$$\delta I = \left(f + (\phi' - y')\frac{\partial f}{\partial y'}\right)_{x=x_2} \delta x_2.$$

For extremum of I we have $\delta I = 0$,

$$\Rightarrow \qquad \left[f + (\phi - y') \frac{\partial f}{\partial y'} \right]_{x=x_2} = 0, \qquad\qquad ...(196)$$

which is the required condition for the free boundary. The condition is known as transversality condition. The conditions (195) and (196) suffice to determine the extremal of the pencil $y = y(x, \alpha)$ on which an extremum is attained. Similarly, if we assume the first boundary point $A(x_1, y_1)$ also moves along another prescribed curve $y = \psi(x)$, then similarly, we obtain another transversality condition as

$$\left[f + (\psi' - y') \frac{\partial f}{\partial y'} \right]_{x=x_1} = 0. \qquad\qquad ...(197)$$

Thus to find the extremal of the functional I with movable boundaries, we must first solve Euler-Lagrange's equation (10) and then use the transversality conditions (196) and (197) to determine the values of two arbitrary constants appearing in the general solution of the Euler-Lagrange's equation.

Note: If δx_2 and δy_2 are independent, then the necessary conditions for the extremum of I are obtained from equation (194) for $\delta I = 0$ in the form

$$\left[f - \frac{\partial f}{\partial y'} y' \right]_{x=x_2} = 0 \text{ and } \left(\frac{\partial f}{\partial y'} \right)_{x=x_2} = 0.$$

Example 37: Find the shortest distance between parabola $y^2 = 4x$ and straight line $x + y = -5$.

Solution: Let one point (x_1, y_1) lies on the straight line

$$y \equiv \phi(x) = -x - 5, \qquad\qquad ...(198)$$

and the other point (x_2, y_2) lies on the parabola

$$y \equiv \psi(x) = \pm 2\sqrt{x}. \qquad\qquad ...(199)$$

Then the shortest distance between the points (x_1, y_1) and (x_2, y_2) is given by the functional

$$I(y(x)) = \int_{x_1}^{x_2} \sqrt{1 + y'^2} \, dx, \qquad\qquad ...(200)$$

and the extremal is the straight line given by

$$y = c_1 x + c_2, \qquad\qquad ...(201)$$

where c_1 and c_2 are arbitrary constants and are to be determined. This shows that the extremal curves are the two parameters family of straight lines. The point (x_1, y_1) lies on the curve (198) and also on the line (201), therefore, we have

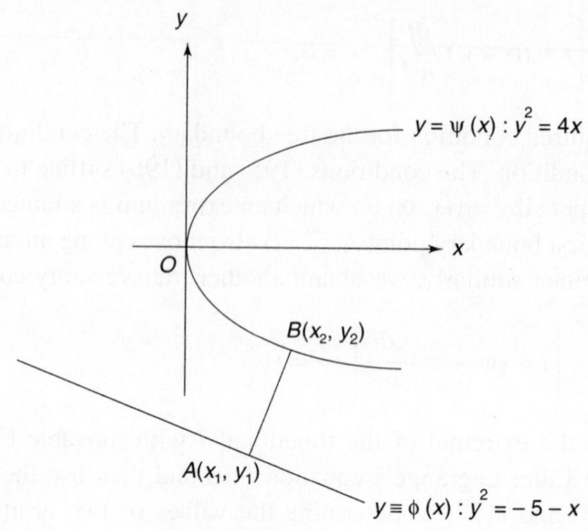

$$y_1 = -x_1 - 5 \quad \text{and} \quad y_1 = c_1 x_1 + c_2$$

$$\Rightarrow \quad c_1 x_1 + c_2 = -x_1 - 5. \qquad \qquad \dots(202)$$

Similarly, the point (x_2, y_2) lies on the parabola (199) and also on the line (201), therefore, we have

$$y_2 = \pm 2\sqrt{x_2} \quad \text{and} \quad y_2 = c_1 x_2 + c_2$$

$$\Rightarrow \quad c_1 x_2 + c_2 = \pm 2\sqrt{x_2}. \qquad \qquad \dots(203)$$

From equations (198), (199) we have $\phi'(x) = -1$, $\psi'(x) = \pm \dfrac{1}{\sqrt{x}}$,

and
$$\frac{\partial f}{\partial y'} = \frac{c_1}{\sqrt{1 + c_1^2}} \quad \text{as} \quad y' = c.$$

Hence from the transversality equations (196) and (197) we have

$$c_1 = 1, \quad \text{and} \quad c_1 = \pm\sqrt{x_2} \Rightarrow \quad x_2 = 1.$$

Now $x_2 = 1$ lies on equation (199) $\Rightarrow y_2 = \pm 2\sqrt{x_2} \Rightarrow \quad y_2 = \pm 2$.

Now using $c_1 = 1$ and $x_2 = 1$ in equation (202), we get $c_2 = -3$ or 1. Hence from equation (202) we have $x_1 = -1$ so that $y_1 = -4$. Thus one end of the extremal (201) on $x + y = -5$ is $A(-1, -4)$ and the other end point on the parabola is $B(1, -2)$ and the extremal joining these two points is the straight line given by $y = x - 3$. Now the shortest distance between the two given curves is the distance between two points on it and is given by either the distance formula between them or from the integral

$$AB = I = \int_{-1}^{1} \sqrt{1 + c_1^2}\ dx = \int_{-1}^{1} \sqrt{2}\ dx$$

$$\Rightarrow \qquad AB = 2\sqrt{2}.$$

Example 38: Find the shortest distance between $y^2 = 4x$ and

$$(x - 9)^2 + y^2 = 4.$$

Solution: Let one point $A(x_1, y_1)$ lies on the circle

$$(x - 9)^2 + y^2 = 4 \qquad\qquad ...(204)$$

and the other point $B(x_2, y_2)$ lies on the parabola

$$y^2 = 4x. \qquad\qquad ...(205)$$

The shortest distance between the points $A(x_1, y_1)$ and $B(x_2, y_2)$ is the solution of the functional (200). It is a two parameter family of straight line given by

$$y = c_1 x + c_2 \qquad\qquad ...(206)$$

where c_1 and c_2 are arbitrary constants and are to be determined. Since the point $A(x_1, y_1)$ lies on the circle (204), this gives $y \equiv \phi(x) = \sqrt{4 - (x - 9)^2}$ and also on the line (206), therefore, we have

$$c_1 x_1 + c_2 = \sqrt{4 - (x_1 - 9)^2}. \qquad\qquad ...(207)$$

Similarly, the other point (x_2, y_2) lies on the parabola $y^2 = 4x$ and also on the line (206), hence we have

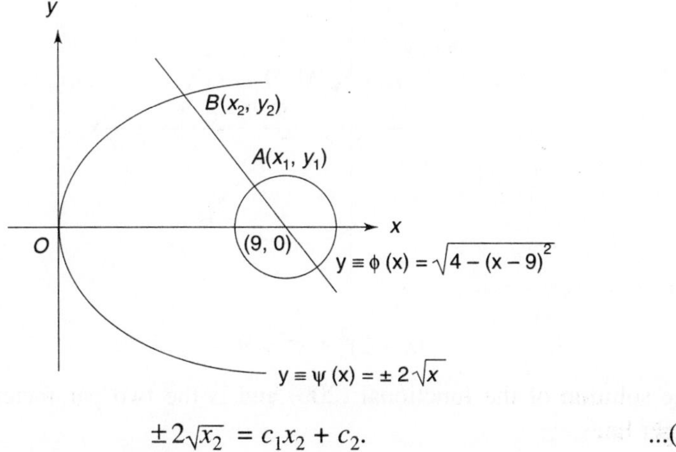

$$\pm 2\sqrt{x_2} = c_1 x_2 + c_2. \qquad\qquad ...(208)$$

The transversality conditions (196) and (197) that the points (x_1, y_1) and (x_2, y_2) move on the curves $y \equiv \phi(x) = \sqrt{4 - (x - 9)^2}$ and $y \equiv \psi(x) = \pm 2\sqrt{x}$ give

$$c_1(x_1 - 9) = \sqrt{4 - (x_1 - 9)^2} \text{ and } c_1 = \pm \sqrt{x_2}. \qquad \qquad ...(209)$$

Now from equations (207) and (209) we obtain $c_2 = -9c_1$. Consequently, the equation (208) gives $x_2 = 7$ or 11. Hence these equations yield $c_1 = -\sqrt{7}$, $c_2 = 9\sqrt{7}$. Now $x_2 = 7$ gives $y_2 = 2\sqrt{7}$. Thus the point is $B(7, 2\sqrt{7})$. Now the equation (207) gives two points on the circle. These points are $A\left(9 - \dfrac{1}{\sqrt{2}}, \sqrt{\dfrac{7}{2}}\right)$ and $C\left(9 + \dfrac{1}{\sqrt{2}}, \sqrt{\dfrac{7}{2}}\right)$. The former point will have the shortest distance from the point $B(7, 2\sqrt{7})$ on the parabola. Thus the shortest distance between the parabola $y^2 = 4x$ and the circle $(x - 9)^2 + y^2 = 4$ is obtained from equation (205) as

$$I = \sqrt{8}\left(2 - \dfrac{1}{\sqrt{2}}\right) \text{ and the line joining two points is}$$

$$y = \sqrt{7}\,(9 - x).$$

Example 39: Find the extremum distance of an interior point $(1, 1)$ from the curve

$$(x - 2)^2 + y^2 = 9.$$

Solution: Let a point on the circle $(x - 2)^2 + y^2 = 9$ be $A(x_1, y_1)$ from which the distance of the point $P(1, 1)$ is extremum. The extremum distance between the point $P(1, 1)$ and the point $A(x_1, y_1)$ on the circle

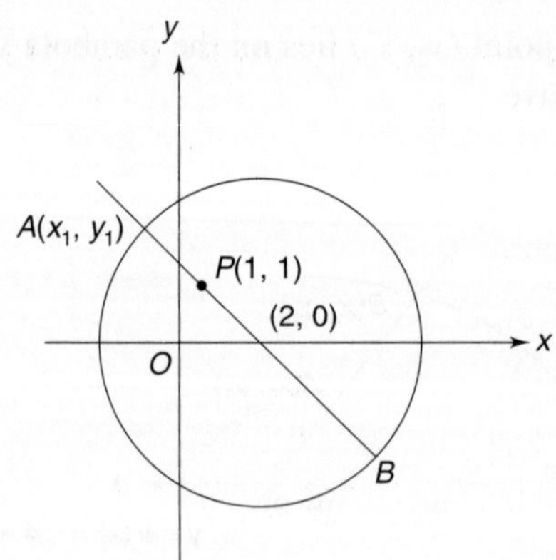

$$(x - 2)^2 + y^2 = 9 \qquad \qquad ...(210)$$

is the solution of the functional (200) and is the two parameter family of the straight lines

$$y = c_1 x + c_2 \qquad \qquad ...(211)$$

where c_1 and c_2 are arbitrary constants and are to be determined. Since this extremal passes through the point $P(1, 1)$ hence we have

$$1 = c_1 + c_2. \qquad \qquad ...(212)$$

Further more the point $A(x_1, y_1)$ lies on the line (211) and also on the circle (210)

$$\Rightarrow \qquad c_1 x_1 + c_2 = \sqrt{9 - (x_1 - 2)^2}. \qquad \qquad ...(213)$$

The transversality condition (196) that the point $A(x_1, y_1)$ moves on the circle gives

$$c_1(x_1 - 2) = \sqrt{9 - (x_1 - 2)^2}. \qquad \qquad ...(214)$$

From equations (213) and (214), we find the relation

$$c_2 = -2c_1. \qquad \qquad ...(215)$$

The equations (212) and (215) yield $c_1 = -1$ and $c_2 = 2$. Hence the required extremal is

$$x + y = 2. \qquad \qquad ...(216)$$

This is the straight-line passes through the center of the circle. Now substituting $c_1 = -1$ and $c_2 = 2$ in equation (213), we obtain $x_1 = 2 \pm \dfrac{3}{\sqrt{2}}$. Consequently, from equation (211) we obtain $y_1 = \pm \dfrac{3}{\sqrt{2}}$. This shows that there are two points on

the circle whose co-ordinates are $A\left(2 - \dfrac{3}{\sqrt{2}}, \dfrac{3}{\sqrt{2}}\right)$ and $B\left(2 + \dfrac{3}{\sqrt{2}}, -\dfrac{3}{\sqrt{2}}\right)$. The

distance of these points from the point $P(1, 1)$ will be the shortest and longest respectively and are given by

$$AP = \int_{2 - \frac{3}{\sqrt{2}}}^{1} \sqrt{1 + c_1^2} \, dx$$

$$AP = 3 - \sqrt{2}.$$

Similarly, we obtain $\qquad BP = 3 + \sqrt{2}.$

Example 40: Find the curve for which the functional $I(y(x)) = \displaystyle\int_{0}^{x_1} \dfrac{\sqrt{1 + y'^2}}{y} \, dx$, $y(0) = 0$ can have extrema if the point (x_1, y_1) can vary along the line $y = x - 5$.

Solution: The integrand f of the given functional does not contain x explicitly, hence the extremals of the functional are the curves obtain as the solution of the equation given by

$$f - y' \frac{\partial f}{\partial y'} = \text{const.} \qquad \qquad ...(217)$$

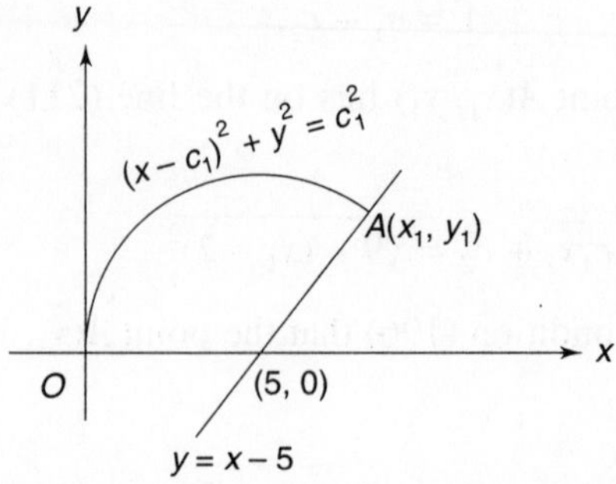

Solving the equation we obtain

$$\frac{y\,dy}{\sqrt{c_1^2 - y^2}} = dx. \qquad \text{...(218)}$$

Integrating equation (218), we get

$$\int \frac{y}{\sqrt{c_1^2 - y^2}}\, dy = x - c_2. \qquad \text{...(219)}$$

Put in the equation (219)

$$y = c_1 \sin \theta, \qquad \text{...(220)}$$

$$\Rightarrow \qquad dy = c_1 \cos \theta\, d\theta.$$

Hence equation (219) becomes

$$\int c_1 \sin \theta\, d\theta = x - c_2,$$

$$\Rightarrow \qquad (x - c_2) = c_1 \cos \theta \qquad \text{...(221)}$$

Squaring and adding equations (220) and (221), we get

$$(x - c_2)^2 + y^2 = c_1^2, \qquad \text{...(222)}$$

where c_1 and c_2 are arbitrary constants and are to be determined. This shows that the extremal curves are the two parameters family of circles. Thus the required extremal will be one of the circles belonging to this family. The boundary condition $y(0) = 0$ gives

$$c_1 = c_2. \qquad \text{...(223)}$$

Hence the extremal is one of the one parameter family of circles

$$x^2 - 2c_1 x + y^2 = 0. \qquad \text{...(224)}$$

Using the transversality condition (196) that the point $A(x_1, y_1)$ moves on the line $y = x - 5$, we obtain

$$y'(x_1) = -1. \qquad \qquad ...(225)$$

Hence the differentiation of the equation (224) at point x_1 gives

$$x_1 - c_1 + y(x_1)\, y'(x_1) = 0.$$

Since, we have $y(x_1) = x_1 - 5$. Using this and the equation (225), we obtain from the above equation $c_1 = 5$. Substitution of this value in equation (224) gives the required arc of the circle $x^2 - 10x + y^2 = 0$.

Example 41: Find the curves for which the functional $I(y(x)) = \displaystyle\int_0^{x_1} \frac{\sqrt{1 + y'^2}}{y}\, dx$, $y(0) = 0$ can have extremal, if the point (x_1, y_1) can vary along the line $(x - 9)^2 + y^2 = 9$.

Solution The integrand f of the given functional does not contain x explicitly, hence the extremals of the functional are the curves obtain as the solutions of the equation (217). Solving equation (217) we obtain the differential equation

$$y' = \frac{\sqrt{c_1^2 - y^2}}{y}. \qquad \qquad ...(226)$$

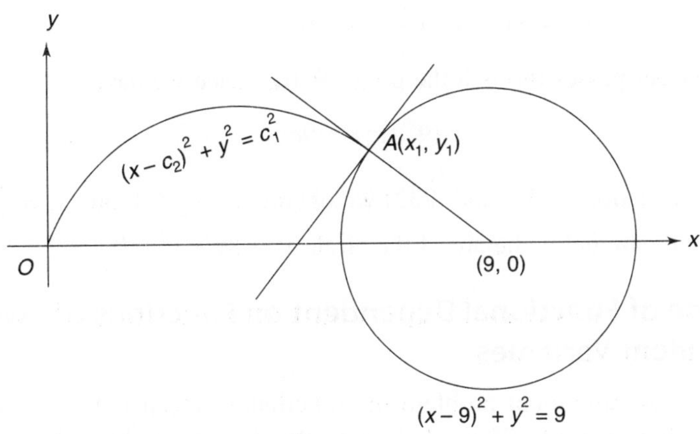

The solution of the equation yields

$$(x - c_2)^2 + y^2 = c_1^2 \qquad \qquad ...(227)$$

where c_1 and c_2 are arbitrary constants and are to be determined. This shows that the extremal curves are the two parameters family of circles. Thus the required extremal will be one of the circles belonging to this family. The boundary condition $y(0) = 0$ gives $c_1 = c_2$. Hence the extremal curves are now becomes one parameter family of circles

$$x^2 - 2c_1x + y^2 = 0. \qquad \qquad ...(228)$$

Since the boundary point $A(x_1, y_1)$ lies on both the circles given by the equation (228) and the circle

$$(x - 9)^2 + y^2 = 9. \qquad \qquad ...(229)$$

This implies that $x_1^2 - 2c_1x_1 + y_1^2 = 0$ and $x_1^2 - 18x_1 + y_1^2 = -72$. Solving these two equations we get

$$x_1(9 - c_1) = 36. \qquad \qquad ...(230)$$

The constant will be determined by using the transversality condition (196) that the point $A(x_1, y_1)$ moves on the circle $y \equiv \phi(x) = \pm \sqrt{9 - (x - 9)^2}$ so that we have

$$\phi'(x_1)\, y'(x_1) = -1. \qquad \qquad ...(231)$$

This is the condition that the two circles (228) and (229) intersect orthogonally. This implies that the tangent to the circle (228) at $A(x_1, y_1)$ passes through the center $(9, 0)$ of the circle (229). The equation of the tangent to the circle (228) is given by

$$xx_1 + yy_1 - c_1(x + x_1) = 0.$$

This tangent passes through the point $(9, 0)$, hence we have

$$x_1(9 - c_1) = 9c_1. \qquad \qquad ...(232)$$

Solving equations (230) and (232) we obtain $c_1 = c_2 = 4$, and $x_1 = \dfrac{36}{5}$. Hence the required extremal is the arc of the circle $x^2 - 8x + y^2 = 0$.

4. A Case of Functional Dependent on Functions of Two Independent Variables

In this type of variational problem the functional, which is to be extremized, involve double integrals. The solutions to this type of problem lead to one or more partial differential equations. The method is illustrated in the following Theorem.

Theorem 7 Find the differential equation for the extremization of the functional

$$I[u(x, y)] = \iint\limits_{G} F(x, y, u, u_x, u_y)\, dx\,dy$$

over a region of integration G, where u is continuous and has continuous derivatives up to the second order and takes on prescribed values on the boundary of G, and F is thrice differentiable.

Proof: Our aim in this case is to extremize the functional given by

$$I[u(x,\ y)] = \iint\limits_{G} F(x,\ y,\ u,\ u_x,\ u_y)\ dxdy \qquad \dots(233)$$

where G is the region of integration. The method is analogous to the method discussed in Theorem 1. Let the surface, which extremizes (233) be given by

$$u = u(x,\ y) \qquad \dots(234)$$

Consider one parameter family of neighboring surfaces given by the equation

$$u(x,\ y,\ \alpha) = u(x,\ y) + \alpha\eta(x,\ y). \qquad \dots(235)$$

Since u is prescribed on the boundary Γ of the surface $G. \Rightarrow \eta(x,\ y) = 0$ on the boundary of G. Consider the value of the functional (233) along the neighboring surface (235) given by

$$I[u(x,\ y,\ \alpha)] = \iint\limits_{G} F\big[x,\ y,\ u(x,\ y) + \alpha\eta(x,\ y),\ u_x(x,\ y) + \alpha\eta_x,\ u_y(x,\ y) + \alpha\eta_y\big]\ dxdy,$$
$$\dots(236)$$

Then the necessary condition for the extremum of the surface $u = u(x,\ y)$, is $(\delta I)_{\alpha=0} = 0,$

$$\Rightarrow \qquad \iint\limits_{G} \left[\frac{\partial F}{\partial u}\ \delta u + \frac{\partial F}{\partial u_x}\ \delta u_x + \frac{\partial F}{\partial u_y}\ \delta u_y \right] dxdy = 0,$$

where
$$\delta u = \frac{\partial u}{\partial \alpha}\ d\alpha = \eta(x,\ y)\ d\alpha$$

$$\Rightarrow \qquad \iint\limits_{G} \left[\frac{\partial F}{\partial u}\ \eta(x,\ y) + \frac{\partial F}{\partial u_x}\ \eta_x(x,\ y) + \frac{\partial F}{\partial u_y}\ \eta_y(x,\ y) \right] d\alpha\, dxdy = 0.$$

Integrating the second and third integrals by parts we get

$$\Rightarrow \qquad \iint\limits_{G} \frac{\partial F}{\partial u}\ \eta(x,\ y)\ d\alpha\ dxdy + \int\limits_{\Gamma} \frac{\partial F}{\partial u_x}\eta(x,\ y)\ d\alpha\, dy -$$

$$-\iint\limits_{G} \frac{\partial}{\partial x}\!\left(\frac{\partial F}{\partial u_x}\right) \eta(x,\ y)\ d\alpha\, dxdy + \int\limits_{\Gamma} \frac{\partial F}{\partial u_y}\eta(x,\ y)\ d\alpha\, dx -$$

$$-\iint\limits_{G} \frac{\partial}{\partial y}\!\left(\frac{\partial F}{\partial u_y}\right) \eta(x,\ y)\ d\alpha\, dxdy = 0.$$

Since $\eta(x,\ y) = 0$ on the boundary Γ of G

$$\Rightarrow \qquad \iint\limits_{G} \left[\frac{\partial F}{\partial u} - \frac{\partial}{\partial x}\!\left(\frac{\partial F}{\partial u_x}\right) - \frac{\partial}{\partial y}\!\left(\frac{\partial F}{\partial u_y}\right) \right] \eta(x,\ y)\ d\alpha\ dxdy = 0.$$

Since $\eta(x, y)$ is arbitrary continuously differentiable function, it follows from the generalization of fundamental basic lemma that

$$\frac{\partial F}{\partial u} - \frac{\partial}{\partial x}\left(\frac{\partial F}{\partial u_x}\right) - \frac{\partial}{\partial y}\left(\frac{\partial F}{\partial u_y}\right) = 0. \qquad \qquad \text{...(237)}$$

The extremization function $u(x, y)$ is determined from the solution of the second order partial differential equation (237). This equation is known as Euler-Ostrogradsky equation.

Note: Similarly, the extremals of the functional

$$I[z(x, y)] = \iint_G F(x, y, z, z_x, z_y, z_{xx}, z_{xy}, z_{yy})\ dxdy$$

are obtained from

$$\frac{\partial F}{\partial z} - \frac{\partial}{\partial x}\left(\frac{\partial F}{\partial z_x}\right) - \frac{\partial}{\partial y}\left(\frac{\partial F}{\partial z_y}\right) + \frac{\partial^2}{\partial x^2}\left(\frac{\partial F}{\partial z_{xx}}\right) + \frac{\partial^2}{\partial x\partial y}\left(\frac{\partial F}{\partial z_{xy}}\right) + \frac{\partial^2}{\partial y^2}\left(\frac{\partial F}{\partial z_{yy}}\right) = 0. \quad \text{...(238)}$$

WORKED EXAMPLES

Example 42: Obtain the differential equation of the surface $z = z(x, y)$ for which the area $\iint_R \left(1 + z_x^2 + z_y^2\right)^{\frac{1}{2}} dxdy$ is minimum.

Solution: The condition that the functional given in the question is minimum provided the integrand F must satisfy the Euler-Ostrogrdsky equation (237)

$$\frac{\partial F}{\partial z} - \frac{\partial}{\partial x}\left(\frac{\partial F}{\partial z_x}\right) - \frac{\partial}{\partial y}\left(\frac{\partial F}{\partial z_y}\right) = 0, \qquad \qquad \text{...(239)}$$

where from the integrand we have

$$\frac{\partial F}{\partial z} = 0,\ \frac{\partial F}{\partial z_x} = \frac{z_x}{F},\ \frac{\partial F}{\partial z_y} = \frac{z_y}{F}.$$

Hence the equation (239) becomes

$$-\frac{\partial}{\partial x}\left(\frac{z_x}{F}\right) - \frac{\partial}{\partial y}\left(\frac{z_y}{F}\right) = 0,$$

$$\Rightarrow \qquad \frac{Fz_{xx} - z_x\dfrac{\partial F}{\partial x}}{F^2} + \frac{Fz_{yy} - z_y\dfrac{\partial F}{\partial y}}{F^2} = 0,$$

$$\Rightarrow \qquad F(z_{xx} + z_{yy}) - z_x\frac{\partial F}{\partial x} - z_y\frac{\partial F}{\partial y} = 0, \qquad \qquad \text{...(240)}$$

where

$$\frac{\partial F}{\partial x} = \frac{1}{F}(z_x z_{xx} + z_y z_{xy}),$$

$$\frac{\partial F}{\partial y} = \frac{1}{F}(z_x z_{xy} + z_y z_{yy}).$$

Hence the equation (240) gives the equation of minimal surfaces in the form

$$z_{xx}\left(1 + z_y^2\right) + z_{yy}(1 + z_x^2) - 2z_x z_y z_{xy} = 0. \qquad ...(241)$$

Example 43:   Find the differential equation for the extremal of

$$\iint_D \left[\left(\frac{\partial u}{\partial x}\right)^2 + \left(\frac{\partial u}{\partial y}\right)^2\right] dxdy$$

where the values of u are prescribed on the boundary of the domain D.

Solution: The functional given in the question is extremum provided the functional must satisfy the Euler-Ostrogrdsky equation (237). The equation (237) reduces to the equation

$$\frac{\partial^2 u}{\partial x^2} + \frac{\partial^2 u}{\partial y^2} = 0. \qquad ...(242)$$

Thus the extremal of the functional is the Laplace equation.

Theorem 8   Find the geodesic on a given surface $G(x, y, z) = 0$ that gives a stationary value to an integral of the form

$$I = \int_{t_0}^{t_1} f(\dot{x}, \dot{y}, \dot{z})\, dt.$$

Proof: Let us assume that the curve lie on the part of the surface $G(x, y, z) = 0$.

$$\Rightarrow \qquad G_x dx + G_y dy + G_z dz = 0$$

Since x, y, z are not all independent but are related by the equation of the surface, this implies that at least one of the derivatives $\dfrac{\partial G}{\partial x}, \dfrac{\partial G}{\partial y}, \dfrac{\partial G}{\partial z}$ is not equal to zero. Let us assume that $\dfrac{\partial G}{\partial z} \neq 0$. Now solving the given equation of the surface for z, the equation of the surface can be expressed as

$$z = g(x, y). \qquad ...(243)$$

Differentiating equation (243) with respect to t, we get

$$\dot{z} = \frac{\partial g}{\partial x}\,\dot{x} + \frac{\partial g}{\partial y}\,\dot{y}. \qquad \text{...(244)}$$

Substituting this value in the given functional we get

$$I = \int_{t_1}^{t_2} f\left(\dot{x},\,\dot{y},\,\frac{\partial g}{\partial x}\,\dot{x} + \frac{\partial g}{\partial y}\,\dot{y}\right) dt. \qquad \text{...(245)}$$

The functional (245) is extremum if the integrand $f = f\left(\dot{x},\,\dot{y},\,\dfrac{\partial g}{\partial x}\,\dot{x} + \dfrac{\partial g}{\partial y}\,\dot{y}\right)$ must satisfy the Euler-Lagrange's differential equations

$$\frac{d}{dt}\left(\frac{\partial f}{\partial \dot{x}}\right) - \frac{\partial f}{\partial x} = 0,\ \frac{d}{dt}\left(\frac{\partial f}{\partial \dot{y}}\right) - \frac{\partial f}{\partial y} = 0. \qquad \text{...(246)}$$

These equations yield

$$\frac{d}{dt}\left(\frac{\partial f}{\partial \dot{x}} + \frac{\partial f}{\partial \dot{z}}\frac{\partial g}{\partial x}\right) - \frac{\partial f}{\partial \dot{z}}\frac{\partial \dot{z}}{\partial x} = 0, \qquad \text{...(247)}$$

and

$$\frac{d}{dt}\left(\frac{\partial f}{\partial \dot{y}} + \frac{\partial f}{\partial \dot{z}}\frac{\partial g}{\partial y}\right) - \frac{\partial f}{\partial \dot{z}}\frac{\partial \dot{z}}{\partial y} = 0, \qquad \text{...(248)}$$

where

$$\frac{\partial \dot{z}}{\partial x} = \frac{\partial^2 g}{\partial x^2}\,\dot{x} + \frac{\partial^2 g}{\partial x \partial y}\,\dot{y} = \frac{d}{dt}\left(\frac{\partial g}{\partial x}\right),\ \frac{\partial \dot{z}}{\partial y} = \frac{\partial^2 g}{\partial x \partial y}\,\dot{x} + \frac{\partial^2 g}{\partial y^2}\,\dot{y} = \frac{d}{dt}\left(\frac{\partial g}{\partial y}\right). \qquad \text{...(249)}$$

Consequently, the Euler's equations (247) and (248) reduce to

$$\frac{d}{dt}\left(\frac{\partial f}{\partial \dot{x}}\right) + \frac{\partial g}{\partial x}\frac{d}{dt}\left(\frac{\partial f}{\partial \dot{z}}\right) = 0 \text{ and } \frac{d}{dt}\left(\frac{\partial f}{\partial \dot{y}}\right) + \frac{\partial g}{\partial y}\frac{d}{dt}\left(\frac{\partial f}{\partial \dot{z}}\right) = 0. \qquad \text{...(250)}$$

Define the function $\lambda(t)$ by the equation

$$\frac{d}{dt}\left(\frac{\partial f}{\partial \dot{z}}\right) = \lambda(t)\,G_z. \qquad \text{...(251)}$$

From the equation of the surface, we have

$$\frac{dG}{dt} = 0 \Rightarrow G_x \dot{x} + G_y \dot{y} + G_z \dot{z} = 0.$$

$$\Rightarrow \qquad \dot{z} = -\frac{(G_x \dot{x} + G_y \dot{y})}{G_z}. \qquad \text{...(252)}$$

Comparing the coefficients of $\dot{x}$ and $\dot{y}$ in equations (244) and (252), we get

$$\frac{\partial g}{\partial x} = -\frac{G_x}{G_z} \text{ and } \frac{\partial g}{\partial y} = -\frac{G_y}{G_z} \qquad \text{...(253)}$$

On using (251) and (253), we write equations (250) as

$$\frac{d}{dt}\left(\frac{\partial f}{\partial \dot{x}}\right) = \lambda(t)\, G_x, \qquad \text{...(254)}$$

and
$$\frac{d}{dt}\left(\frac{\partial f}{\partial \dot{y}}\right) = \lambda(t)\, G_y. \qquad \text{...(255)}$$

Thus the necessary condition for a stationary value of I is the existence of a function $\lambda(t)$ satisfying equations (251), (254) and (255). Eliminating $\lambda(t)$ between (251), (254), and (255), we obtain the equations

$$\frac{\dfrac{d}{dt}\left(\dfrac{\partial f}{\partial \dot{x}}\right)}{G_x} = \frac{\dfrac{d}{dt}\left(\dfrac{\partial f}{\partial \dot{y}}\right)}{G_y} = \frac{\dfrac{d}{dt}\left(\dfrac{\partial f}{\partial \dot{z}}\right)}{G_z} \qquad \text{...(256)}$$

These equations together with the equation of the surface $G(x, y, z) = 0$ determine the extremal of the given problem and the extremal is the required geodesic.

Note: Equations (251), (254) and (255) can be regarded as the Euler equations for the problem of finding unconstrained stationary functions for the integral

$$\int_{t_1}^{t_2} \left[f(\dot{x}, \dot{y}, \dot{z}) + \lambda(t)\, G(x, y, z) \right] dt.$$

For this integrand if we solve the Euler-Lagrange's equation we immediately obtain the equation (256).

WORKED EXAMPLES

Example 44: Find the extremal of the functional $\int_{t_1}^{t_2} \sqrt{\dot{x}^2 + \dot{y}^2 + \dot{z}^2}\; dt$ subject to $x^2 + y^2 + z^2 = a^2$.

Solution: The surface is a sphere of radius a. Denote this surface by the equation

$$G(x, y, z) \equiv x^2 + y^2 + z^2 - a^2 = 0. \qquad \text{...(257)}$$

The given functional is extremal provided the integrand $f = \sqrt{\dot{x}^2 + \dot{y}^2 + \dot{z}^2}$ must satisfy the following Euler-Lagrange's conditions (256). This reduces to

$$\frac{f\ddot{x} - \dot{x}\dot{f}}{2xf^2} = \frac{f\ddot{y} - \dot{y}\dot{f}}{2yf^2} = \frac{f\ddot{z} - \dot{z}\dot{f}}{2zf^2}. \qquad \qquad ...(258)$$

Considering the first two and the second two ratios of equations (258) we obtain

$$\frac{\dot{f}}{f} = \frac{x\ddot{y} - y\ddot{x}}{x\dot{y} - y\dot{x}} \quad \text{and} \quad \frac{\dot{f}}{f} = \frac{z\ddot{y} - y\ddot{z}}{z\dot{y} - y\dot{z}}.$$

This gives $\qquad \dfrac{x\ddot{y} - y\ddot{x}}{x\dot{y} - y\dot{x}} = \dfrac{z\ddot{y} - y\ddot{z}}{z\dot{y} - y\dot{z}}.$

The equation can be written as

$$\frac{\dfrac{d}{dt}(x\dot{y} - y\dot{x})}{x\dot{y} - y\dot{x}} = \frac{\dfrac{d}{dt}(z\dot{y} - y\dot{z})}{z\dot{y} - y\dot{z}}.$$

Integrating we get $\log(x\dot{y} - y\dot{x}) = \log(y\dot{z} - z\dot{y}) + \log c_1.$

$$\Rightarrow \qquad y(c_1\dot{z} + \dot{x}) = \dot{y}(c_1 z + x),$$

$$\frac{(c_1\dot{z} + \dot{x})}{(c_1 z + x)} = \frac{\dot{y}}{y}.$$

Integrating we get

$$(c_1 z + x) = c_2 y. \qquad \qquad ...(259)$$

This is the equation of a plane passing through the origin, so the geodesics on a sphere are arcs of great circles.

Example 45: Find the geodesic on the sphere $x^2 + y^2 + z^2 = a^2$.

Solution: Define the curve $x = x(t)$, $y = y(t)$, $z = z(t)$ on the surface $G(x, y, z) = 0$, where $G(x, y, z) = x^2 + y^2 + z^2 - a^2$. Let P and Q be two points on the curve.

The metric on this surface has the form $ds = (\dot{x}^2 + \dot{y}^2 + \dot{z}^2)^{\frac{1}{2}} \, dt$. Hence the total length of the curve on the surface becomes

$$I = \int_{t_1}^{t_2} (\dot{x}^2 + \dot{y}^2 + \dot{z}^2)^{\frac{1}{2}} dt. \qquad \qquad ...(260)$$

The curve is a geodesic provided that the integrand of the functional satisfies the equation (256). The equation gives

$$\frac{f\ddot{x} - \dot{x}\dot{f}}{2xf^2} = f\ddot{y} - \frac{\dot{y}\dot{f}}{2yf^2} = \frac{f\ddot{z} - \dot{z}\dot{f}}{2zf^2}. \qquad \qquad ...(261)$$

This is same as that of equation (258). Hence continuing in the same way one can establish that the geodesic is the arc of the great circle.

Theorem 9   Find the Euler – Lagrange's condition that is to be satisfied by the function, which extremizes the curve $v = v(u)$ on the surface

$$x = x(u, v),\ y = y(u, v),\ z = z(u, v).$$

Proof: Define curve on the given surface by the equation $v = v(u)$. The metric on the surface along the curve becomes

$$ds^2 = Edu^2 + 2Fdudv + Gdv^2, \qquad \text{...(262)}$$

where the coefficients of the fundamental form are

$$E = x_u \cdot x_u + y_u \cdot y_u + z_u \cdot z_u,$$

$$G = x_v \cdot x_v + y_v \cdot y_v + z_v \cdot z_v,$$

$$F = x_u \cdot x_v + y_u \cdot y_v + z_u \cdot z_v, \qquad \text{...(263)}$$

Hence the total distance between two points $P(u_1, v_1)$ and $Q(u_2, v_2)$ along the

curve is given by
$$I = \int_{u_1}^{u_2} (Edu^2 + 2Fdudv + Gdv^2)^{\frac{1}{2}},$$

$$I = \int_{u_1}^{u_2} (E + 2Fv' + Gv'^2)^{\frac{1}{2}}\, du,\ v' = \frac{dv}{du}. \qquad \text{...(264)}$$

Thus our problem is to find the function $v(u)$ that renders the integral (264) minimum. This is true if the integrand of the functional $f = (E + 2Fv' + Gv'^2)^{\frac{1}{2}}$ satisfies the Euler-Lagrange's equation which reduces on simplifying to

$$\frac{E_v + 2v'F_v + v'^2 G_V}{2f} - \frac{d}{du}\left(\frac{F + Gv'}{f}\right) = 0. \qquad \text{...(265)}$$

Special Cases:

Case 1:   If the coefficients of the fundamental form E, F and G are functions of u alone, then from equation (265), we have

$$\frac{F + Gv'}{f} = c_1. \qquad \text{...(266)}$$

If in particular $F = 0$, then

$$\frac{Gv'}{f} = c_1,$$

$$\Rightarrow \qquad v'G = c_1\sqrt{E + Gv'},$$

$$\Rightarrow \qquad v' = \frac{c_1\sqrt{E}}{\sqrt{G^2 - c_1^2\,G}},$$

$$v = c_1\int \frac{\sqrt{E}\ du}{\sqrt{G^2 - c_1^2\,G}} + c_2. \qquad \qquad \text{...(267)}$$

Case (2): Let $F = 0$ and E and G are function of v alone. In this case the integrand f is explicitly independent of the independent variable u and we have

$$v'\frac{\partial f}{\partial v'} - f = c_1,$$

$$\Rightarrow \qquad \frac{v'^2 G}{f} - f = c_1,$$

$$\Rightarrow \qquad v'^2 G - f^2 = c_1 f$$

$$\Rightarrow \qquad -E = c_1\sqrt{E + Gv'^2}$$

$$\Rightarrow \qquad v' = \frac{\sqrt{E^2 - c_1^2\,E}}{c_1\sqrt{G}}$$

$$du = \frac{c_1\sqrt{G}}{\sqrt{E^2 - c_1^2\,E}}\ dv.$$

Integrating we get

$$u = c_1\int \frac{\sqrt{G}}{\sqrt{E_2 - c_1^2\,E}}\ dv + c_2. \qquad \qquad \text{...(268)}$$

Solving the integral we will get the required geodesic on the surface.

WORKED EXAMPLES

Example 46: Find the equation of geodesic on the surface of a sphere of radius r, whose parametric equations are $x = r\sin\theta\cos\phi$, $y = r\sin\theta\sin\phi$, $z = r\cos\theta$.

Solution: Here the parameters of the surface are θ and ϕ. Hence from the equation of the surface, we have

$$x_\theta = r \cos \theta \cos \phi, \; x_\phi = -r \sin \theta \sin \phi,$$

$$y_\theta = r \cos \theta \sin \phi, \; y_\phi = r \sin \theta \cos \phi,$$

$$z_\theta = -r \sin \theta, \; z_\phi = 0.$$

The coefficients of the first fundamental form become

$$E = r^2, \; F = 0, \; G = r^2 \sin^2 \theta.$$

Now the total length of the curve $\phi = \phi(\theta)$ between two points becomes

$$I = \int_{\theta_1}^{\theta_2} (r^2 + r^2 \sin^2 \theta \dot\phi^2)^{\frac{1}{2}} \, d\theta. \qquad \qquad ...(269)$$

We see that the coefficients of the fundamental form are functions of a single parameter θ. Thus from the condition (267), the curve be a geodesic if

$$\phi = c_1 \int \frac{r d\theta}{\sqrt{r^4 \sin^4 \theta - c_1^2 r^2 \sin^2 \theta}},$$

$$\phi = \int \frac{k \cos ec^2\theta d\theta}{\sqrt{1 - k^2 \cot^2 \theta}}.$$

The solution of this integral is an arc of the great circle (refer Example 3).

Note: In the Theorem 9 the equation of the curve used was a single valuedness of v as a function of u. But instead if we use the parametric representation $u = u(t)$, $v = v(t)$, then the Euler-Lagrange's conditions for the curve to be a geodesic are obtained in the following theorem. However, we will show that (Refer Theorem 12) the solution of the Euler-Lagrange's equation derived on the basis of the assumption of the single valuedness of v as a function of u also satisfies the system of Euler-Lagrange's equations derived on the basis of the parametric relationship $u = u(t)$, $v = v(t)$, between v and u.

Theorem 10 Find the Euler–Lagrange's conditions that are to be satisfied by the function, which extremize the curve $u = u(t)$, $v = v(t)$ on the surface

$$x = x(u, v), \; y = y(u, v), \; z = z(u, v).$$

Proof: The total distance between two points $P(u_1, v_1)$ and $Q(u_2, v_2)$ along the curve defined on the surface in question is given by

$$I = \int_{u_1}^{u_2} (Eu'^2 + 2Fu'v' + Gv'^2)^{\frac{1}{2}} \, dt, \; u' = \frac{du}{dt}, \qquad ...(270)$$

where

$$E = x_u \cdot x_u + y_u \cdot y_u + z_u \cdot z_u,$$

$$G = x_v \cdot x_v + y_v \cdot y_v + z_v \cdot z_v,$$

$$F = x_u \cdot x_v + y_u \cdot y_v + z_u \cdot z_v,$$

Solving the Euler-Lagrange's equations we obtain

$$\frac{E_u u'^2 + 2u'v'F_u + v'^2 G_u}{f} - \frac{d}{dt}\left(\frac{2(Eu' + Fv')}{f}\right) = 0,$$

$$\frac{E_v u'^2 + 2u'v'F_v + v'^2 G_v}{f} - \frac{d}{dt}\left(\frac{2(Fu' + Gv')}{f}\right) = 0 \qquad \ldots(271)$$

These are the required Euler-Lagrange's conditions.

Remark: If the parametric curves $u = $ const. and $v = $ const. are orthogonal, then the tangent vectors x_u and x_v to the respective curves are orthogonal and $F = x_u \cdot x_v = 0$. In this case the Euler-Lagrange's conditions (271) become

$$\frac{E_u u'^2 + v'^2 G_u}{f} - \frac{d}{dt}\left(\frac{2Eu'}{f}\right) = 0, \quad \frac{E_v u'^2 + v'^2 G_v}{f} - \frac{d}{dt}\left(\frac{2Gv'}{f}\right) = 0. \quad \ldots(272)$$

WORKED EXAMPLES

Example 47: Find the geodesic on the surface of a cylinder.

Solution: The surface of a circular cylinder is characterized by the equations

$$x = a \cos \phi, \; y = a \sin \phi, \; z = z.$$

Here the parameters of the surface are ϕ and z. Hence from the equations of the surface, we have

$$x_\phi = -a \sin \phi, \; x_z = 0, \; y_\phi = a \cos \phi, \; y_z = 0,$$

$$z_z = 1, \; z_\phi = 0.$$

Hence the coefficients of the first fundamental form become $E = a^2$, $F = 0$, $G = 1$. The total length of the curve between two given points is given by the integral

$$I = \int_{\theta_1}^{\theta_2} (a^2 \dot{\phi}^2 + \dot{z}^2)^{\frac{1}{2}} \, dt. \qquad \ldots(273)$$

The curve is a geodesic if equations (271) are satisfied. These conditions reduce to

$$\frac{d}{dt}\left(\frac{a^2 \dot{\phi}}{\sqrt{a^2 \dot{\phi}^2 + \dot{z}^2}}\right) = 0, \; \frac{d}{dt}\left(\frac{\dot{z}}{\sqrt{a^2 \dot{\phi}^2 + \dot{z}^2}}\right) = 0.$$

Integrating these equations we get

$$a^2\dot\phi = c_1\sqrt{a^2\dot\phi^2 + \dot z^2} \text{ and } \dot z = c_2\sqrt{a^2\dot\phi^2 + \dot z^2}.$$

Taking the ratio of these equations we get

$$\frac{\dot z}{\dot\phi} = c_1 \Rightarrow \frac{dz}{d\phi} = c_1$$

Integrating we obtain $z = c_1\phi + c_2$. This equation represents two parameters family of helix, which lies on the cylinder. Thus the geodesic on the surface of a cylinder is the helix.

Theorem 11 Find the Euler-Lagrange's condition to be satisfied by the function for a curve be a geodesic on the surface obtained by revolving the curve $y = g(x)$ about x-axis.

Proof: The surface generated by revolving the curve $y = g(x)$ about x-axis is given by

$$y^2 + z^2 = [g(x)]^2. \qquad \qquad ...(274)$$

The parametric representation of this surface is

$$x = u, \; y = g(u)\cos v, \; z = g(u)\sin v. \qquad ...(275)$$

Let $\qquad\qquad\qquad v = v(u) \qquad\qquad\qquad\qquad\qquad ...(276)$

be a curve on the surface (275). The square of the infinitesimal distance between two neighboring points on the curve (276) is given by

$$ds^2 = Edu^2 + 2Fdudv + Gdv^2, \qquad ...(277)$$

where

$$E = (1 + g'^2(u)), \; F = 0, \; G = g^2(u). \qquad ...(278)$$

Hence the total distance between two points $P(u_1)$ and $Q(u_2)$ along the curve becomes

$$I = \int_{u_1}^{u_2} \left[(1 + g'^2) + g^2 v'^2\right]^{\frac{1}{2}} du, \; v' = \frac{dv}{du} \qquad ...(279)$$

The curve is of shortest length if

$$\Rightarrow \qquad\qquad \frac{d}{du}\left(\frac{g^2 v'}{f}\right) = 0,$$

$$\Rightarrow \qquad\qquad \frac{g^2 v'}{f} = c_1,$$

$$\Rightarrow \qquad\qquad v' = \frac{c_1\sqrt{1 + g'^2}}{g(u)\sqrt{g^2 - c_1^2}},$$

$$\Rightarrow \qquad v = c_1 \int \frac{\sqrt{1 + g'^2}}{g(u)\sqrt{g^2 - c_1^2}}\, du + c_2. \qquad \qquad \text{...(280)}$$

This is the required condition.

Example 48: Find the geodesic on the surface obtained by revolving the line $y = x$ about x-axis.

Solution: The revolution of the line $y = x$ about x-axis generates a surface given by the equation

$$y^2 + z^2 = x^2. \qquad \qquad \text{...(281)}$$

The surface is a cone with semi vertical angle $\dfrac{\pi}{4}$ whose parametric representation is

$$x = x,\; y = x \cos \phi,\; z = x \sin \phi. \qquad \qquad \text{...(282)}$$

Combing these equations with the equations (275), we have

$$x = x,\; g(u) = x,\; g'(x) = 1.$$

Hence the Euler-Lagrange's condition (280) for the curve on the surface (282) to be geodesic becomes

$$\phi = c_1 \int \frac{\sqrt{2}}{x\sqrt{x^2 - c_1^2}}\, dx + \alpha,$$

$$\phi = \frac{\sqrt{2}}{c_1} \int \frac{dx}{\left(\dfrac{x}{c_1}\right)\sqrt{\left(\dfrac{x}{c_1}\right)^2 - 1}} + \alpha,$$

$$\phi = \sqrt{2}\, \sec^{-1}\left(\frac{x}{c_1}\right) + \alpha,$$

$$\Rightarrow \qquad x = c_1 \sec\left(\frac{\phi - \alpha}{\sqrt{2}}\right). \qquad \qquad \text{...(283)}$$

This is the required geodesic on the surface of the cone.

Example 49: Show that the geodesics on the space characterized by the metric $ds^2 = \dfrac{1}{t^2}(dx^2 - dt^2)$ are the hyperbolas.

Solution: The curve $x = x(t)$ on this space is geodesic provided

$$\delta \int (1 - \dot{x}^2)^{\frac{1}{2}}\, \frac{dt}{t} = 0. \qquad \qquad \text{...(284)}$$

The Euler-Lagrange's differential equation gives

$$\Rightarrow \qquad \frac{d}{dt}\left(\frac{\dot{x}}{t\sqrt{1-\dot{x}^2}}\right) = 0$$

$$\dot{x} = \frac{kt}{\sqrt{1+k^2t^2}}.$$

On integrating we obtain

$$(x - x_0)^2 = t^2 + a^2, \qquad\qquad ...(285)$$

where x_0, a are some constant. This shows that the required geodesic curves are the hyperbolas.

Theorem 12   Show that the solution of the Euler-Lagrange's equation derived on the basis of the assumption of the single valuedness of y as a function of x also satisfies the system of Euler-Lagrange's equations derived on the basis of the parametric relationship between x and y.

Proof: Let $y = y(x)$ be a single valued function of x. The integral to be extremized is given by

$$I = \int_{x_0}^{x_1} f(x, y, y')dx \qquad\qquad ...(286)$$

with the conditions $y(x_1) = y_1$, $y(x_2) = y_2$. The corresponding Euler-Lagrange's equation is

$$\frac{\partial f}{\partial y} - \frac{d}{dx}\left(\frac{\partial f}{\partial y'}\right) = 0. \qquad\qquad ...(287)$$

Now if instead we use the parametric representation $x = x(t)$, $y = y(t)$ for the curve, then we write

$$y' = \frac{dy}{dx} = \frac{dy/dt}{dx/dt} = \frac{\dot{y}}{\dot{x}} \text{ and } dx = \dot{x}dt.$$

Hence the integral (286) becomes

$$I = \int_{t_0}^{t_1} f\left(x, y, \frac{\dot{y}}{\dot{x}}\right) \dot{x}dt. \qquad\qquad ...(288)$$

Let $g(x, y, \dot{x}, \dot{y}) = f(x, y, y')\dot{x}$, for $y' = \frac{\dot{y}}{\dot{x}}$. $\qquad\qquad ...(289)$

Now to find the associated Euler-Lagrange's conditions, we find

$$\frac{\partial g}{\partial x} = \frac{\partial f}{\partial x}\dot{x}, \qquad\qquad ...(290)$$

$$\frac{\partial g}{\partial \dot{x}} = \frac{\partial f}{\partial y'}\frac{\partial y'}{\partial \dot{x}}\dot{x} + f,$$

$$\frac{\partial g}{\partial \dot{x}} = \dot{x}\,\frac{\partial f}{\partial y'}\left(-\frac{\dot{y}}{\dot{x}^2}\right) + f.$$

$$\Rightarrow \qquad \frac{\partial g}{\partial \dot{x}} = f - y'\,\frac{\partial f}{\partial y'}. \qquad\qquad ...(291)$$

Thus

$$\frac{d}{dt}\left(\frac{\partial g}{\partial \dot{x}}\right) = \frac{d}{dx}\left(f - y'\,\frac{\partial f}{\partial y'}\right)\dot{x},$$

$$\frac{d}{dt}\left(\frac{\partial g}{\partial \dot{x}}\right) = \dot{x}\left[\frac{\partial f}{\partial x} + y'\left\{\frac{\partial f}{\partial y} - \frac{d}{dx}\left(\frac{\partial f}{\partial y'}\right)\right\}\right]. \qquad\qquad ...(292)$$

Similarly, we obtain from equation (289) that

$$\frac{\partial g}{\partial y} = \frac{\partial f}{\partial y}\,\dot{x}$$

$$\frac{\partial g}{\partial \dot{y}} = \frac{\partial f}{\partial y'}\,\frac{\partial y'}{\partial \dot{y}}\,\dot{x} = \frac{\partial f}{\partial y'}.$$

Hence

$$\frac{d}{dt}\left(\frac{\partial g}{\partial \dot{y}}\right) = \frac{d}{dt}\left(\frac{\partial f}{\partial y'}\right) = \frac{d}{dx}\left(\frac{\partial f}{\partial y'}\right)\dot{x} \qquad\qquad ...(293)$$

Thus we have

$$\frac{\partial g}{\partial x} - \frac{d}{dt}\left(\frac{\partial g}{\partial \dot{x}}\right) = \dot{x}\,\frac{\partial f}{\partial x} - \dot{x}\left[\frac{\partial f}{\partial x} + y'\left\{\frac{\partial f}{\partial y} - \frac{d}{dx}\left(\frac{\partial f}{\partial y'}\right)\right\}\right]$$

$$\frac{\partial g}{\partial x} - \frac{d}{dt}\left(\frac{\partial g}{\partial \dot{x}}\right) = -\dot{y}\left[\frac{\partial f}{\partial y} - \frac{d}{dx}\left(\frac{\partial f}{\partial y'}\right)\right], \qquad\qquad ...(294)$$

and

$$\frac{\partial g}{\partial y} - \frac{d}{dt}\left(\frac{\partial g}{\partial \dot{y}}\right) = \dot{x}\left[\frac{\partial f}{\partial y} - \frac{d}{dx}\left(\frac{\partial f}{\partial y'}\right)\right]. \qquad\qquad ...(295)$$

Equations (294) and (295) show that any single valued relationship $y = y(x)$ that satisfies the Euler-Lagrange's equation (287) also satisfies the system of equations

$$\frac{\partial g}{\partial x} - \frac{d}{dt}\left(\frac{\partial g}{\partial \dot{x}}\right) = 0, \quad \frac{\partial g}{\partial y} - \frac{d}{dt}\left(\frac{\partial g}{\partial \dot{y}}\right) = 0.$$

Exercise

1. Find the shortest distance between circle $x^2 + y^2 = 4$ and straight line $2x + y = 6$.

2. Find the shortest distance of an interior point $(1, 1)$ from the curve $\dfrac{x^2}{9} + \dfrac{y^2}{4} = 1$.

3. Find the extremum distances of an interior point $(1, 1)$ from the curve $x^2 + y^2 = 4$.

4. Find the shortest distance between parabola $y = x^2$ and straight line $x - y = 5$.

5. Find the shortest distance of an exterior point $(-1, 4)$ from the curve $4x^2 + 9y^2 = 36$. Show that the shortest distance between two points along the curve $x = x(t)$, $y = y(t)$ in a Euclidean plane is a straight line.

6. Show that the geodesic defined by $r = r(t)$, $\theta = \theta(t)$ in a plane is a straight line.

7. Show that the stationary (extremum) distance between two points along the curve $\theta = \theta(t)$, $\phi = \phi(t)$ on the sphere
$$x = r \sin \theta \cos \phi, \ y = r \sin \theta \sin \phi, \ z = r \cos \theta$$
is an arc of the great circle.

8. Find the geodesic on the surface obtained by generating the parabola $y^2 = 4ax$ about x-axis.

   Ans: The surface of revolution is $y^2 + z^2 = 4ax$, whose parametric representation is
   $$x = au^2, \ y = 2au \sin v, \ z = 2au \cos v$$
   The geodesic on this surface is obtain by solving the integral
   $$v = c_1 \int \frac{\sqrt{1 + u^2}}{u\sqrt{u^2 - c_1^2}} \, du + c_2.$$

9. The time of travel for photon from one point to another in a medium is given by
   $$\int_1^2 \frac{ds}{v(x, y)},$$
   where s is the arc length and $v(x, y)$ is the velocity of photon in a medium, is a minimum. Show that the path travel is given by
   $$v \frac{d^2y}{dx^2} + \left(1 + y'^2\right) \frac{\partial v}{\partial y} - \frac{dy}{dx} \left(1 + y'^2\right) \frac{\partial v}{\partial x} = 0.$$

10. Find the curve which corresponds to the extreme value of the integral

$$I = \int_{x_1}^{x_2} x^n \left(\frac{dy}{dx}\right)^2 dx$$ and passes through two fixed points $x = x_1$, $x = x_2$.

Ans: $$y = \frac{cx^{1-n}}{1-n} + b.$$

11. Derive the Euler-Lagrange's equations that are to be satisfied by twice differential functions $x(t)$, $y(t)$, ..., $z(t)$, that extremize the integral

$$I = \int_{t_1}^{t_2} f(x, y, ..., z, \dot{x}, \dot{y}, ..., \dot{z}, t)dt, \quad \dot{x} = \frac{dx}{dt}$$

which achieve prescribed values at the fixed points t_1, t_2.

Ans: $$\frac{\partial f}{\partial x} - \frac{d}{dt}\left(\frac{\partial f}{\partial \dot{x}}\right) = 0, \quad \frac{\partial f}{\partial y} - \frac{d}{dt}\left(\frac{\partial f}{\partial \dot{y}}\right) = 0, \quad ..., \quad \frac{\partial f}{\partial z} - \frac{d}{dt}\left(\frac{\partial f}{\partial \dot{z}}\right) = 0.$$

12. Find the curve which generates a surface of revolution of minimum area when it is revolved about x-axis.

Ans: Area of revolution of a curve about x-axis is $I = \int_{x_0}^{x_1} y\sqrt{1 + y'^2}\, dx.$

The curve is a catenary given by $y = c \sec \psi$, or $y = a \cosh\left(\frac{x-b}{a}\right)$.

13. Find the function on which the functional can be extremized

$$I[y(x)] = \int_0^1 \left(y''^2 - 2xy\right)dx, \quad y(0) = 0, \quad y(1) = \frac{1}{120},$$

and y' is not prescribed at both the ends.

Ans: $$y = \frac{x^5}{120} - \frac{x^3}{36} + \frac{x}{36}.$$

14. Find the stationary function of $\int_0^4 (xy' - y'^2)dx$, which is determined by the boundary conditions $y(0) = 0$, $y(4) = 3$.

Ans: $$y = \frac{x^2}{4} - \frac{x}{4}.$$

15. Find the extremum of $I[y(x)] = \int_1^2 \frac{x^3}{y'^2}\, dx$, $y(1) = 1$, $y(2) = 4$.

Ans: $$y = x^2.$$

16. Find the extremal of $I[y(x)] = \int_{x_0}^{x_1} \frac{y'^2}{x^3}\, dx.$

Ans: $\qquad y = c_1 \dfrac{x^4}{4} + c_2.$

17. Find the extremal of $\int_0^{\pi} (y'^2 - y^2 + 4y \cos x)dx$, $y(0) = 0$, $y(\pi) = 0$.

    Ans: $\qquad y = (c_2 + x \sin x).$

18. Find the extremal of the functional $\int_0^1 (y'^2 - 12xy)dx$,

    $$y(0) = 1, \ y(1) = 2.$$

    Ans: $\qquad y = -x^3 + 2x + 1.$

19. Determine the curve $z = z(x)$ for which the functional

    $$I = \int_1^2 (z'^2 - 2xz)dx, \ z(1) = 0, \ z(2) = -1 \text{ is extremal.}$$

    Ans: $\qquad z = -\dfrac{x^3}{3} + \dfrac{x}{6}.$

20. Determine the curve $z = z(x)$ for which the functional

    $$I = \int_0^{\frac{\pi}{2}} (z'^2 - z^2)dx, \ z(0) = 0, \ z\left(\frac{\pi}{2}\right) = 1 \text{ is extremal.}$$

    Ans: $\qquad z = \sin x.$

21. Find the extremal of the functional

    $$I[y(x)] = \int_0^{\frac{\pi}{2}} (y''^2 - y^2 + x^2)dx, \ y(0) = 1, \ y\left(\frac{\pi}{2}\right) = 0, \ y'(0) = 0, \ y'\left(\frac{\pi}{2}\right) = -1.$$

    Ans: $\qquad y = \cos x.$

22. Find the differential equation of extremals for an isoperimetric problem of extremal of the functional $I(y(x)) = \int_{x_1}^{x_1} [p(x)y'^2 + q(x)y^2]dx$ subject to the conditions that $\int_0^{x_1} r(x)y^2 dx = 1$, $y(0) = 0$, $y(x_1) = 0$.

    Ans: $\qquad p(x)y'' + p'(x)\,y' - [q(x) + \lambda r(x)]y = 0.$

23. Obtain the differential equation in which the extremizing function makes the integral

    $$I(y(x)) = \int_{x_1}^{x_2} f(x, y, y')dx$$

extremum subject to the conditions $y(x_1) = y_1$, $y(x_2) = y_2$, and

$$J_k = \int_{x_1}^{x_2} g_k(x, y, y')dx = \text{const.} \ k = 1, 2, \dots, n$$

Ans: The problem would be extrmization of $I^* = \int_{x_1}^{x_2} f^*(x, y, y')dx$,

where $I^* = I + \sum_{k=1}^{n} \lambda_k J_k$ and $f^* = f + \sum_{k=1}^{n} \lambda_k g_k$.

The required differential equation is $\dfrac{\partial f^*}{\partial y} - \dfrac{d}{dx}\left(\dfrac{\partial f^*}{\partial y'}\right) = 0$.

24. Find the extremals of the isoperimetric problem $I = \int_{x_0}^{x_1} y'^2 dx$, such that

$$\int_{x_0}^{x_1} y\,dx = c.$$

Ans: $\qquad y = \lambda\dfrac{x^2}{4} + ax + b.$

25. Find the extremal of the functional

$$I[y(x)] = \int_{0}^{1}(1 + y''^2)\,dx, \ y(0) = 0, \ y(1) = 1, \ y'(0) = 1, \ y'(1).$$

Ans: $\qquad y = x.$

26. Find the extremal of the functional

$$I[y(x), z(x)] = \int_{x_0}^{x_1}\left(2yz - 2y^2 + y'^2 - z'^2\right)dx.$$

Ans: $y = (c_1 x + c_2)\cos x + (c_3 x + c_4)\sin x$, $z = 2y + y''$.

27. Find the extremal of the functional $I[y(x)] = \int_{x_0}^{x_1}\left(y^2 + y'^2 + 2ye^x\right)dx$.

Ans: $y = \dfrac{xe^x}{2} + c_1 e^x + c_2 e^{-x}.$

28. Find the extremal of the functional

$$I[y(x), z(x)] = \int_{x_0}^{x_1}\left(y'^2 + z'^2 + y'z'\right)dx.$$

Ans: $y = c_1 x + c_2$, $z = c_3 x + c_4$.

29. If $x(t)$ is an extremal of the functional $\int\limits_a^b \left(\dfrac{1}{2}\,m\dot{x}^2 - cx^2\right)dt$, where a, b, c are arbitrary constants and $\dot{x} = \dfrac{dx}{dt}$, then the function $x(t)$ satisfies

    $x(t) = k_1 \sin\left(\sqrt{\dfrac{2c}{m}}\,t + k_2\right)$, k_1, k_2 are constants. Show also that the energy of the particle is conserved.

    Hint: The Euler-Lagrange's equation is $m\ddot{x} + 2cx = 0$. The solution of this equation is the answer to the question and the first integral of the equation of motion represents the energy of the particle and is given by $m\dot{x}^2 + 2cx^2$ and is conserved.

30. If $u(x)$ and $v(x)$ satisfying $u(0) = 1$, $v(0) = -1$, $u\left(\dfrac{\pi}{2}\right) = 0$, $v\left(\dfrac{\pi}{2}\right) = 0$ are the extremals of the functional $\int\limits_0^{\frac{\pi}{2}} [\dot{u}^2 + \dot{v}^2 + 2uv]dx$, $\dot{u} = \dfrac{du}{dx}$, then show that $u(x) = -v(x) = \cos x$.

31. Show that the variational problem of extremizing the functional

    $$\int\limits_0^{2\pi} [y'^2 + y^2]dx, \quad y' = \frac{dy}{dx}, \quad y(0) = 1, \quad y(2\pi) = 1$$

    has an infinite number of solutions.

Hamilton's Principle and Hamilton's Formulation

INTRODUCTION

In the Chapter 2, we used the techniques of variational principles of Calculus of Variation and found the stationary path between two points, the curve of quickest decent and the curve for which the area of surface of revolution is minimum when revolved about a co-ordinate axis. Hamilton's principle is one of the variational principles in mechanics. It is the most fundamental and important principle in the sense that all laws of mechanics can be derived from it. In this chapter, we first derive this principle from D'Alembert's principle, for both conservative and non-conservative systems and then derive the Hamilton's canonical equations of motion and also the Lagrange's equation of motion from it. Accordingly we define

Hamilton's Principle (for non-conservative system):

Hamilton's principle for non-conservative systems states that "The motion of a dynamical system between two points at time intervals t_0 to t_1 is such that the line integral $I = \int_{t_0}^{t_1} (T + W)\, dt$ is extremum for the actual path followed by the system", where T is the kinetic energy of the system and W the total work done.

It is equivalent to say that δ variation in the value of the line integral $I = \int_{t_0}^{t_1} (T + W)\, dt$ is zero for the actual path.

Mathematically, it means that

$$\delta I = \delta \int_{t_0}^{t_1} (T + W)\, dt = 0 \qquad \qquad ...(1)$$

for the actual path.

Hamilton's Principle (for conservative system):

"Of all possible paths between two points along which a dynamical system may move from one point to another within a given time interval from t_0 to t_1, the actual path followed by the system is the one which minimizes the line integral of Lagrangian."

This means that the motion of a dynamical system from t_0 to t_1 is such that the line integral $\int_{t_0}^{t_1} L dt$ is extremum for actual path. This means that a small δ variation in the actual path followed by the system is zero.

Mathematically, we express this as $\delta \int_{t_0}^{t_1} L dt = 0$ for the actual path.

Note: We will show below in the Theorem 2 that the Hamilton's principle $\delta \int_{t_0}^{t_1} L dt = 0$ also holds good for the non-conservative system.

Action in Mechanics:

Let $L = L(q_j, \dot{q}_j, t)$ be the Lagrangian for the conservative system. Then the integral

$$I = \int_{t_0}^{t_1} L dt$$

is called the action of the system.

Hence we can also define the Hamilton's principle as "Out of all possible paths of a dynamical system between the time instants t_0 and t_1, the actual path followed by the system is one for which the action has a stationary value",

$$\Rightarrow \qquad \delta I = \delta \int_{t_0}^{t_1} L dt = 0 \qquad \qquad ...(2)$$

for the actual path.

Theorem 1   Derive Hamilton's principle for non-conservative system from D'Alembert's principle and hence deduce from it the Hamilton's principle for conservative system.

Proof: We start with D'Alembert's principle (refer equation (47) of Chapter 1) which states that

$$\sum_i \left(\overline{F}_i - \dot{p}_i \right) \delta r_i = 0.$$

Note that in this principle the knowledge of force whether it is conservative or non-conservative and also the requirement of holonomic or non-holonomic constraints does not arise. We write the principle in the form

$$\sum_i F_i \delta r_i = \sum_i \dot{p}i \delta r_i,$$

$$\Rightarrow \qquad \delta W = \sum_i \dot{p}_i \delta r_i \qquad\qquad ...(3)$$

where $\delta W = \sum_i F_i \delta r_i$ is the virtual work. Now consider the term on the right hand side of the equation (3) and write

$$\sum_i \dot{p}_i \delta r_i = \sum_i m_i \ddot{r} \delta r_i,$$

$$= \sum_i \frac{d}{dt}\left(m_i \dot{r}_i \delta r_i\right) - \sum_i m_i \dot{r}_i \frac{d}{dt}(\delta r_i)$$

Since we have $\delta \dot{r}_i = \dfrac{d}{dt}\,\delta r_i$, therefore, we write

$$\sum_i \dot{p}_i \delta r_i = \frac{d}{dt}\left[\sum_i m_i \dot{r}_i \delta r_i\right] - \delta\left(\sum_i \frac{1}{2} m_i \dot{r} i^2\right).$$

$$\sum_i \dot{p}_i \delta r_i = \frac{d}{dt}\left[\sum_i m_i \dot{r}_i \delta r_i\right] - \delta T, \qquad\qquad ...(4)$$

where $T = \sum_i \dfrac{1}{2} m_i \dot{r}_i^2$ is the kinetic energy of the system. Substituting this in equation (3) we get

$$\delta W = \frac{d}{dt}\left[\sum_i m_i \dot{r}_i \delta r_i\right] - \delta T,$$

$$\Rightarrow \qquad \delta(W + T) = \frac{d}{dt}\left[\sum_i m_i \dot{r}_i \delta r_i\right].$$

Integrating the above equation with respect to t between t_0 to t_1 we get

$$\int_{t_0}^{t_1} \delta(W + T)\, dt = \left[\sum_i m_i \dot{r}_i \delta r_i\right]_{t_0}^{t_1}.$$

Since, there is no variation in co-ordinates along any paths at the end points. i.e., $(\delta r_i)_{t_0}^{t_1} = 0$, hence from above equation we have

$$\delta \int_{t_0}^{t_1} (W + T)\, dt = 0.$$

This is known as Hamilton's principle for non-conservative systems. If however, the system is conservative, then the forces are derivable from potential V which involves only the co-ordinates. In this case the expression for virtual work becomes

$$\delta W = \sum_i F_i \delta r_i = -\sum_i \frac{\partial V}{\partial r_i} \delta r_i = -\,\delta V.$$

Hence the Hamilton's principle for non-conservative system becomes

$$\delta \int_{t_0}^{t_1} (T - V)\, dt = 0,$$

$$\Rightarrow \qquad \delta \int_{t_0}^{t_1} L\, dt = 0,$$

where $L = T - V$ is the Lagrangian of the conservative system. This is the required Hamilton's principle for conservative system.

Theorem 2   Show that the Hamilton's principle $\delta \int_{t_0}^{t_1} L\, dt = 0$ also holds for the non-conservative system.

Proof: We recall here the expression (63) of the Chapter 1 for the virtual work, given by

$$\delta W = \sum_j Q_j\, \delta q_j, \qquad\qquad \dots(5)$$

where $Q_j = \sum_i F_i \dfrac{\partial r_i}{\partial q_j}$ are the components of generalized forces. In the case of non-conservative system the potential energy depends on velocity of the particle called the velocity dependent potential or generalized potential. The corresponding components of generalized force are given in the equation (66) of the Chapter 1. Using this equation in the equation (5) and integrating it between the limits t_0 to t_1, we find

$$\int_{t_0}^{t_1} \delta W\, dt = \int_{t_0}^{t_1} \sum_i Q_j \delta q_j dt = \int_{t_0}^{t_1} \left\{ \sum_j \left[-\frac{\partial U}{\partial q_j} + \frac{d}{dt}\left(\frac{\partial U}{\partial \dot{q}_j}\right) \right] \delta q_j \right\} dt,$$

Substituting this in the Hamilton's principle for non-conservative system given in the equation (1), we get

$$\int_{t_0}^{t_1} \delta T dt = \int_{t_0}^{t_1} \sum_j \frac{\partial U}{\partial q_j}\, \delta q_j dt - \int_{t_0}^{t_1} \left\{ \sum_j \left[\frac{d}{dt}\left(\frac{\partial U}{\partial \dot{q}_j}\right) \right] \delta q_j \right\} dt.$$

Integrating the second integral by parts we obtain

$$\int_{t_0}^{t_1} \delta T dt = \int_{t_0}^{t_1} \sum_j \frac{\partial U}{\partial q_j}\, \delta q_j dt - \sum_j \left(\frac{\partial U}{\partial \dot{q}_j}\, \delta q_j \right)\Bigg|_{t_0}^{t_1} + \int_{t_0}^{t_1} \left\{ \sum_j \left(\frac{\partial U}{\partial \dot{q}_j} \right) \frac{d}{dt}\, (\delta q_j) \right\} dt.$$

Since change in co-ordinates at the end point is zero. This implies $\left(\delta q_j\right)_{t_0}^{t_1} = 0$ and also

$\dfrac{d}{dt}\left(\delta q_j\right) = \delta \dot{q}_j$, then we have

$$\int_{t_0}^{t_1} \delta T dt = \int_{t_0}^{t_1}\left\{\sum_j\left(\frac{\partial U}{\partial \dot{q}_j}\delta \dot{q}_j + \frac{\partial U}{\partial q_j}\delta q_j\right)\right\} dt.$$

Since in Hamilton's variational principle, time t is fixed along any path hence, there is no variation in time along any path therefore change in time along any path is zero. *i.e.*, $\delta t = 0$.

Hence we write above equation as

$$\int_{t_0}^{t_1}\delta T dt = \int_{t_0}^{t_1}\left[\sum_j\left(\frac{\partial U}{\partial \dot{q}_j}\delta \dot{q}_j + \frac{\partial U}{\partial q_j}\delta q_j\right) + \frac{\partial U}{\partial t}\delta t\right] dt.$$

$$\int_{t_0}^{t_1}\delta T dt = \int_{t_0}^{t_1}\delta U dt,$$

$$\Rightarrow \qquad \int_{t_0}^{t_1}\delta(T - U)\,dt = 0,$$

$$\Rightarrow \qquad \delta \int_{t_0}^{t_1} L\,dt = 0,$$

where $L = T - U$ is a Lagrangian of the non-conservative system. This proves that the Hamilton's principle holds good even for non-conservative systems.

Theorem 3   Deduce from D'Alembert's principle the Hamilton's principle for conservative system.

Proof: Using equation (4) in the D'Alembert's principle, we get

$$\sum_j F_i \delta r_i = \frac{d}{dt}\left[\sum_i m_i \dot{r}_i \delta r_i\right] - \delta T. \qquad\qquad ...(6)$$

Since the forces are conservative $\Rightarrow F_i = -\dfrac{\partial V}{\partial r_i}$.

$$\Rightarrow \qquad \frac{d}{dt}\left[\sum_i m_i \dot{r}_i \delta r_i\right] = \delta T - \sum_i \frac{\partial V}{\partial r_i}\delta r_i,$$

$$\frac{d}{dt}\left[\sum_i m_i \dot{r}_i \delta r_i\right] = \delta T - \delta V = \delta(T - V) = \delta L,$$

where $L = T - V$ is a Lagrangian for the conservative system. Integrating the above equation with respect to t between the limits t_0 to t_1, we get

$$\left[\sum_i m_i \dot{r}_i \delta r_i\right]_{t_0}^{t_1} = \delta \int_{t_0}^{t_1} L\,dt.$$

Since the coordinates are fixed at the end points of the path, hence there is no variation in coordinates along any paths at the end points. *i.e.*, $(\delta r_i)_{t_0}^{t_1} = 0$. Hence from above equation we have

$$\delta \int_{t_0}^{t_1} L\,dt = 0.$$

This is the required Hamilton's principle for conservative systems.

Theorem 4   State Hamilton's principle for non-conservative system and hence derive from it the Lagrange's equations of motion for non-conservative holonomic system.

Proof: Consider a non-conservative holonomic dynamical system whose configuration at any instant t is specified by n generalized co-ordinates q_1, q_2, q_3, ..., q_n. The kinetic energy of the system of particles is a function of generalized coordinates q_j, generalized velocities $\dot{q}_j$ and time t, *i.e.*, $T = T(q_j, \dot{q}_j, t)$. From this we find

$$\delta T = \sum_j \frac{\partial T}{\partial q_j} \delta q_j + \sum_j \frac{\partial T}{\partial \dot{q}_j} \delta \dot{q}_j + \frac{\partial T}{\partial t} \delta t. \qquad \text{...(7)}$$

In Hamilton's principle, the variation in time along any path is always zero. $\Rightarrow \delta t = 0$. This implies that

$$\delta T = \sum_j \frac{\partial T}{\partial q_j} \delta q_j + \sum_j \frac{\partial T}{\partial \dot{q}_j} \delta \dot{q}_j. \qquad \text{...(8)}$$

Integrating equation (8) between the limits t_0 to t_1, we get

$$\int_{t_0}^{t_1} \delta T\,dt = \int_{t_0}^{t_1} \sum_j \frac{\partial T}{\partial q_j} \delta q_j\,dt + \int_{t_0}^{t_1} \sum_j \frac{\partial T}{\partial \dot{q}_j} \delta \dot{q}_j\,dt.$$

Since we have $\dfrac{d}{dt}\delta q_j = \delta \dot{q}_j$. Therefore we write the above integral as

$$\int_{t_0}^{t_1} \delta T\,dt = \int_{t_0}^{t_1} \sum_j \frac{\partial T}{\partial q_j} \delta q_j\,dt + \int_{t_0}^{t_1} \sum_j \frac{\partial T}{\partial \dot{q}_j} \frac{d}{dt}(\delta q_j)\,dt.$$

Integrating the second integral by parts, we obtain

$$\int_{t_0}^{t_1} \delta T\,dt = \int_{t_0}^{t_1} \sum_j \frac{\partial T}{\partial q_j} \delta q_j\,dt + \left[\sum_j \frac{\partial T}{\partial \dot{q}_j} \delta q_j\right]_{t_0}^{t_1} - \int_{t_0}^{t_1} \sum_j \frac{d}{dt}\left(\frac{\partial T}{\partial \dot{q}_j}\right) \delta q_j\,dt.$$

Since in δ variation there is no change in the co-ordinates at the end points,

$$\Rightarrow \left(\delta q_j\right)_{t_0}^{t_1} = 0.$$

Hence

$$\int_{t_0}^{t_1} \delta T dt = \int_{t_0}^{t_1} \sum_j \left[\frac{\partial T}{\partial q_j} - \frac{d}{dt}\left(\frac{\partial T}{\partial \dot{q}_j} \right) \right] \delta q_j\, dt. \qquad \ldots(9)$$

Using equations (5) and (9) in the Hamilton's principle for non-conservative system (1), we get

$$\int_{t_0}^{t_1} \sum_j \left[\frac{\partial T}{\partial q_j} - \frac{d}{dt}\left(\frac{\partial T}{\partial \dot{q}_j} \right) + Q_j \right] \delta q_j\, dt = 0. \qquad \ldots(10)$$

If the constraints are holonomic then δq_j are independent. (Note that if the constraints are non-holonomic, then δq_j are not all independent. In this case vanishing of the integral (10) does not imply the coefficient vanish separately) Hence the integral (10) vanishes if and only if the coefficient must vanish separately. This gives

$$\frac{d}{dt}\left(\frac{\partial T}{\partial \dot{q}_j} \right) - \frac{\partial T}{\partial q_j} = Q_j, j = 1, 2, 3, \ldots, n. \qquad \ldots(11)$$

These are the Lagrange's equations of motion for non-conservative holonomic system.

Configuration space and Phase space: In the absence of constraints the configuration space of a system is described by n generalized coordinates q_j. The n dimensional space spanned by n generalized coordinates is called the configuration space. However, in the presence of k constraints, the system of n particles has $3n - k$ independent generalized coordinates and $3n - k$ dimension space is called configuration space.

In the Hamiltonian formulation n generalized coordinates q_j and n generalized momenta p_j are taken as the independent coordinates. The $2n$ dimensional space spanned by n generalized coordinates q_j and n generalized momenta p_j is known as phase space.

WORKED EXAMPLES

Example 1:   Deduce Newton's second law of motion from Hamilton's principle.

Solution: Let a particle of mass m be moving under the action of a conservative force $\overline{F}$. If $\overline{r}$ is the position vector of the particle, then its kinetic energy is $T = \frac{1}{2}\, m\dot{\overline{r}}^2$, while the force is $\overline{F} = -\frac{\partial V}{\partial r}$. This gives $\delta V = -\overline{F}\,\delta\overline{r}$. Using these equations in the Hamilton's principle, we get

$$\int_{t_0}^{t_1} \left(m\dot{r}\, \delta\dot{r} + \overline{F}\, \delta\overline{r} \right) dt = 0$$

Interchanging the time derivative and δ variation as they are commutative and then integrating the first term of the integral by parts, we obtain

$$\left(m\dot{r}\, \delta\overline{r} \right)_{t_0}^{t_1} - \int_{t_0}^{t_1} \left(m\ddot{r} - \overline{F} \right) \delta\overline{r}\, dt = 0. \qquad \qquad ...(12)$$

Since the end points are fixed hence there is no variation of co-ordinates at the end points, therefore, $\delta\overline{r}$ vanishes at the end points. This reduces the equation (12) in to

$$\int_{t_0}^{t_1} \left(m\ddot{r} - \overline{F} \right) \delta\overline{r}\, dt = 0. \qquad \qquad ...(13)$$

This equation is true for every virtual displacement $\delta\overline{r}$ and hence the integrand must vanish.

$$\Rightarrow \qquad \qquad \overline{F} = m\ddot{r}. \qquad \qquad ...(14)$$

This is the Newton's second law of motion.

Example 2:   A particle moves on a smooth surface under gravity. Use Hamilton's principle to find the equation of motion.

Solution: Let (x, y, z) be the co-ordinates of a particle at some instant moving on a smooth surface under gravity. If z is the vertical, then the kinetic and potential energies of the particle are given by respectively,

$$T = \frac{1}{2} m\left(\dot{x}^2 + \dot{y}^2 + \dot{z}^2 \right), \ V = mgz.$$

Hence the Lagrangian of the particle becomes

$$L = \frac{1}{2} m\left(\dot{x}^2 + \dot{y}^2 + \dot{z}^2 \right) - mgz.$$

The Hamilton's Principle (2) yields

$$\Rightarrow \qquad \qquad \int_{t_0}^{t_1} \left[m(\dot{x}\delta\dot{x} + \dot{y}\delta\dot{y} + \dot{z}\delta\dot{z}) - mg\delta z \right] dt = 0.$$

Since $\delta \dfrac{d}{dt} = \dfrac{d}{dt} \delta$. We therefore write the above equation as

$$\int_{t_0}^{t_1} \left[m\left(\dot{x} \frac{d}{dt} (\delta x) + \dot{y} \frac{d}{dt} (\delta y) + \dot{z} \frac{d}{dt} (\delta z) \right) - mg\delta z \right] dt = 0.$$

Integrating the first three terms by parts we get

$$m(\dot{x}\delta x)_{t_0}^{t_1} + m(\dot{y}\delta y)_{t_0}^{t_1} + m(\dot{z}\delta z)_{t_0}^{t_1} - \int_{t_0}^{t_1}\left[m(\ddot{x}\delta x + \ddot{y}\delta y + (\ddot{z} + g)\,\delta z)\right] dt = 0.$$

Since the co-ordinates at the end points are fixed, hence

$$(\delta x)_{t_0}^{t_1} = (\delta y)_{t_0}^{t_1} = (\delta z)_{t_0}^{t_1} = 0.$$

Consequently, we have

$$\int_{t_0}^{t_1}\left[m(\ddot{x}\delta x + \ddot{y}\delta y + (\ddot{z} + g)\,\delta z)\right] dt = 0.$$

Since δx, δy, δz are all independent, we therefore obtain the equations of motion as

$$\ddot{x} = 0,\ \ddot{y} = 0,\ \ddot{z} = -g.$$

Example 3:   Use Hamilton's principle to find the equation of motion of a simple pendulum.

Solution: In case of a simple pendulum, the only generalized co-ordinate is θ, and the Lagrangian is given by (refer Example 26 of Chapter 1)

$$L = \frac{1}{2}\,ml^2\dot{\theta}^2 - mgl(1 - \cos\theta).$$

The Hamilton's Principle (2) states that "the path followed by the pendulum is one along which the line integral of Lagrangian is extremum". This gives

$$\int_{t_0}^{t_1} \delta\left[\frac{1}{2}\,ml^2\dot{\theta}^2 - mgl(1 - \cos\theta)\right] dt = 0$$

$$\int_{t_0}^{t_1}\left[ml^2\dot{\theta}\delta\dot{\theta} - mgl\,\sin\theta\delta\theta\right] dt = 0.$$

Since the time derivative and δ variation is commutative, we have therefore,

$$\int_{t_0}^{t_1}\left[ml^2\dot{\theta}\,\frac{d}{dt}(\delta\theta) - mgl\,\sin\theta\delta\theta\right] dt = 0.$$

Integrating the first integral by parts we get

$$ml^2(\dot{\theta}\delta\theta)_{t_0}^{t_1} - \int_{t_0}^{t_1} m[l^2\ddot{\theta} + gl\,\sin\theta]\,\delta\theta\,dt = 0.$$

Since $(\delta\theta)_{t_0}^{t_1} = 0$, we have therefore,

$$\int_{t_0}^{t_1} m\left[l^2\ddot{\theta} + gl\sin\theta\right]\delta\theta\, dt = 0.$$

As $\delta\theta$ is arbitrary, we have

$$\ddot{\theta} + \frac{g}{l}\sin\theta = 0.$$

This is the required equation of motion of the simple pendulum.

Example 4:   A particle of mass m is moving on the surface of the sphere of radius r in the gravitational field. Use Hamilton's principle to show the equation of motion is given by

$$\ddot{\theta} - \frac{p_\phi^2\cos\theta}{m^2 r^4\sin^3\theta} - \frac{g}{r}\sin\theta = 0,$$

where p_ϕ is the constant of angular momentum.

Solution: Let a particle of mass m be moving on the surface of a sphere of radius r. The particle has two degrees of freedom and hence two generalized co-ordinates θ, ϕ. The Lagrangian of the motion is (refer Example 38 of Chapter 1) given by

$$L = \frac{1}{2}\,mr^2\left(\dot{\theta}^2 + \sin^2\theta\,\dot{\phi}^2\right) - mgr\cos\theta$$

The Hamilton's Principle (2) gives

$$\int_{t_0}^{t_1}\delta\left[\frac{1}{2}\,mr^2(\dot{\theta}^2 + \sin^2\theta\,\dot{\phi}^2) - mgr\cos\theta\right] dt = 0$$

$$\int_{t_0}^{t_1}\left[mr^2(\dot{\theta}\,\delta\dot{\theta} + \sin^2\theta\,\dot{\phi}\,\delta\dot{\phi} + \sin\theta\cos\theta\,\dot{\phi}^2\,\delta\theta) + mgr\sin\theta\,\delta\theta\right] dt = 0$$

Since, $\delta\dfrac{d\theta}{dt} = \dfrac{d}{dt}\,\delta\theta$, we have therefore,

$$\int_{t_0}^{t_1}\left[mr^2\left(\dot{\theta}\,\frac{d}{dt}(\delta\theta) + \sin^2\theta\,\dot{\phi}\,\frac{d}{dt}(\delta\phi)\right) + \left(mr^2\sin\theta\cos\theta\,\dot{\phi}^2 + mgr\sin\theta\right)\delta\theta\right] dt = 0$$

Integrating the first two integrals by parts we get

$$mr^2(\dot{\theta}\,\delta\theta)_{t_0}^{t_1} + mr^2\sin^2\theta(\dot{\phi}\,\delta\phi)_{t_0}^{t_1} - \left[\int_{t_0}^{t_1} mr^2\left(\ddot{\theta} - \sin\theta\cos\theta\,\dot{\phi}^2 - \frac{g}{r}\sin\theta\right)\delta\theta\right] dt -$$

$$-\int_{t_0}^{t_1} mr^2 \frac{d}{dt}(\sin^2 \theta \dot\phi)\, \delta\phi\, dt = 0.$$

Since $(\delta\theta)_{t_0}^{t_1} = 0 = (\delta\phi)_{t_0}^{t_1}$, we have therefore,

$$\left[\int_{t_0}^{t_1} mr^2\left(\ddot\theta - \sin\theta\cos\theta\dot\phi^2 - \frac{g}{r}\sin\theta\right)\delta\theta\right] dt - \int_{t_0}^{t_1} mr^2 \frac{d}{dt}(\sin^2\theta\dot\phi)\,\delta\phi\, dt = 0$$

Since θ and ϕ are independent and arbitrary, hence we have

$$\ddot\theta - \sin\theta\cos\theta\dot\phi^2 - \frac{g}{r}\sin\theta = 0,$$

$$mr^2 \frac{d}{dt}(\sin^2\theta\dot\phi) = 0 \Rightarrow mr^2 \sin^2\theta\dot\phi = p_\phi \text{ (const.)}$$

Eliminating $\dot\phi$ we obtain differential equation of motion for spherical pendulum

$$\ddot\theta - \frac{P_\phi^2 \cos\theta}{m^2 r^4 \sin^3\theta} - \frac{g}{r}\sin\theta = 0.$$

Example 5:   A particle of mass m moves in a space under the action of a force

$$\overline{F} = (y^2\cos x + z^3)\, i + (2y\sin x - 4)\, j + (3xz^2 + 2)k$$

Find the Lagrangian and total energy of the particle and hence the equation of motion of the particle.

Solution: One can easily check that $\nabla \times \overline{F} = 0$, hence the force in question is conservative. The conservative force is given by

$$F = -\nabla V = -\left(i\frac{\partial V}{\partial x} + j\frac{\partial V}{\partial y} + k\frac{\partial V}{\partial z}\right)$$

$$\overline{F}d\overline{r} = -\left(\frac{\partial V}{\partial x}dx + \frac{\partial V}{\partial y}dy + \frac{\partial V}{\partial z}dz\right) = -dV$$

Integrating we get

$$V = -\int (y^2\cos x + z^3)\, dx + (2y\sin x - 4)dy + (3xz^2 + 2)\, dz$$

$$V = -\int d(y^2\sin x) + d(xz^3) + d(-4y + 2z),$$

$$V = -\int d(y^2\sin x + xz^3 - 4y + 2z),$$

$$V = -(y^2\sin x + xz^3 - 4y + 2z),$$

If T is the kinetic energy of the particle then we have $T = \dfrac{1}{2} m(\dot{x}^2 + \dot{y}^2 + \dot{z}^2)$. The Lagrangian function is defined by

$$L = \dfrac{1}{2} m(\dot{x}^2 + \dot{y}^2 + \dot{z}^2) + (y^2 \sin x + xz^3 - 4y + 2z).$$

The total energy of the particle is given by

$$E = \sum_j \dot{q}_j \dfrac{\partial L}{\partial \dot{q}_j} - L.$$

This becomes

$$E = \dfrac{1}{2} m(\dot{x}^2 + \dot{y}^2 + \dot{z}^2) - (y^2 \sin x + xz^3 - 4y + 2z).$$

The equations of motion are

$$m\ddot{x} - y^2 \cos x - z^3 = 0, \quad m\ddot{y} - (2y \sin x - 4) = 0, \quad m\ddot{z} - (3xz^2 + 2) = 0.$$

Example 6: Show that the action for the actual path followed by a particle of unit mass in a uniform gravitational field is smaller than any other path.

Solution: Consider a single particle of unit mass in a uniform gravitational field. The path followed by the particle in time t is given by the equation

$$x = ut + \dfrac{1}{2} gt^2. \qquad \qquad \text{...(15)}$$

Since gravitational field is conservative, the potential energy of unit mass particle is

$$V = -gx, \qquad \qquad \text{...(16)}$$

where x is vertical. The Lagrangian of the particle becomes

$$L = \dfrac{1}{2} \dot{x}^2 + gx,$$

$$L = \dfrac{1}{2} (u + gt)^2 + g\left(ut + \dfrac{1}{2} gt^2\right). \qquad \qquad \text{...(17)}$$

Therefore the action of the particle along the actual path (15) becomes

$$I = \int_0^t \left[\dfrac{1}{2} (u + gt)^2 + g\left(ut + \dfrac{1}{2} gt^2\right)\right] dt. \qquad \qquad \text{...(18)}$$

Consider the varied path given by the equation

$$x = ut + \dfrac{1}{2} gt^2 + \eta(t), \qquad \qquad \text{...(19)}$$

where $\eta(t)$ is a continuous function vanishing at the end points. *i.e.,*

$$\eta(o) = \eta(t) = 0. \qquad \ldots(20)$$

Hence action along the varied path becomes

$$I' = \int_0^t \left[\frac{1}{2}(u + gt + \dot{\eta})^2 + g\left(ut + \frac{1}{2}gt^2 + \eta\right) \right] dt. \qquad \ldots(21)$$

Change in the action is obtained taking the difference between the equations (21) and (20). Thus, we have

$$\delta I = I' - I =$$

$$= \int_0^t \left[\frac{1}{2}(u + gt + \dot{\eta})^2 + g\left(ut + \frac{1}{2}gt^2 + \eta\right) - \frac{1}{2}(u + gt)^2 - g\left(ut + \frac{1}{2}gt^2\right) \right] dt$$

$$\delta I = \int_0^t \frac{1}{2}(\dot{\eta})^2 \, dt + \int_0^t (u + gt)\, \dot{\eta}\, dt + \int_0^t g\eta\, dt.$$

Integrating the second integral by parts and using boundary conditions (20) we get

$$\delta I = \int_0^t \frac{1}{2}(\dot{\eta})^2 \, dt. \qquad \ldots(22)$$

Since $\dot{\eta}^2 > 0 \Rightarrow \delta I > 0 \Rightarrow I < I'$. This shows that the action along actual path is smaller than the action along any other path.

Lagrange's equations of motion from Hamilton's Principle

We have derived in the Chapter 1 the Lagrange's equations of motion from D'Alembert's principle. The same set of equations can also be derived from the Hamilton's variational principle. The method is illustrated in the following theorem.

Theorem 5   Show that the Lagrange's equations are necessary conditions for the action to have a stationary value.

Proof: Let $L = L(q_j, \dot{q}_j, t)$ be a Lagrangian of a particle. Then from the action function, we obtain

$$\delta I = \delta \int_{t_0}^{t_1} L\, dt = \int_{t_0}^{t_1} \left[\sum_j \frac{\partial L}{\partial q_j}\delta q_j + \sum_j \frac{\partial L}{\partial \dot{q}_j}\delta \dot{q}_j \right] dt.$$

As there is no variation in time along any path, hence the term containing δt is absent in the above equation. Time derivative and δ variation are commutative, therefore, we write the second integral on the right hand side as

$$\delta \int_{t_0}^{t_1} L\, dt = \int_{t_0}^{t_1} \sum_j \frac{\partial L}{\partial q_j}\delta q_j\, dt + \int_{t_0}^{t_1} \sum_j \frac{\partial L}{\partial \dot{q}_j}\frac{d}{dt}(\delta q_j)\, dt.$$

Integrating the second integral on the right hand side of above equation, we get

$$\delta \int_{t_0}^{t_1} L dt = \int_{t_0}^{t_1} \sum_j \frac{\partial L}{\partial q_j} \delta q_j dt + \left[\sum_j \frac{\partial L}{\partial \dot{q}_j} \delta q_j \right]_{t_0}^{t_1} - \int_{t_0}^{t_1} \sum_j \frac{d}{dt}\left(\frac{\partial L}{\partial \dot{q}_j} \right) \delta q_j dt.$$

Since there is no variation in the co-ordinates along any path at the end points, hence change in the co-ordinates at the end points is zero. *i.e.*, $(\delta q_j)_{t_0}^{t_1} = 0$. This gives

$$\delta \int_{t_0}^{t_1} L dt = \int_{t_0}^{t_1} \sum_j \left[\frac{\partial L}{\partial q_j} - \frac{d}{dt}\left(\frac{\partial L}{\partial \dot{q}_j} \right) \right] \delta q_j dt.$$

$$\delta \int_{t_0}^{t_1} L dt = 0 \Leftrightarrow \int_{t_0}^{t_1} \sum_j \left[\frac{\partial L}{\partial q_j} - \frac{d}{dt}\left(\frac{\partial L}{\partial \dot{q}_j} \right) \right] \delta q_j dt = 0.$$

If the system is holonomic, then all the generalized co-ordinates are linearly independent, consequently, this yields

$$\delta \int_{t_0}^{t_1} L dt = 0 \Leftrightarrow \frac{\partial L}{\partial q_j} - \frac{d}{dt}\left(\frac{\partial L}{\partial \dot{q}_j} \right) = 0.$$

These are the required Lagrange's equations of motion derived from the Hamilton's principle. The equation shows that the Lagrange's equations of motion for holonomic system are necessary and sufficient conditions for action to have a stationary value.

Hamiltonian Formulation

Introduction: We have developed in the Chapter 1 Lagrangian formulation as a description of mechanics in terms of the generalized co-ordinates q_j, generalized velocities $\dot{q}_j$ and time t as a parameter and the equations of motion were used to solve some problems. We now introduce another powerful formulation known as Hamilton's formulation in which the independent variables are the generalized co-ordinates q_j, the generalized momenta p_j and time t. This formulation is an alternative to the Lagrangian formulation but proved to be more convenient and useful, particularly in dealing with problems of modern physics. Hence all the examples solved in the Chapter 1 can also be solved by the Hamiltonian procedure. As an illustration some of them are solved in this chapter by Hamilton's procedure and physical meaning to the Hamiltonian H is given.

The Hamiltonian Function

The quantity $\sum_j P_j \dot{q}_j - L$ when expressed in terms of generalized coordinates q_1, q_2, ..., q_n, generalized momenta p_1, p_2, ...p_n, and time t is called Hamiltonian and it is denoted by H. Thus, we have

$$H = H(q_j, p_j, t) = \sum_j p_j \dot{q}_j - L. \qquad \ldots(23)$$

Hamilton's Canonical Equations of Motion

Theorem 6 Derive the Hamilton's canonical equations of motion from Hamiltonian function.

Proof: From the Hamiltonian function H defined in (23), we find

$$dH = \sum_j \frac{\partial H}{\partial p_j} dp_j + \sum_j \frac{\partial H}{\partial q_j} dq_j + \frac{\partial H}{\partial t} dt. \qquad \ldots(24)$$

Now from the second half equation (23) we also find

$$dH = \sum_j \dot{q}_j dp_j + \sum_j d\dot{q}_j p_j - dL,$$

$$\Rightarrow \quad dH = \sum_j \dot{q}_j dp_j + \sum_j d\dot{q}_j p_j - \sum_j \frac{\partial L}{\partial q_j} dq_j - \sum_j \frac{\partial L}{\partial \dot{q}_j} d\dot{q}_j - \frac{\partial L}{\partial t} dt. \qquad \ldots(25)$$

This equation, on using the expression for generalized momentum $p_j = \dfrac{\partial L}{\partial \dot{q}_j}$, reduces to

$$dH = \sum_j \dot{q}_j dp_j - \sum_j \frac{\partial L}{\partial q_j} dq_j - \frac{\partial L}{\partial t} dt. \qquad \ldots(26)$$

Now comparing the coefficients of dp_j, dq_j in equations (24) and (26) and using the Lagrange's equations of motion, we obtain the required Hamilton's canonical equations of motion

$$\dot{q}_j = \frac{\partial H}{\partial p_j}, \; \dot{p}_j = -\frac{\partial H}{\partial q_j}, \; j = 1, 2, \ldots, n. \qquad \ldots(27)$$

·These are the set of $2n$ first order differential equations of motion in place of n Lagrange's second order differential equations of motion.

Note: In the derivation of the equations (27), the knowledge of force whether conservative or non-conservative does not arise. Hence the set of equations (27) are usually referred as the Hamilton's canonical equations of motion for both conservative and non-conservative systems. The difference lies only in the definition of Lagrangian L. In case of non-conservative system L contains the potential of non-conservative forces, while for conservative system it involves the potential of conservative forces. However, for partially conservative and partially non-conservative systems, the Hamilton's canonical equations of motion are different and are derived in the following.

Hamilton's Canonical Equations of Motion for partially conservative and partially non-conservative system

The Lagrange's equations of motion for partially conservative and partially non-conservative system are given in equation (83) of Chapter 1. We recall it for the use in the sequel

$$\frac{d}{dt}\left(\frac{\partial L}{\partial \dot{q}_j}\right) - \frac{\partial L}{\partial q_j} = Q_j^{(d)},$$

where $Q_j^{(d)}$ represents the forces not arising from a potential and L contains the potential of the conservative forces. From this equation we obtain

$$\frac{\partial L}{\partial q_j} = \dot{p}_j - Q_j^{(d)}.$$

Putting this value in the equation (26), we get

$$dH = \sum_j \dot{q}_j dp_j - \sum_j \left(\dot{p}_j - Q_j^{(d)}\right) dq_j - \frac{\partial L}{\partial t} dt.$$

Equating the coefficients of dp_j and dq_j in the above equation and the equation (24) we obtain the Hamilton's canonical equations of motion for the partially conservative and partially non-conservative system in the form

$$\dot{q}_j = \frac{\partial H}{\partial p_j} \text{ and } \dot{p}_j = -\frac{\partial H}{\partial q_j} + Q_j^{(d)}. \qquad \qquad \dots(28)$$

Derivation of Hamilton's equations of motion from Hamilton's Principle

The above set of Hamilton's canonical equations of motion for conservative/non-conservative system can also be obtained from the Hamilton's principle. The method is illustrated in the following theorem.

Theorem 7 Obtain Hamilton's equations of motion from the Hamilton's principle.

Proof: Replace the Lagrangian function L by using the definition of Hamiltonian (23) in the Action function, we get

$$\delta \int_{t_0}^{t_1} L\,dt = \delta \int_{t_0}^{t_1} \left[\sum_j p_j \dot{q}_j - H\right] dt$$

$$\delta \int_{t_0}^{t_1} L\,dt = 0 \Rightarrow \delta \int_{t_0}^{t_1} \left[\sum_j p_j \dot{q}_j - H\right] dt = 0.$$

This is known as the modified Hamilton's principle. Consider

$$\delta \int_{t_0}^{t_1} L\, dt = \int_{t_0}^{t_1}\left[\sum_j \delta p_j \dot{q}_j + \sum_j p_j \delta \dot{q}_j - \sum_j \frac{\partial H}{\partial q_j}\delta q_j - \sum_j \frac{\partial H}{\partial p_j}\delta p_j - \frac{\partial H}{\partial t}\delta t \right] dt.$$

Since time is fixed along any path, hence change in time along any path is zero. This reduces to

$$\delta \int_{t_0}^{t_1} L\, dt = \int_{t_0}^{t_1}\left[\sum_j \left(\dot{q}_j - \frac{\partial H}{\partial p_j}\right)\delta p_j + \sum_j p_j \delta \dot{q}_j - \sum_j \frac{\partial H}{\partial q_j}\delta q_j \right] dt. \qquad \ldots(29)$$

Now consider the integral

$$\int_{t_0}^{t_1} \sum_j p_j \delta \dot{q}_j\, dt = \int_{t_0}^{t_1} \sum_j p_j \frac{d}{dt}\left(\delta q_j\right) dt.$$

Integrating the integral on the right hand side by parts, we get

$$\int_{t_0}^{t_1} \sum_j p_j \delta \dot{q}_j\, dt = \left(\sum_j p_j \delta q_j\right)_{t_0}^{t_1} - \int_{t_0}^{t_1} \sum_j \dot{p}_j \delta q_j\, dt.$$

Since $\left(\delta q_j\right)_{t_0}^{t_1} = 0$, we have therefore,

$$\int_{t_0}^{t_1} \sum_j p_j \delta \dot{q}_j\, dt = -\int_{t_0}^{t_1} \sum_j \dot{p}_j \delta q_j\, dt.$$

Substituting this in equation (29), we get

$$\delta \int_{t_0}^{t_1} L\, dt = \int_{t_0}^{t_1}\left[\sum_j \left(\dot{q}_j - \frac{\partial H}{\partial p_j}\right)\delta p_j - \sum_j \left(\dot{p}_j + \frac{\partial H}{\partial q_j}\right)\delta q_j \right] dt.$$

Now we see that

$$\delta \int_{t_0}^{t_1} L\, dt = 0 \Leftrightarrow \int_{t_0}^{t_1}\left[\sum_j \left(\dot{q}_j - \frac{\partial H}{\partial p_j}\right)\delta p_j - \sum_j \left(\dot{p}_j + \frac{\partial H}{\partial q_j}\right)\delta q_j \right] dt = 0.$$

For holonomic system we have q_j, p_j are independent, hence, we have

$$\delta \int_{t_0}^{t_1} L\, dt = 0 \Leftrightarrow \dot{q}_j - \frac{\partial H}{\partial p_j} = 0,\ \dot{p}_j + \frac{\partial H}{\partial q_j} = 0.$$

$$\Rightarrow \qquad \delta \int_{t_0}^{t_1} L\, dt = 0 \Leftrightarrow \dot{q}_j = \frac{\partial H}{\partial p_j},\ \dot{p}_j = -\frac{\partial H}{\partial q_j}.$$

These are the Hamilton's canonical equations of motion.

Remark: We see from the above equation that the Hamilton's canonical equations of motion are the necessary and sufficient conditions for the action to have stationary value.

WORKED EXAMPLES

Example 7: Use Hamilton's principle to find the equations of motion of a particle of unit mass moving on a plane in a conservative force field.

Solution: Let the force $\overline{F}$ be conservative and under the action of which the particle of unit mass be moving on the xy plane. Let $P(x, y)$ be the position of the particle. We write the force $\overline{F} = iF_x + jF_y$. Since $\overline{F}$ is conservative, we have therefore,

$$F_x = -\frac{\partial V}{\partial x}, \; F_y = -\frac{\partial V}{\partial y}.$$

The kinetic energy of the particle is given by

$$T = \frac{1}{2}\left(\dot{x}^2 + \dot{y}^2\right).$$

Hence the Lagrangian of the particle becomes

$$L = \frac{1}{2}\left(\dot{x}^2 + \dot{y}^2\right) - V(x, y).$$

The Hamilton's principle gives

$$\int_{t_0}^{t_1}\left[\frac{\partial L}{\partial x}\,\delta x + \frac{\partial L}{\partial y}\,\delta y + \frac{\partial L}{\partial \dot{x}}\,\delta \dot{x}\,\frac{\delta L}{\delta \dot{y}}\,\delta \dot{y}\right] dt = 0,$$

$$\Rightarrow \qquad \int_{t_0}^{t_1}\left[(\dot{x}\delta \dot{x} + \dot{y}\delta \dot{y}) - \frac{\partial V}{\partial x}\,\delta x - \frac{\partial V}{\partial y}\,\delta y\right] dt = 0.$$

The role of time derivative and δ variation is interchanged and then integrating the first two terms by parts, we obtain

$$\left(\dot{x}\delta x\right)_{t_0}^{t_1} = \left(\dot{y}\delta y\right)_{t_0}^{t_1} - \int_{t_0}^{t_1}\left[\left(\ddot{x} + \frac{\partial V}{\partial x}\right)\delta x + \left(\ddot{y} + \frac{\partial V}{\partial y}\right)\delta y\right] dt = 0.$$

Since $\delta x = 0$, $\delta y = 0$ at both the ends t_0 and t_1 along any path, therefore, we have

$$\int_{t_0}^{t_1}\left[\left(\ddot{x} + \frac{\partial V}{\partial x}\right)\delta x + \left(\ddot{y} + \frac{\partial V}{\partial y}\right)\delta y\right] dt = 0.$$

Since δx and δy are independent and arbitrary, therefore, we have

$$\ddot{x} + \frac{\partial V}{\partial x} = 0, \ddot{y} + \frac{\partial V}{\partial y} = 0.$$

$$\ddot{x} = -\frac{\partial V}{\partial x} = F_x, \ddot{y} = -\frac{\partial V}{\partial y} = F_y.$$

These are the equations of motion of a particle of unit mass moving under the action of the conservative force field.

Example 8: Show that addition of the total time derivative of any function of the form $f(q_j, t)$ to the Lagrangian of a holonomic system, the generalized momentum and the Jacobi integral are respectively given by $p_i + \dfrac{\partial f}{\partial q_i}$ and $H - \dfrac{\partial f}{\partial t}$. Does the new Lagrangian L' unchange the Hamilton's principle? Justify your claim.

Solution: Let the new Lagrangian function obtained after addition of the time derivative of any function of the form $f(q_j, t)$ to the Lagrangian L be denoted by L'. Thus we have

$$L' = L + \frac{df}{dt}. \qquad \ldots(30)$$

Thus the generalized momentum corresponding to the new Lagrangian L' is defined by

$$p_j' = \frac{\partial L'}{\partial \dot{q}_j}.$$

This gives

$$p_j' = \frac{\partial L}{\partial \dot{q}_j} + \frac{\partial}{\partial \dot{q}_j}\left(\frac{df}{dt}\right).$$

$$\Rightarrow \qquad p_j' = p_j + \frac{\partial}{\partial \dot{q}_j}\left(\sum_k \frac{\partial f}{\partial q_k}\dot{q}_k + \frac{\partial f}{\partial t}\right).$$

As f involves q_j and time t, its derivative functions will at the most involve q_j and t only and do not involve $\dot{q}_j$. We have therefore,

$$p_j' = \frac{\partial L'}{\partial \dot{q}_j} = p_j + \frac{\partial f}{\partial q_j}. \qquad \ldots(31)$$

This is the required generalized momentum corresponding to the new Lagrangian L'. Similarly, the Hamiltonian corresponding to the new function L' is given by

$$H' = \sum_j p_j'\dot{q}_j - L',$$

$$H' = \sum_j \frac{\partial L'}{\partial \dot{q}_j} \, \dot{q}_j - \left(L + \frac{df}{dt} \right)$$

On using equation (31), we get

$$H' = \sum_j \left(p_j + \frac{\partial f}{\partial q_j} \right) \dot{q}_j - \left(L + \frac{df}{dt} \right)$$

$$H' = \sum_j (p_j \dot{q}_j - L) - \frac{\partial f}{\partial t} \, ,$$

$$\Rightarrow \qquad H' = H - \frac{\partial f}{\partial t} \, . \qquad \qquad \qquad \qquad ...(32)$$

This is called the Jacobi integral for the new Lagrangian L'. Now we show that the new Lagrangian L' also satisfies the Hamilton's principle. Consider therefore,

$$\delta \int_1^2 L' dt = \delta \int_1^2 L dt + \delta \int_1^2 \frac{df}{dt} \, dt,$$

$$\delta \int_1^2 L' dt = \delta \int_1^2 L dt + \delta \int_1^2 df,$$

$$\delta \int_1^2 L' dt = \delta \int_1^2 L dt + (\delta f)_1^2,$$

$$\delta \int_1^2 L' dt = \delta \int_1^2 L dt + \left[\sum_j \frac{\partial f}{\partial q_j} \, \delta q_j + \frac{\partial f}{\partial t} \, \delta t \right]_1^2 .$$

But in δ variation time is held fixed along any path and hence $\delta t = 0$ along any path. Further, co-ordinates at the end points are held fixed. This gives $(\delta q_j)_1^2$. Hence the above equation yields

$$\delta \int_1^2 L' dt = \delta \int_1^2 L dt.$$

The Hamilton's Principle implies

$$\delta \int_1^2 L dt = 0 \Leftrightarrow \delta \int_1^2 L' dt = 0.$$

This shows that the new Lagrangian L' also satisfies the Hamilton's principle.

Example 9: Obtain Lagrangian L from Hamiltonian H and show that it satisfies Lagrange's equations of motion. Prove also that the Hamiltonian H thus defined also satisfies the Hamilton's canonical equations of motion.

Solution: As pointed earlier we will show here that Lagrangian formulation and Hamiltonian formulation are not different but are interdependent. Let H be the Hamiltonian defined in equation (23) satisfying the Hamilton's canonical equations of motion (27). From the Hamiltonian function (23), the Lagrangian function L is given by

$$L = \sum_j p_j \dot{q}_j - H. \qquad \qquad ...(33)$$

It is evident from equation (33) that

$$\frac{\partial L}{\partial q_j} = -\frac{\partial H}{\partial q_j}, \text{ and } \frac{\partial L}{\partial \dot{q}_j} = p_j.$$

Now consider the equation

$$\frac{\partial L}{\partial q_j} - \frac{d}{dt}\left(\frac{\partial L}{\partial \dot{q}_j}\right) = -\frac{\partial H}{\partial q_j} - \frac{d}{dt}(p_j).$$

Using Hamilton's equations of motion, we obtain

$$\frac{\partial L}{\partial q_j} - \frac{d}{dt}\left(\frac{\partial L}{\partial \dot{q}_j}\right) = \dot{p}_j - \dot{p}_j,$$

$$\Rightarrow \qquad \frac{\partial L}{\partial q_j} - \frac{d}{dt}\left(\frac{\partial L}{\partial \dot{q}_j}\right) = 0.$$

This shows that the Lagrangian obtained from the Hamiltonian satisfies the Lagrange's equations of motion.

Now consider the Hamiltonian H in terms of Lagrangian L defined as

$$H = \sum_j p_j \dot{q}_j - L,$$

where L satisfies the Lagrange's equations of motion. Therefore, we have

$$\frac{\partial L}{\partial q_j} = \frac{d}{dt}\left(\frac{\partial L}{\partial \dot{q}_j}\right),$$

$$\frac{\partial L}{\partial q_j} = \frac{d}{dt}(p_j),$$

$$\Rightarrow \qquad \frac{\partial L}{\partial q_j} = \dot{p}_j.$$

Now differentiating the Hamiltonian H with respect to q_j and p_j we find the required Hamilton's canonical equations as

$$\frac{\partial H}{\partial q_j} = -\frac{\partial L}{\partial q_j} = -\dot{p}_j \text{ and } \frac{\partial H}{\partial p_j} = \dot{q}_j.$$

Physical Meaning of the Hamiltonian

In the physical world there exist a number of conservation laws. For example, conservation law of total energy, conservation law of momentum, conservation law of charge and various other physical quantities. Conservation laws proved to be very powerful techniques in solving mechanical problems. They tell us a great deal about the motion and provide a convenient method of finding the motion of a particle or a system of particles. Conservation of a physical quantity depends upon the set of generalized coordinates employed. In this chapter we pay attention to the conservation of Hamiltonian and it's representation as total energy. The Hamiltonian H of a system can represent a constant of motion or total energy of the system or both depending upon the type of a dynamical system whether conservative or non-conservative or partially conservative and partially non-conservative involving scleronomic or rheonomic constraints. It should be emphasized that the conditions for conservation of the Hamiltonian H are quite distinct from those that H represents total energy. One can have a set of generalized coordinates such that for a particular dynamical system H is conserved but it cannot represent the total energy. On the other hand, there can exist a dynamical system for which H can represent total energy but is not conserved. Sometimes, under certain circumstances conservation of H is equivalent to the conservation of the total energy, while in some other circumstances, neither it is conserved nor it represents total energy. However, for the partially conservative and partially non-conservative system where some forces are derivable from dissipative function, the physical interpretation of H is quite reverse. To understand what these circumstances are we prove the following theorem.

Theorem 8   Prove that

(a) for conservative scleronomic system the Hamiltonian H represents a constant ofmotion. It also represents total energy of the system, provided kinetic energy is a homogeneous quadratic function of generalized velocities.

(b) for conservative rheonomic system the Hamiltonian H may represent a constant of motion but does not represent the total energy.

(c) for non-conservative system whether constraints are rheonomic or scleronomic the Hamiltonian H never represent total energy, but H is conserved whenever the Lagrangian L is independent of time.

(d) for partially conservative and partially non-conservative scleronomic system the Hamiltonian H does not represent a constant of motion even if the Lagrangian L does not contain time t explicitly but represents the total energy.

Proof: Let $H(p_j, q_j, t)$ and $L(q_j, \dot{q}_j, t)$ be the Hamiltonian and the Lagrangian describing the motion of a dynamical system. These functions are interdependent by the equation

$$H = \sum_j p_j \dot{q}_j - L. \qquad \qquad ...(34)$$

Differentiating equation (34) with respect to time t, we get

$$\frac{dH}{dt} = \sum_j \dot{p}_j \dot{q}_j + \sum_j p_j \ddot{q}_j - \sum_j \frac{\partial L}{\partial q_j} \dot{q}_j - \sum_j \frac{\partial L}{\partial \dot{q}_j} \ddot{q}_j - \frac{\partial L}{\partial t}. \qquad ...(35)$$

Since $p_j = \dfrac{\partial L}{\partial \dot{q}_j}$ is the generalized momentum. From the Lagrange's equation

of motion we have $\dot{p}_j = \dfrac{d}{dt}\left(\dfrac{\partial L}{\partial \dot{q}_j}\right) = \dfrac{\partial L}{\partial q_j}.$ On using these equations we see that the

coefficients of $\dot{q}_j$ and $\ddot{q}_j$ get cancelled with each other and we have

$$\frac{dH}{dt} = -\frac{\partial L}{\partial t}, \qquad \qquad ...(36)$$

where L is a Lagrangian of a conservative/non-conservative system. We see from

the equation (36) that if L does not contain time t explicitly, then $\dfrac{dH}{dt} = 0$. This

shows that whether the system is conservative or non-conservative, H is always conserved whenever L is independent of time t. However, the condition that L does not contain time t explicitly will be satisfied by neither the kinetic energy nor the potential energy involves time t explicitly. There are two cases that the kinetic energy T does not involve time t explicitly.

(a) The system is conservative and scleronomic

In the case of a conservative system when the constraints are scleronomic, whether transformation equations involve time t or not, the kinetic energy T is independent of time t and the potential energy V is only a function of co-ordinates. Consequently, the Lagrangian L does not involve time t explicitly and hence from the equation (36) the Hamiltonian H always represents a constant of motion. Further, for conservative scleronomic system, when kinetic energy is a homogeneous quadratic function of generalized velocities, then we have from the Theorem 6 of the Chapter 1,

$$\sum_j \dot{q}_j \frac{\partial T}{\partial \dot{q}_j} = 2T. \qquad \qquad ...(37)$$

Substituting this in the equation (34), the Hamiltonian H of the system becomes

$$H = 2T - (T - V)$$

$$H = T + V = E,$$

where E is the total energy of the system, showing that the Hamiltonian H represents total energy.

However, for conservative scleronomic system when the kinetic energy is not a homogeneous quadratic function of generalized velocities because of the rotation of the system or due to any other reason, then we have from the equation (139a) of the Chapter 1

$$\sum_j \dot{q}_j \frac{\partial T}{\partial \dot{q}_j} = 2T + \text{terms containing } T_1 \text{ and/or } T_0,$$

where T_1 is a homogeneous function of generalized velocities and T_0 is a term not involving generalized velocities arise due to the rotation of the system. Consequently, from the equation (34), we get $H = T + V +$ terms containing $T_1, T_0 \neq E$. This proves that the Hamiltonian does not represent the total energy of the system.

(b) The system is conservative and rheonomic

In the case of conservative rheonomic system, the transformation equations do involve time t explicitly, though some times the kinetic energy may not involve time t explicitly. Consequently, neither the kinetic energy T nor the potential energy V involves time t, and hence the Lagrangian L, in these cases, will be independent of time t. Thus in all such cases the Hamiltonian H represents a constant of motion. However, in all other cases where kinetic energy involves time t because of the presence of rheonomic constraints in the system, the Lagrangian function of the system involves time t explicitly. In all those cases the Hamiltonian H does not represent a constant of motion. However, for the conservative rheonomic system, whether the Lagrangian L involves time t or not, the kinetic energy T is no more a homogeneous quadratic function of generalized velocities. In this case we have from the equation (142) of the Chapter 1 that

$$\sum_j \dot{q}_j \frac{\partial T}{\partial \dot{q}_j} = 2T_2 + T_1. \qquad \text{...(38)}$$

Substituting this in the equation (34), we get

$$H = T_2 - T_0 + V,$$

showing that the Hamiltonian H does not represent total energy. Thus for the conservative rheonomic systems H may or may not represent the constant of motion but it never represents total energy. Some examples involving scleoronomic and rheonomic constraints are illustrated below and physical meaning of the Hamiltonian is explored. Readers can refer Examples (27), (28), (29), (31) and (43) solved below for understanding these concepts.

(c) The system is non-conservative and rheonomic

For the non-conservative system, potential energy is the velocity dependent potential and the Lagrangian function contains the potential of non-conservative forces. The Lagrangian of the system in this case is given by $L = T - U$, where U is the generalized potential energy function. In this case, from the equation (444) of the Chapter 1, we know the components of the generalized momentum are given by

$$p_j = \frac{\partial T}{\partial \dot{q}_j} - V_j.$$

The Hamiltonian function H in this case becomes

$$H = \sum_j \left(\frac{\partial T}{\partial \dot{q}_j} - V_j \right) \dot{q}_j - (T - U),$$

$$H = \sum_j \left(\dot{q}_j \frac{\partial T}{\partial \dot{q}_j} - V_j \dot{q}_j \right) - (T - U).$$

Since, for the non-conservative system, the generalized potential energy function U in a large number of realistic situations is given by $U = \sum_j V_j \dot{q}_j + V$. Using this we obtain

$$H = \sum_j \left(\dot{q}_j \frac{\partial T}{\partial \dot{q}_j} \right) - (U - V) - (T - U).$$

On using the equation (38), we obtain from this

$$H = 2T_2 + T_1 - T + V,$$
$$H = 2T - T_1 - 2T_0 - T + V,$$
$$H = T + V - T_1 - 2T_0.$$

This shows that for the non-conservative rheonomic system, the Hamiltonian H does not represent total energy. Even if, the system is scleronomic, then we have $T_1 = T_0 = 0$. Thus when the kinetic energy is a homogeneous quadratic function of generalized velocities, then we have from above equation

$$H = T + V \neq T + U = E.$$

This shows that for non-conservative system whether the constraints are rheonomic or scleronomic, H never represents total energy. Nevertheless, even in the case of non-conservative system, it is evident from the equation (36) that H may or may not be conserved depending upon whether the Lagrangian L is independent of time t or not.

(d) The system is partially conservative, partially non-conservative and scleronomic

This case arises when some dissipative forces are present in the system. For such systems we obtain after differentiating the Hamiltonian function

$H = H(p_j, q_j, t)$ with respect to time t, the expression

$$\frac{dH}{dt} = \sum_j \frac{\partial H}{\partial p_j} \dot{p}_j + \sum_j \frac{\partial H}{\partial q_j} \dot{q}_j + \frac{\partial H}{\partial t}.$$

We use the Hamilton's canonical equations of motion (28) for the partially conservative and partially non-conservative system in the above equation and on simplifying we obtain

$$\frac{dH}{dt} = \sum_j Q_j^{(d)} \dot{q}_j + \frac{\partial H}{\partial t}. \qquad \qquad ...(39)$$

We have seen that if the Lagrangian L does not contain time t explicitly then Hamiltonian H also does not contain time t, then we have $\dfrac{\partial H}{\partial t} = 0$. This gives

$$\frac{dH}{dt} = \sum_j Q_j^{(d)} \dot{q}_j \Rightarrow \frac{dH}{dt} \neq 0.$$

Hence H is not conserved. Thus when dissipative forces are present in the system then Hamiltonian H will never represent a constant of motion even if the Lagrangian L is independent of time t. However, the Lagrangian involves the potential of the conservative forces and the forces which are not derivable from

the potential are represented in $Q_j^{(d)}$. In this case we have therefore, $\dfrac{\partial L}{\partial \dot{q}_j} = \dfrac{\partial T}{\partial \dot{q}_j}$,

and hence the equation (34) yields $H = T + V$. This proves that the Hamiltonian H represents total energy.

To illustrate the case (d), we consider the Example (23) on page 1.64 of the Chapter 1. In the Example (23), a particle is falling vertically under gravity when the frictional forces are present in the system and are obtainable from the dissipation function $R = \dfrac{1}{2} k\dot{z}^2$. In this case the Lagrangian function is given by $L = \dfrac{1}{2} m\dot{z}^2 + mgz$, and the corresponding Lagrange's equation of motion is $m\ddot{z} + k\dot{z} - mg = 0$. The dissipative force is represented by $Q_j^{(d)}$, where

$Q^{(d)} = -\dfrac{\partial R}{\partial \dot{z}} = -k\dot{z}$. We see that the dissipative force does not enter in to the Lagrangian function but enters in to the system through the equation of motion.

Now the Hamiltonian function $H = \sum\limits_{j} p_j \dot{q}_j - L$ gives $H = \dfrac{p_z^2}{2m} - mgz$. Hence the Hamilton's canonical equations of motion for the partially conservative and partially non-conservative system give $\dot{z} = \dfrac{\partial H}{\partial p_z} \Rightarrow \dot{z} = \dfrac{p_z}{m}$, and $\dot{p}_z = -\dfrac{\partial H}{\partial z} + Q^{(d)}$. This gives $\dot{p}_z = mg - k\dot{z}$. Eliminating p_z between these two equations we get the same equation of motion obtained by Lagrangian procedure. However, we see

that $\dfrac{dH}{dt} = \dfrac{p_z}{m}(m\ddot{z} - mg)$. Using the equation of motion we get $\dfrac{dH}{dt} = \dfrac{p_z}{m}(-k\dot{z}) \neq 0$.

This shows that because of the presence of dissipative forces in the system the Hamiltonian H does not represent a constant of motion even if the Lagrangian L does not involve time t explicitly. Further, the total energy E of the system is given

by $E = T + V = \dfrac{1}{2}m\dot{z}^2 - mgz$. Eliminating $\dot{z}$ we obtain $E = T + V = \dfrac{p_z^2}{2m} - mgz$. This

is same as the Hamiltonian of the system. Hence H represents total energy.

Thus we summarize the results below:

(1) The Hamiltonian H does not represent the constant of motion in spite of the fact that the Lagrangian L does not contain time t explicitly.

(2) The Hamiltonian H represents the total energy of the system though some dissipative forces are present in the system.

The presence of dissipative nature of forces in the system exhibits the amazing behavior of the Hamiltonian. This happens mainly because the dissipative forces enter in to the system through the equations which describe the motion of the particle and not through the definitions of the Lagrangian and the Hamiltonian functions. Some more examples involving dissipative nature of forces are given in the exercise of the Chapters 1 and 3. Readers can refer to the example (8) in the exercise of the Chapter 3 and the Examples (15) and (16) in the exercise of the Chapter 1.

Cyclic Co-ordinates in Hamiltonian

Theorem 9 Prove that a co-ordinate which is cyclic in the Lagrangian is also cyclic in the Hamiltonian.

Solution: It has been proved in the Chapter 1 that corresponding to the cyclic coordinate the generalized momentum is conserved. Thus if q_j is cyclic in L then $\dot{p}_j = 0$. However, from Hamilton's canonical equations of motion we have $\dot{p}_j = -\dfrac{\partial H}{\partial q_j} = 0$. This shows that the co-ordinate q_j is also absent in the Hamiltonian, and consequently, it is also cyclic in H. Thus a co-ordinate which is cyclic in the Lagrangian is also cyclic in the Hamiltonian.

WORKED EXAMPLES

Example 10: Describe by Hamilton's procedure the motion of a particle of mass m moving near the surface of the Earth under the Earth's constant gravitational field.

Solution: Consider a particle of mass m moving near the surface of the Earth under the Earth's constant gravitational field. Let (x, y, z) be the Cartesian co-ordinates of the projectile, z being vertical. Then the Lagrangian of the projectile is given by

$$L = \frac{1}{2}\, m(\dot{x}^2 + \dot{y}^2 + \dot{z}^2) - mgz. \qquad \ldots(40)$$

We see that the generalized co-ordinates x and y are absent in the Lagrangian, hence they are the cyclic co-ordinates. This implies that any change in these co-ordinates can not affect the Lagrangian. Consequently, horizontal components of the generalized momenta are conserved. *i.e.*, $p_x = m\dot{x} = const.$, $p_y = m\dot{y} = const$ and $p_z = m\dot{z}$. The Hamiltonian of the particle is defined in equation (23) becomes

$$H = p_x\,\dot{x} + p_y\,\dot{y} + p_z\,\dot{z} - \frac{1}{2}\,m\big(\dot{x}^2 + \dot{y}^2 + \dot{z}^2\big) + mgz. \qquad \ldots(41)$$

Eliminating $\dot{x}$, $\dot{y}$, $\dot{z}$ from equation (41), we get

$$H = \frac{1}{2m}\big(p_x^2 + p_y^2 + p_z^2\big) + mgz. \qquad \ldots(42)$$

The Hamilton's equations of motion give

$$\dot{p}_x = -\frac{\partial H}{\partial x} = 0,\; \dot{p}_y = -\frac{\partial H}{\partial y} = 0,\; \dot{p}_z - \frac{\partial H}{\partial z} = -mg.$$

and

$$\dot{x} = \frac{\partial H}{\partial p_x} = \frac{p_x}{m},\; \dot{y} = \frac{\partial H}{\partial p_y} = \frac{p_y}{m},\; \dot{z} = \frac{\partial H}{\partial p_z} = \frac{p_z}{m}.$$

From these set of equations we obtain the equations of motion of the projectile near the surface of the Earth in the form

$$\ddot{x} = 0,\; \ddot{y} = 0,\; \ddot{z} = -g.$$

Routh's Procedure

Introduction: The presence of cyclic co-ordinates in the Lagrangian L is not much profitable because even if the co-ordinate q_j does not appear in L, the corresponding generalized velocity $\dot{q}_j$ generally does, so that one has to deal the problem with all variables and the system has n degrees of freedom. However, if q_j is cyclic in the Hamiltonian then p_j is constant and then one has to deal with the problem involving only $2n - 2$ variables, *i.e.*, only $n - 1$ degrees of freedom. Hence the presence of cyclic coordinate is profitable in the Hamiltonian procedure. Hence Hamiltonian procedure is especially adapted to the problems involving cyclic co-ordinates. The advantage of Hamiltonian formulation in handling with cyclic co-ordinates is utilized by Routh and devised a method by combining with the Lagrangian procedure and the method is known as Routh's Procedure. The Method is described in the following theorem.

Theorem 10   Describe the Routh's procedure to solve the problem involving cyclic and non-cyclic co-ordinates.

Proof: Consider a system of particles involving both cyclic and non-cyclic co-ordinates. Let q_1, q_2, q_3, ..., q_s of q_1, q_2, q_3, ..., q_n are cyclic co-ordinates, then a new function R, known as the Routhian is defined as

$$R(q_1, q_2, ..., q_n; p_1, p_2, ..., p_s; \dot{q}_{s+1}, \dot{q}_{s+2}, ..., \dot{q}_n, t) = \sum_{j=1}^{s} p_j \dot{q}_j - L(q_j, \dot{q}_j, t) \quad ...(43)$$

One can see that the Routhian R is obtained by modifying the Lagrangian L so that it is no longer a function of the generalized velocities corresponding to the cyclic co-ordinates, but instead involves only its conjugate momentum. The advantage in doing so is that p_j can then be considered one of the constants of integration and the remaining integrations involve only the non-cyclic co-ordinates. Now we take

$$R = R(q_1, q_2, ..., q_n; p_1, p_2, ..., p_s; \dot{q}_{s+1}, \dot{q}_{s+2}, ..., \dot{q}_n, t),$$

and find the total differential dR as

$$dR = \sum_{j=1}^{n} \frac{\partial R}{\partial q_j} dq_j + \sum_{j=1}^{s} \frac{\partial R}{\partial p_j} dp_j + \sum_{j=s+1}^{n} \frac{\partial R}{\partial \dot{q}_j} d\dot{q}_j - \frac{\partial R}{\partial t} dt. \quad ...(44)$$

Now we consider second half of the equation (43)

$$R = \sum_{j=1}^{s} p_j \dot{q}_j - L(q_j, \dot{q}_j, t)$$

and find the total differential dR as

$$dR = \sum_{j=1}^{s} p_j d\dot{q}_j + \sum_{j=1}^{s} \dot{q}_j dp_j - \sum_{j=1}^{n} \frac{\partial L}{\partial q_j} dq_j - \sum_{j=1}^{n} \frac{\partial L}{\partial \dot{q}_j} d\dot{q}_j - \frac{\partial L}{\partial t} dt, \quad ...(45)$$

The third and the fourth sums on the right hand side of the above equation can be split up into two sums as.

$$dR = \sum_{j=1}^{s} p_j \, d\dot{q}_j + \sum_{j=1}^{s} \dot{q}_j \, dp_j - \left(\sum_{j=1}^{s} \frac{\partial L}{\partial q_j} \, dq_j + \sum_{j=s+1}^{n} \frac{\partial L}{\partial q_j} \, dq_j \right) -$$

$$- \left(\sum_{j=1}^{s} \frac{\partial L}{\partial \dot{q}_j} \, d\dot{q}_j + \sum_{j=s+1}^{n} \frac{\partial L}{\partial \dot{q}_j} \, d\dot{q}_j \right) - \frac{\partial L}{\partial t} \, dt.$$

The coefficients of $d\dot{q}_j$ term get cancelled. Hence we have

$$dR = \sum_{j=1}^{s} \dot{q}_j \, dp_j - \sum_{j=1}^{s} \frac{\partial L}{\partial q_j} \, dq_j - \sum_{j=s+1}^{n} \frac{\partial L}{\partial q_j} \, dq - \sum_{j=s+1}^{n} \frac{\partial L}{\partial \dot{q}_j} \, d\dot{q}_j - \frac{\partial L}{\partial t} \, dt. \qquad \dots(46)$$

Now equating the corresponding coefficients of the equations (45) and (46) we obtain

$$\frac{\partial R}{\partial p_j} = \dot{q}_j, \, j = 1, \, 2, \, \dots, \, s, \qquad \dots(47)$$

$$\frac{\partial R}{\partial q_j} = -\frac{\partial L}{\partial q_j} = -\dot{p}_j, \, j = 1, \, 2, \, \dots, \, s, \qquad \dots(48)$$

$$\frac{\partial R}{\partial q_j} = -\frac{\partial L}{\partial q_j} = -\dot{p}_j, \, j = s + 1, \, s + 2, \, \dots, \, n, \qquad \dots(49)$$

and
$$\frac{\partial R}{\partial \dot{q}_j} = -\frac{\partial L}{\partial \dot{q}_j} = -p_j, \, j = s + 1, \, s + 2, \, \dots, \, n,. \qquad \dots(50)$$

We see that for cyclic co-ordinates $q_1, q_2, \dots, q_s$ equations (47) and (48) represent Hamilton's canonical equations of motion with R as the Hamiltonian, while equations (49) and (50) for the non-cyclic co-ordinates q_j ($j = s + 1, s + 2, \dots, n$) represent Lagrange's equations of motion with R as the Lagrangian function. Thus by Routhian procedure a problem involving cyclic and non-cyclic co-ordinates can be solved by solving Lagrange's equations for non-cyclic co-ordinates with Routhian R as the Lagrangian function and solving Hamiltonian equations for the given cyclic co-ordinates with R as the Hamiltonian function. In this way The Routhian has a dual character of Hamiltonian H and the Lagrangian L.

WORKED EXAMPLES

Example 11: Find Lagrangian L, Hamiltonian H and the Routhian R in spherical polar co-ordinates for a particle moving in space under the action of conservative force.

Solution: Let a particle of mass m be moving in a space under the action of conservative force. If (x, y, z) and (r, θ, ϕ) are the Cartesian co-ordinates and spherical co-ordinates of the particle, then the relations between them are given by

$$x = r \sin \theta \cos \phi, \; y = r \sin \theta \sin \phi, \; z = r \cos \theta.$$

The kinetic energy $T = \dfrac{1}{2} m\left(\dot{x}^2 + \dot{y}^2 + \dot{z}^2\right)$ of the particle, in spherical polar co-ordinates becomes

$$T = \frac{1}{2} m\left(\dot{r}^2 + r^2\dot{\theta}^2 + r^2 \sin^2 \theta\dot{\phi}^2\right).$$

The force is conservative therefore; the potential energy of the particle is the function of position only. Hence the Lagrangian function of the particle becomes

$$L = \frac{1}{2} m\left(\dot{r}^2 + r^2\dot{\theta}^2 + r^2 \sin^2 \theta\dot{\phi}^2\right) - V. \qquad \text{...(51)}$$

We notice that ϕ is cyclic in the Lagrangian L, hence the corresponding generalized momentum is conserved. *i.e.*,

$$p_\phi = \frac{\partial L}{\partial \dot{\phi}} = mr^2 \sin^2 \theta\dot{\phi} = const. \qquad \text{...(52)}$$

Similarly, the generalized momenta corresponding to the generalized coordinates r and θ give respectively the linear momentum and angular momentum.

$$p_r = \frac{\partial L}{\partial \dot{r}} = m\dot{r}, \; p_\theta = \frac{\partial L}{\partial \dot{\theta}} = mr^2\dot{\theta}. \qquad \text{...(53)}$$

Thus the Hamiltonian function (23) becomes

$$H = p_r\dot{r} + p_\theta\dot{\theta} + p_\phi\dot{\phi} - \frac{1}{2} m\left(\dot{r}^2 + r^2\dot{\theta}^2 + r^2 \sin^2 \theta\dot{\phi}^2\right) + V \qquad \text{...(54)}$$

Eliminating the generalized velocities $\dot{r}$, $\dot{\theta}$, $\dot{\phi}$ between equations (52), (53) and (54) we obtain the Hamiltonian function.

$$H = \frac{1}{2m}\left(p_r^2 + \frac{1}{r^2} p_\theta^2 + \frac{1}{r^2 \sin^2 \theta} p_\phi^2 \right) + V. \qquad \text{...(55)}$$

Now the Routhian function R defined in equation (43) becomes

$$R = p_\phi\dot{\phi} - L, \qquad \text{...(56)}$$

This becomes after eliminating $\dot{r}$, $\dot{\theta}$, $\dot{\phi}$ between (52), (53) and (56), we get

$$R\left(r, \theta, \phi, \dot{r}, \dot{\theta}, t\right) = \frac{p_\phi^2}{2mr^2 \sin^2 \theta} - \frac{1}{2} m\left(\dot{r}^2 + r^2\dot{\theta}^2\right) + V. \qquad \text{...(57)}$$

Example 12: A planet moves under the inverse square law of attractive force, Find Lagrangian L, Hamiltonian H, and the Routhian R for the planet.

Solution: Let a planet of mass m move under the action of an inverse square law of attractive force. The force lies along the radius vector and hence the angular momentum of the planet is always zero. This implies that the motion of the planet is a motion in the plane. Let therefore, (r, θ) be the generalized co-ordinates of the planet. The kinetic and potential energies of the planet are respectively given by

$$T = \frac{1}{2} m\left(\dot{r}^2 + r^2\dot{\theta}^2\right), \ V = -\frac{K}{r}.$$

Hence the Lagrangian function becomes

$$L = \frac{1}{2} m\left(\dot{r}^2 + r^2\dot{\theta}^2\right) + \frac{K}{r}. \qquad \qquad ...(58)$$

One can see that θ is the cyclic co-ordinate in L. Hence the corresponding angular momentum of the planet is conserved.

$$p_\theta = \frac{\partial L}{\partial \dot{\theta}} = mr^2\dot{\theta} = const. \Rightarrow \dot{\theta} = \frac{p_\theta}{mr^2}. \qquad ...(59)$$

Similarly, the linear momentum corresponding to the coordinate r is

$$p_r = \frac{\partial L}{\partial \dot{r}} = m\dot{r} \Rightarrow \dot{r} = \frac{p_r}{m}. \qquad \qquad ...(60)$$

The Hamiltonian function (23) yields

$$H = p_r\dot{r} + p_\theta\dot{\theta} - \frac{1}{2} m\left(\dot{r}^2 + r^2\dot{\theta}^2\right) - \frac{K}{r}.$$

This on eliminating generalized velocities, we obtain the Hamiltonian function

$$H = \frac{1}{2m}\left(p_r^2 + \frac{p_\theta^2}{r^2}\right) - \frac{K}{r}. \qquad \qquad ...(61)$$

As θ is the only cyclic coordinate, therefore the Routhian is defined as

$$R(r, \theta, p_\theta, \dot{r}, t) = p_\theta\dot{\theta} - L$$

$$R(r, \theta, p_\theta, \dot{r}, t) = p_\theta\dot{\theta} - \frac{1}{2} m\left(\dot{r}^2 + r^2\dot{\theta}^2\right) - \frac{K}{r}$$

Eliminating $\dot{\theta}$ we get

$$R(r, \theta, p_\theta, \dot{r}, t) = \frac{p_\theta^2}{2mr^2} - \frac{1}{2} m\dot{r}^2 - \frac{K}{r} \qquad \qquad ...(62)$$

Example 13: Set up the Hamiltonian for the Lagrangian

$$L(q, \dot{q}, t) = \frac{m}{2}\left[\dot{q}^2 \sin^2 \omega t + q\dot{q} \sin 2\omega t + q^2\omega^2\right]$$

Derive the Hamilton's equations of motion. Reduce the equations in to a single second order differential equation.

Solution: The given Lagrangian function shows that the system has only one degree of freedom and hence only one generalized co-ordinate q. The corresponding generalized momentum becomes

$$p = \frac{\partial L}{\partial \dot{q}} = \frac{m}{2}(2\dot{q} \sin^2 \omega t + q\omega \sin 2\omega t). \qquad \text{...(63)}$$

Solving this equation for $\dot{q}$ we get

$$\dot{q} = \frac{1}{\sin^2 \omega t}\left[\frac{p}{m} - \frac{q}{2}\,\omega \sin 2\omega t\right]. \qquad \text{...(64)}$$

Hence the Hamiltonian function H can be found as

$$H = p\dot{q} - \frac{m}{2}\left(\dot{q}^2 \omega t + q\dot{q}\omega \sin 2\omega t + q^2\omega^2\right). \qquad \text{...(65)}$$

Substituting the value of $\dot{q}$ from equation (64) in (65) and on simplifying we obtain the expression for the Hamiltonian for the system as

$$H = \frac{p^2}{2m \sin^2 \omega t} - pq\omega \cot \omega t + \frac{q^2 m\omega^2}{2} \cos^2 \omega t - \frac{m}{2} q^2 \omega^2. \qquad \text{...(66)}$$

The Hamilton's canonical equations of motion give

$$\dot{q} = \frac{\partial H}{\partial p} = \frac{p}{m \sin^2 \omega t} - q\omega \cot \omega t. \qquad \text{...(67)}$$

and
$$\dot{p} = p\omega \cot \omega t - q\omega^2 m \cos^2 \omega t + mq\omega^2. \qquad \text{...(68)}$$

From equation (6), we find

$$p = \frac{m}{2}\left[2\dot{q} \sin^2 \omega t + q\omega \sin 2\omega t\right]$$

Differentiating equation (63) with respect to t, we get

$$\dot{p} = \frac{m}{2}\left[2\ddot{q} \sin^2 \omega t + 4\dot{q}\omega \sin \omega t \cos \omega t + \dot{q}\omega \sin 2\omega t + 2q\omega^2 \cos 2\omega t\right] \text{...(69)}$$

Equating equations (68) and (69) we obtain the equation of motion

$$\ddot{q} + 2\omega\dot{q} \cot \omega t - 2q\omega^2 = 0. \qquad \text{...(70)}$$

Example 14: A Lagrangian of a system is given by

$$L(x, y, \dot{x}, \dot{y}) = \frac{m}{2}\left(a\dot{x}^2 + 2b\dot{x}\dot{y} + c\dot{y}^2\right) - \frac{k}{2}\left(ax^2 + 2bxy + cy^2\right),$$

where a, b, c, k, m are constants and $b^2 - ac \neq 0$. Find the Hamiltonian and equations of motion. Examine the particular cases (i) $a = c = 0$ and (ii) $b = 0$, $c = -a$.

Solution: We see from the Lagrangian that the system has two degrees of freedom and hence two generalized co-ordinates x and y. The corresponding generalized momenta are

$$p_x = \frac{\partial L}{\partial \dot{x}} = m(a\dot{x} + b\dot{y}) \text{ and } p_y = \frac{\partial L}{\partial \dot{y}} = m(b\dot{x} + c\dot{y}).$$

Solving these equations for $\dot{x}$ and $\dot{y}$ we get

$$\dot{x} = -\frac{cp_x - bp_y}{m\left(b^2 - ac\right)}, \quad \dot{y} = \frac{bp_x - ap_y}{m\left(b^2 - ac\right)}. \qquad \ldots(71)$$

The Hamiltonian function H defined in the equation (23) becomes

$$H = p_x\dot{x} + p_y\dot{y} - \frac{m}{2}\left(a\dot{x}^2 + 2b\dot{x}\dot{y} + c\dot{y}^2\right) + \frac{k}{2}\left(ax^2 + 2bxy + cy^2\right). \qquad \ldots(72)$$

Using equations (71) in (72) we obtain after simplifying the Hamiltonian of the system

$$H = \frac{1}{m\left(b^2 - ac\right)}\left[bp_xp_y - \frac{a}{2}p_y^2 - \frac{c}{2}p_x^2\right] + \frac{k}{2}\left(ax^2 + 2bxy + cy^2\right). \qquad \ldots(73)$$

The Hamilton's equations of motion corresponding to two generalized co-ordinates x, y are

$$\dot{p}_x = -\frac{\partial H}{\partial x} = -k(ax + by), \quad \dot{p}_y = -\frac{\partial H}{\partial y} = -k(bx + cy).$$

$$\dot{x} = \frac{\partial H}{\partial p_x} = \frac{1}{m\left(b^2 - ac\right)}(bp_y - cp_x), \quad \dot{y} = \frac{\partial H}{\partial p_y} = \frac{1}{m\left(b^2 - ac\right)}(bp_x - cp_y).$$

Solving these equations we obtain the equations of motion in the form

$$m(a\ddot{x} + b\ddot{y}) + k\,(ax + by) = 0, \quad m(b\ddot{x} + c\ddot{y}) + k\,(bx + cy) = 0. \qquad \ldots(74)$$

Solving these equations for $\ddot{x}$ and $\ddot{y}$ we obtain respectively

$$m\ddot{x} + kx = 0, \quad m\ddot{y} + ky = 0. \qquad \ldots(75)$$

The solutions of these equations yield the relations

$$x = c_1 \cos\left(\frac{k}{m}\right)t + c_2 \sin\left(\frac{k}{m}\right)t \text{ and } y = d_1 \cos\left(\frac{k}{m}\right)t + d_2 \sin\left(\frac{k}{m}\right)t.$$

Now the cases $a = c = 0$ and $b = 0$, $c = -a$ yield from the above equations of motion (74) the same set of equations (75).

Example 15: Find the Routhian for the Lagrangian

$$L = \frac{1}{2} I_3(\dot{\psi} + \dot{\phi} \cos \theta)^2 + \frac{1}{2} I_1(\dot{\theta} + \dot{\phi}^2 \sin^2 \theta) - mgl \cos \theta,$$

where I_1, I_2, I_3, m, g, l are all constants.

Solution: We notice from the Lagrangian function in the question that θ and ψ are cyclic co-ordinates. Hence the corresponding generalized momenta are conserved. Thus,

$$p_\phi = \frac{\partial L}{\partial \dot{\phi}} = I_3(\dot{\psi} + \dot{\phi} \cos \theta) \cos \theta + I_1 \dot{\phi} \sin^2 \theta = \text{const.} \quad \ldots(76)$$

and

$$p_\psi = \frac{\partial L}{\partial \dot{\psi}} = I_3(\dot{\psi} + \dot{\phi} \cos \theta) = \text{const.} \quad \ldots(77)$$

Eliminating $(\dot{\psi} + \dot{\phi} \cos \theta)$ between (76) and (77) we obtain

$$\dot{\phi} \sin \theta = \frac{p_\phi - p_\psi \cos \theta}{I_1 \sin \theta}. \quad \ldots(78)$$

By definition, the Routhian function R is given by

$$R\left(\theta, \dot{\theta}, p_\theta, p_\phi\right) = p_\phi \dot{\phi} + p_\psi \dot{\psi} - L$$

$$R = p_\phi \dot{\phi} + p_\psi \dot{\psi} - \frac{1}{2} I_3(\dot{\psi} + \dot{\phi} \cos \theta)^2 - \frac{1}{2} I_1(\dot{\theta} + \dot{\phi}^2 \sin^2 \theta) + mgl \cos \theta, \quad \ldots(79)$$

Using (76) and (77) in the equation (79) we obtain the required Routhian R as

$$R(\theta, \dot{\theta}, p_\theta, p_\psi) = \frac{p_\psi^2}{2I_3} + \frac{(p_\phi - p_\psi \cos \theta)^2}{2I_1 \sin^2 \theta} - \frac{1}{2} I_1 \dot{\theta}^2 + mgl \cos \theta.$$

Example 16: A Hamiltonian of one degree of freedom has the form

$$H = \frac{p^2}{2\alpha} - bpqe^{-\alpha t} + \frac{ab}{2} q^2 e^{-\alpha t}(\alpha + be^{-\alpha t}) + \frac{kq^2}{2},$$

where a, b, α and k are constants.

1. Find a Lagrangian corresponding to this Hamiltonian
2. Find an equivalent Lagrangian that is not explicitly dependent of time.
3. What is the Hamiltonian corresponding to this second Lagrangian and what is the relationship between the two Hamiltonians?

Solution: We see from the given Hamiltonian function that the system has one degree of freedom and hence one generalized co-ordinate q. The Hamilton's canonical equations give

$$\dot{q} = \frac{\partial H}{\partial p} = \frac{p}{\alpha} - bqe^{-\alpha t},$$

$$\Rightarrow \qquad p = \alpha(\dot{q} + bqe^{-\alpha t}). \qquad \qquad ...(80)$$

1. The Lagrangian L through H is defined by

$$L(q, \dot{q}, t) = p\dot{q} - H(p, q, t)$$

$$\Rightarrow \qquad L(q, \dot{q}, t) = p\dot{q} - \frac{p^2}{2\alpha} + bqpe^{-\alpha t} - \frac{ab}{2} q^2 e^{-\alpha t}(\alpha + be^{-\alpha t}) - \frac{kq^2}{2}. \qquad ...(81)$$

Eliminating p from equation (81), we get the required Lagrangian as

$$L = \alpha\dot{q}(\dot{q} + bqe^{-\alpha t}) - \frac{1}{2} \alpha(\dot{q} + bqe^{-\alpha t})^2 + bq\alpha e^{-\alpha t}(\dot{q} + bqe^{-\alpha t}) -$$

$$- \frac{ab}{2} q^2 e^{-\alpha t}(\alpha + be^{-\alpha t}) - \frac{kq^2}{2}. \qquad ...(82)$$

2. Lagrangian independent of time: To find the Lagrangian that is not explicit dependent of time, we simplify equation (82) and write

$$L = \frac{1}{2} \alpha\dot{q}^2 + \frac{1}{2} \alpha b^2 q^2 e^{-2\alpha t} + bq\dot{q}\alpha e^{-\alpha t} - \frac{ab}{2} \alpha q^2 e^{-\alpha t} - \frac{ab^2}{2} q^2 e^{-2\alpha t} - \frac{kq^2}{2}. \qquad ...(83)$$

Now the Lagrangian that is not explicitly dependent of time is obtained by dropping the terms containing time t explicitly from equation (83), we get

$$L = \frac{\alpha}{2} \dot{q}^2 - \frac{kq^2}{2}. \qquad \qquad ...(84)$$

The generalized momentum corresponding to the generalized co-ordinates q is given by

$$p = \frac{\partial L}{\partial \dot{q}} = \alpha\dot{q}, \Rightarrow \dot{q} = \frac{p}{\alpha}.$$

3. Now the Hamiltonian say H' corresponding to the Lagrangian (84) is defined by

$$H'(q, p, t) = p\dot{q} - L(q, \dot{q}, t),$$

$$H'(q, p, t) = p\dot{q} - \frac{\alpha}{2}\dot{q}^2 + \frac{kq^2}{2}. \qquad ...(85)$$

Eliminating $\dot{q}$ from equation (85), we get

$$H' = \frac{p^2}{2\alpha} + \frac{kq^2}{2}. \qquad ...(86)$$

The relation between the two Hamiltonians is obtained as

$$H = H' - bpqe^{-\alpha t} + \frac{ab}{2}q^2 e^{-\alpha t}(\alpha + be^{-\alpha t}). \qquad ...(87)$$

Example 17: The Lagrangian for a system can be written as

$$L = a\dot{x}^2 + b\frac{\dot{y}}{x} + c\dot{z}\dot{y} + fy^2\dot{x}\dot{z} + g\dot{y} - k\sqrt{x^2 + y^2},$$

where a, b, c, f, g and k are constants. What is Hamiltonian? What quantities are conserved?

Solution: From the given Lagrangian L, we see that the system has three degrees of freedom and has three generalized co-ordinates (x, y, z) of which z is cyclic. This implies the corresponding generalized momentum p_z is conserved.

$$\Rightarrow \qquad p_z = \frac{\partial L}{\partial \dot{z}} = c\dot{y} + fy^2\dot{x} = \text{const.} \qquad ...(88)$$

Similarly, we find

$$p_x = \frac{\partial L}{\partial \dot{x}} = 2a\dot{x} + fy^2\dot{z} \text{ and } p_y = \frac{\partial L}{\partial \dot{y}} = \frac{b}{x} + c\dot{z} + g. \qquad ...(89)$$

Solving these equations for $\dot{x}$, $\dot{y}$, $\dot{z}$ we get

$$\dot{x} = \frac{1}{2a}\left[p_x - \frac{fy^2}{c}\left(p_y - \frac{b}{x} - g\right)\right], \ \dot{y} = \frac{1}{c}\left[p_z - \frac{fy^2}{2a}\left(p_x - \frac{fy^2}{c}\left(p_y - \frac{b}{x} - g\right)\right)\right],$$

and $\qquad \dot{z} = \frac{1}{c}\left(p_y - \frac{b}{x} - g\right). \qquad ...(90)$

The Hamiltonian of the system is defined as

$$H = p_x\dot{x} + p_y\dot{y} + p_z\dot{z} - L,$$

$$H = p_x\dot{x} + p_y\dot{y} + p_z\dot{z} - a\dot{x}^2 - b\frac{\dot{y}}{x} - c\dot{x}\dot{y} - fy^2\dot{x}\dot{z} - g\dot{y} + k\sqrt{x^2 + y^2}. \qquad ...(91)$$

The required Hamiltonian is obtained by eliminating $\dot{x}$, $\dot{y}$, $\dot{z}$ from equation (91).

Example 18:   A dynamical system has the Lagrangian

$$L = \dot{q}_1^2 + \frac{\dot{q}_2^2}{a + bq_1^2} + k_1 q_1^2 + k_2 \dot{q}_1 \dot{q}_2, \text{ where } a, b, k_1, k_2 \text{ are constants.}$$

Solution: The system has two degrees of freedom and hence two generalized co-ordinates q_1 and q_2. The corresponding generalized momenta are given by

$$p_1 = \frac{\partial L}{\partial \dot{q}_1} = 2\dot{q}_1 + k_2 \dot{q}_2, \quad p_2 = \frac{\partial L}{\partial \dot{q}_2} = \frac{2\dot{q}_2}{a + bq_1^2} + k_2 \dot{q}_1. \qquad \text{...(92)}$$

Solving these equations for $\dot{q}_1$ and $\dot{q}_2$ we get

$$\dot{q}_1 = \frac{\left[-\dfrac{2p_1}{(a + bq_1^2)} + k_2 p_2 \right]}{\left(k_2^2 - \dfrac{4}{a + bq_1^2} \right)}, \quad \dot{q}_2 = \frac{k_2 p_1 - 2p_2}{\left(k_2^2 - \dfrac{4}{a + bq_1^2} \right)}. \qquad \text{...(93)}$$

The Hamiltonian of the system is defined by

$$H = p_1 \dot{q}_1 + p_2 \dot{q}_2 - \dot{q}_1^2 - \frac{\dot{q}_2^2}{a + bq_1^2} - k_1 q_1^2 - k_2 \dot{q}_1 \dot{q}_2. \qquad \text{...(94)}$$

Eliminating $\dot{q}_1$ and $\dot{q}_2$ from equation (94), we get the required Hamiltonian.

Example 19:   For a particle the kinetic energy and potential energy are respectively given by,

$$T = \frac{1}{2} m\dot{r}^2, \quad U = \frac{1}{r}\left(1 + \frac{\dot{r}^2}{c^2} \right).$$

Find the Hamiltonian H and determine whether (i) $H = T + U$,

(ii) $\dfrac{dH}{dt} = 0$.

Solution: The Lagrangian function of the system is therefore given by

$$L = \frac{1}{2} m\dot{r}^2 - \frac{1}{r}\left(1 + \frac{\dot{r}^2}{c^2} \right). \qquad \text{...(95)}$$

We see that the particle has only one degree of freedom and hence it has only one generalized co-ordinate and that is r. We also notice that the potential U contains the generalized velocity, hence the system is non-conservative. The Lagrangian is independent of time t hence we infer from the Theorem 8 that the Hamiltonian H will be a constant of motion but will not represent total energy. We will justify below this conclusion mathematically. Therefore, the generalized momentum is given by

$$p_r = \frac{\partial L}{\partial \dot{r}} = \dot{r}\left(m - \frac{2}{rc^2}\right),$$

$$\Rightarrow \qquad \dot{r} = \frac{p_r}{\left(m - \dfrac{2}{rc^2}\right)}. \qquad\qquad \text{...(96)}$$

Thus the corresponding Hamiltonian function is defined by

$$H = p_r \dot{r} - \frac{1}{2}\, m\dot{r}^2 + \frac{1}{r}\left(1 + \frac{\dot{r}^2}{c^2}\right). \qquad\qquad \text{...(97)}$$

Eliminating $\dot{r}$ between (96) and (97) we obtain the Hamiltonian H in the form

$$H = \frac{1}{2}\frac{p_r^2}{\left(m - \dfrac{2}{rc^2}\right)} + \frac{1}{r}. \qquad\qquad \text{...(98)}$$

(i) Now the sum of the kinetic and potential energies is given by

$$T + U = \frac{1}{2}\, m\dot{r}^2 + \frac{1}{r}\left(1 + \frac{\dot{r}^2}{c^2}\right). \qquad\qquad \text{...(99)}$$

Eliminating $\dot{r}$ between (96) and (99), we get

$$T + U = \frac{1}{2}\frac{p_r^2\left(m + \dfrac{2}{rc^2}\right)}{\left(m - \dfrac{2}{rc^2}\right)^2} + \frac{1}{r}. \qquad\qquad \text{...(100)}$$

We see from equations (98) and (100) that the Hamiltonian H does not represent the total energy.

$$\Rightarrow \qquad T + U \neq H.$$

(ii) Now differentiating equation (98) with respect to time t we get

$$\frac{dH}{dt} = \frac{p_r \dot{p}_r}{\left(m - \dfrac{2}{rc^2}\right)} - \frac{p_r^2 \dot{r}}{\left(m - \dfrac{2}{rc^2}\right)^2 r^2 c^2} - \frac{\dot{r}}{r^2}, \qquad\qquad \text{...(101)}$$

where from equation (96) we have $\dot{p}_r = \ddot{r}\left(m - \dfrac{2}{rc^2}\right) + \dfrac{2\dot{r}^2}{r^2 c^2}.$ Substituting this in the equation (101) and simplifying we get

$$\frac{dH}{dt} = \frac{p_r}{\left(m - \dfrac{2}{rc^2}\right)}\left[\dot{p}_r - \frac{p_r^2}{\left(m - \dfrac{2}{rc^2}\right)^2 r^2 c^2} - \frac{1}{r^2}\right].$$

$$\frac{dH}{dt} = \frac{p_r}{\left(m - \dfrac{2}{rc^2}\right)}\left[\ddot{r}\left(m - \frac{2}{rc^2}\right) + \frac{\dot{r}^2}{r^2 c^2} - \frac{1}{r^2}\right].$$

$\Rightarrow \dfrac{dH}{dt} = 0$ due to the equation of motion. This shows that H is a constant of motion.

Example 20: A particle is thrown horizontally from the top of a building of height h with an initial velocity u. Write down the Hamiltonian of the problem. Show that H represents both a constant of motion and the total energy.

Solution: Let a particle be thrown horizontally from the top of a building of height h with an initial velocity u. The motion of the particle is in a plane. Let P be the position of the particle at any instant t latter. The position co-ordinates $P(x, y)$ of the particle are the generalized coordinates. The kinetic energy and the potential energy of the particle are respectively given by

$$T = \frac{1}{2}m\left(\dot{x}^2 + \dot{y}^2\right), \qquad\qquad ...(102)$$

$$V = -mg(h - y). \qquad\qquad ...(103)$$

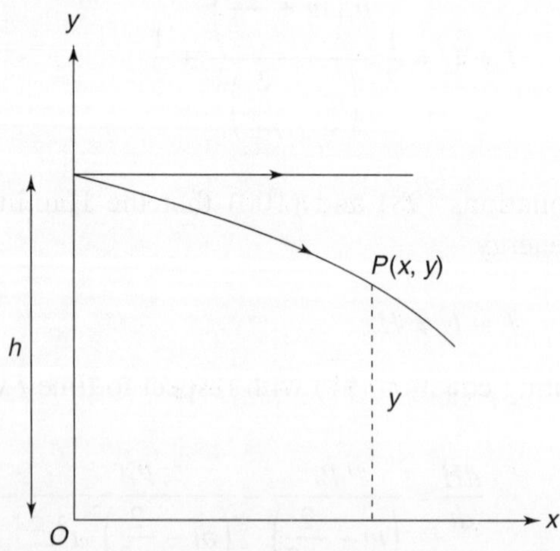

Hence the Lagrangian of the particle becomes

$$L = \frac{1}{2}m\left(\dot{x}^2 + \dot{y}^2\right) + mg(h - y). \qquad\qquad ...(104)$$

We see from equations (102) and (104) that the kinetic energy is a homogeneous quadratic function of generalized velocities and the Lagrangian L is independent of time t. Hence we infer from the Theorem 8 that the Hamiltonian will be a constant of motion as well as H will represent the total energy of the system. We will illustrate below the conclusion mathematically. We see that the generalized co-ordinate x is cyclic in L, hence the corresponding generalized momentum is conserved.

$$\Rightarrow \qquad p_x = \frac{\partial L}{\partial \dot{x}} m\dot{x} = \text{const.}, \ p_y = \frac{\partial L}{\partial \dot{y}} = m\dot{y}. \qquad \qquad \text{...(105)}$$

The Hamiltonian function H becomes

$$H = p_x\dot{x} + p_y\dot{y} - \frac{1}{2} m\left(\dot{x}^2 + \dot{y}^2\right) - mh(h - y). \qquad \text{...(106)}$$

Eliminating the generalized velocities from equations (105) and (106), we obtain the Hamiltonian of motion as

$$H = \frac{1}{2m}\left(p_x^2 + p_y^2\right) - mg(h - y). \qquad \text{...(107)}$$

The corresponding Hamilton's canonical equations of motion are

$$\dot{x} = \frac{\partial H}{\partial p_x} = \frac{p_x}{m}, \ \dot{y} = \frac{\partial H}{\partial p_y} = \frac{p_y}{m},$$

and $\qquad \qquad p_x = -\frac{\partial H}{\partial x} = 0, \ \dot{p}_y = -\frac{\partial H}{\partial y} = -mg.$

Solving these equations we get the equations of motion as

$$\ddot{x} = 0, \ \ddot{y} = -g. \qquad \qquad \text{...(108)}$$

Now differentiating equation (107) with respect to t we get

$$\frac{dH}{dt} = \frac{1}{m}\left(p_x\dot{p}_x + p_y\dot{p}_y\right) + mg\dot{y}$$

$$\Rightarrow \qquad \qquad \frac{dH}{dt} = 0.$$

This proves that H is a constant of motion. Now to see whether H represents total energy or not, we consider

$$T + V = \frac{1}{2} m\left(\dot{x}^2 + \dot{y}^2\right) - mg \ (h - y)$$

Putting the values of $\dot{x}$ and $\dot{y}$ we obtain the expression for the Hamiltonian as

$$T + V = \frac{1}{2m}\left(p_x^2 + p_y^2\right) - mg(h - y). \qquad \text{...(109)}$$

From equations (107) and (109) that the Hamiltonian H represents the total energy of the particle.

Example 21: A particle is constrained to move on the arc of a parabola $x^2 = 4ay$ under the action of gravity. Show that the Hamiltonian of the system is

$$H = \frac{2a^2 p_x^2}{m\left(4a^2 + x^2\right)} + \frac{mg}{4a}\,x^2.$$

Does the Hamiltonian of the particle represent total energy? Is it a constant of motion? Justify your claim.

Solution: The particle is constrained to move on the arc of the parabola $x^2 = 4ay$, where y is vertical axis, under the action of gravity. The kinetic energy and the potential energy of the particle are given by

$$T = \frac{1}{2}\,m(\dot{x}^2 + \dot{y}^2) \text{ and } V = mgy. \qquad \text{...(110)}$$

However, x and y are not the generalized co-ordinates, because they are related by the constraint equation $x^2 = 4ay$. Eliminating y from equations (110), we obtain

$$T = \frac{1}{2}\,m\dot{x}^2\left(1 + \frac{x^2}{4a^2}\right), \quad V = \frac{x^2}{4a}\,mg.$$

Hence the Lagrangian function is

$$L = \frac{1}{2}\,m\dot{x}^2\left(1 + \frac{x^2}{4a^2}\right) - \frac{x^2}{4a}\,mg. \qquad \text{...(111)}$$

The kinetic energy of the system is a homogeneous quadratic function of generalized velocities and the Lagragian is independent of time t. Hence from the Theorem 8 we infer that the Hamiltonian of the system will represent a constant of motion as well as the total energy. These conclusions are proved below mathematically. Now we see that the system has one degree of freedom and only one generalized co-ordinate x. Furthermore, the generalized momentum gives

$$\Rightarrow \qquad p_x = \frac{\partial L}{\partial \dot{x}} = m\dot{x}\left(1 + \frac{x^2}{4a^2}\right)$$

$$\Rightarrow \qquad \dot{x} = \frac{4a^2 p_x}{m\left(4a^2 + x^2\right)}. \qquad \text{...(112)}$$

Now the Hamiltonian H becomes

$$H = \dot{x}p_x - \frac{1}{2} m\dot{x}^2\left(1 + \frac{x^2}{4a^2}\right) + \frac{x^2}{4a} mg. \qquad ...(113)$$

On using (112) we write equation (113) as

$$H = \frac{2a^2 p_x^2}{m(4a^2 + x^2)} + \frac{mgx^2}{4a}. \qquad ...(114)$$

This is the required Hamiltonian function. Now to see whether this H represents total energy or not, we consider

$$T + V = \frac{1}{2} m\dot{x}^2\left(1 + \frac{x^2}{4a^2}\right) + \frac{x^2}{4a} mg. \qquad ...(115)$$

Eliminating $\dot{x}$ from equation (115) we obtain

$$T + V = \frac{2a^2 p_x^2}{m(4a^2 + x^2)} + \frac{mgx^2}{4a}.$$

Which is the Hamiltonian of the motion, showing that it represents the total energy of the particle. Now to show that the Hamiltonian H represents constant of motion, we first find the equation of motion. From equation (111), we have

$$\frac{\partial L}{\partial x} = \frac{m}{4a^2} x\dot{x}^2 - \frac{mgx}{2a}, \quad \frac{\partial L}{\partial \dot{x}} = \left(1 + \frac{x^2}{4a^2}\right) m\dot{x}.$$

Hence the Lagrange's equation of motion becomes

$$\frac{d}{dt}\left[\left(1 + \frac{x^2}{4a^2}\right) m\dot{x}\right] - \frac{m}{4a^2} x\dot{x}^2 + \frac{mgx}{2a} = 0.$$

This gives on simplification the equation of motion

$$\left(4a^2 + x^2\right) \ddot{x} + x\dot{x}^2 + 2agx = 0. \qquad ...(116)$$

Differentiation of equation (114) with respect to t yields

$$\frac{dH}{dt} = \frac{4a^2}{m}\left[\frac{p_x \dot{p}_x}{(4a^2 + x^2)} - \frac{x\dot{x}p_x^2}{(4a^2 + x^2)^2}\right] + \frac{mg}{2a} x\dot{x}.$$

Eliminating p_x, $\dot{p}_x$ we obtain

$$\frac{dH}{dt} = \frac{m}{4a^2}\left[\left(4a^2 + x^2\right) \ddot{x} + x\dot{x}^2 + 2agx\right] \dot{x}.$$

This implies from equation (116) that $\dfrac{dH}{dt} = 0$. This shows that the Hamiltonian H is a constant of motion.

Example 22: For the Lagrangian $L = \dot{q}_1^2 + \dot{q}_2^2 - \dot{q}_1\dot{q}_2 - q_1^2 - q_2^2$. Find the conjugate momenta p_1, p_2 and obtain the Hamiltonian. Write down the Hamilton's canonical equations of motion.

Solution: We see that the system has two degrees of freedom and hence two generalized co-ordinates q_1 and q_2. The corresponding conjugate momenta are given by

$$p_1 = \frac{\partial L}{\partial \dot{q}_1} = 2\dot{q}_1 - \dot{q}_2, \quad p_2 = \frac{\partial L}{\partial \dot{q}_2} = 2\dot{q}_2 - \dot{q}_1. \qquad \text{...(117)}$$

Solving these equations for $\dot{q}_1, \dot{q}_2$ we obtain

$$\dot{q}_1 = \frac{1}{3}(2p_1 + p_2), \quad \dot{q}_2 = \frac{1}{3}(p_1 + 2p_2). \qquad \text{...(118)}$$

The Hamiltonian of the system gives

$$H = p_1\dot{q}_1 + p_2\dot{q}_2 - \dot{q}_1^2 - \dot{q}_2^2 + \dot{q}_1\dot{q}_2 + q_1^2 + q_2^2$$

Eliminating the generalized velocities we obtain the Hamiltonian function as

$$H = \frac{1}{3}\left(p_1^2 + p_2^2 + p_1p_2\right) + q_1^2 + q_2^2. \qquad \text{...(119)}$$

The Hamilton's canonical equations of motion are $\dot{p}_1 = -2q_1$, $\dot{p}_2 = -2q_2$, and the other set of equations are given in (118). From these set of equations we find

$$\ddot{q}_1 = -\frac{2}{3}(2q_1 + q_2), \quad \ddot{q}_2 = -\frac{2}{3}(q_1 + 2q_2).$$

Example 23: The Hamiltonian for 3-dimensional isotropic harmonic oscillator is given by

$$H = \frac{1}{2} \sum_{i=1}^{3} (p_i^2 + \mu q_i^2), \quad \mu = \text{const.}$$

Show that $F_1 = q_2p_3 - q_3p_2$, $F_2 = q_3p_1 - q_1p_3$, $F_3 = -q_1p_2 - q_2p_1$ are constants of motion.

Solution: For the given Hamiltonian, the Hamilton's canonical equations of motion give

$$\dot{q}_j = \frac{\partial H}{\partial p_j} = p_j, \quad \dot{p}_j = -\frac{\partial H}{\partial q_j} = -\mu q_j. \qquad \text{...(120)}$$

Now consider

$$\frac{dF_1}{dt} = \dot{p}_3 q_2 + p_3 \dot{q}_2 - \dot{p}_2 q_3 - p_2 \dot{q}_3.$$

On using (120) we see that $\frac{dF_1}{dt} = 0 \Rightarrow F_1 = \text{const.}$ Similarly, we prove that $F_2 = \text{const.}$, $F_3 = \text{const.}$

Example 24: Obtain the Hamiltonian H and the Hamilton's equations of motion of a simple pendulum. Prove or disprove that H represents the constant of motion and total energy.

Solution: The Example is solved earlier by various methods. The Lagrangian of the pendulum is given by (refer Example 26 of Chapter 1)

$$L = \frac{1}{2} ml^2 \dot{\theta}^2 - mgl(1 - \cos \theta), \qquad \qquad ...(121)$$

where the generalized momentum is given by

$$p_\theta = \frac{\partial L}{\partial \dot{\theta}} = ml^2 \dot{\theta} \Rightarrow \dot{\theta} = \frac{p_\theta}{ml^2}. \qquad \qquad ...(122)$$

The Hamiltonian of the system is given by

$$H = p_\theta \dot{\theta} - \frac{1}{2} ml^2 \dot{\theta}^2 + mgl(1 - \cos \theta).$$

Eliminating $\dot{\theta}$ we obtain

$$H = \frac{p_\theta^2}{2ml^2} + mgl(1 - \cos \theta). \qquad \qquad ...(123)$$

Hamilton's canonical equations of motion yield

$$\dot{\theta} = \frac{p_\theta}{ml^2}, \ \dot{p}_\theta = -mgl \sin \theta. \qquad \qquad ...(124)$$

Now eliminating p_θ from these equations we get

$$\ddot{\theta} + \frac{g}{l} \sin \theta = 0. \qquad \qquad ...(125)$$

We observe from the Lagrangian (121) that it is independent of time t, and the constraints involved are scleronomic hence H represents the constant of motion and represents total energy of the system. The proof is left as an exercise for the readers.

Example 25: The Lagrangian for a particle moving on a surface of a sphere of radius r is given by $L = \frac{1}{2} mr^2 (\dot{\theta}^2 + \sin^2 \theta \dot{\phi}^2) - mgr \cos \theta$. Find the Hamiltonian

H and show that it is constant of motion. Prove or disprove that *H* represents the total energy. Is the energy of the particle constant? Justify your claim.

Solution: The Lagrangian in the question is the Lagragian for spherical pendulum and is derived in Example 38 of Chapter 1. We see that ϕ is a cyclic co-ordinate. Hence the corresponding generalized momentum is conserved. *i.e.,*

$$p_\phi = \frac{\partial L}{\partial \dot\phi} = mr^2 \sin^2 \theta \dot\phi = \text{const.} \qquad \text{...(126)}$$

Similarly,
$$p_\theta = \frac{\partial L}{\partial \dot\theta} = mr^2 \dot\theta. \qquad \text{...(127)}$$

The Hamiltonian of the particle is defined as

$$H = \dot\theta p_\theta + \dot\phi p_\phi - \frac{1}{2} mr^2(\dot\theta^2 + \sin^2 \theta \dot\phi^2) + mgr \cos \theta.$$

Eliminating $\dot\theta$ and $\dot\phi$ from the equation we obtain the expression for the Hamilton in the form

$$H = \frac{1}{2}\left(\frac{p_\theta^2}{mr^2} + \frac{p_\phi^2}{mr^2 \sin^2 \theta} \right) + mgr \cos \theta. \qquad \text{...(128)}$$

The Hamilton's canonical equations of motion give

$$\dot{p}_\theta = \frac{\cos \theta \, p_\phi^2}{mr^2 \sin^3 \theta} + mgr \sin \theta, \quad \dot{p}_\phi = 0. \qquad \text{...(129)}$$

Eliminating p_θ from equation (129), we get the equation of motion of spherical pendulum as

$$mr^2\ddot\theta - \frac{\cos \theta \, p_\phi^2}{mr^2 \sin^3 \theta} - mgr \sin \theta = 0. \qquad \text{...(130)}$$

(i) We notice that the Lagragian *L* is independent of time *t*, hence one can prove that the Hamiltonian *H* is a constant of motion.

(ii) Similarly, we see that the kinetic energy of the particle is homogeneous quadratic function of generalized velocities; consequently by using the Theorem 8, we conclude that the Hamiltonian *H* represents the total energy of the system.

(iii) To find the first integral of equation of motion, multiply equation (130) by $\dot\theta$ to get

$$mr^2\ddot{\theta}\dot{\theta} - \frac{p_\phi^2 \cos\theta\dot{\theta}}{mr^2 \sin^3\theta} - mgr\sin\theta\dot{\theta} = 0.$$

This can also be written as

$$\frac{d}{dt}\left(\frac{1}{2}mr^2\dot{\theta}^2\right) + \frac{d}{dt}\left(\frac{p_\phi^2}{2mr^2\sin^2\theta}\right) + \frac{d}{dt}(mgr\cos\theta) = 0.$$

Integrating we get

$$\left(\frac{1}{2}mr^2\dot{\theta}^2\right) + \left(\frac{p_\phi^2}{2mr^2\sin^2\theta}\right) + (mgr\cos\theta) = \text{const.}$$

Eliminating $\dot{\theta}$ on using equation (127), we get

$$\frac{1}{2}\left(\frac{p_\theta^2}{mr^2} + \frac{p_\phi^2}{mr^2\sin^2\theta}\right) + mgr\cos\theta = \text{const.} \qquad ...(131)$$

This shows that the first integral of the equation of motion represents the total energy (H) of the system.

Example 26: Two mass points of mass m_1 and m_2 are connected by a string passing through a hole in a smooth table so that m_1 rests on the table surface and m_2 hangs suspended. Assuming m_2 moves only in a vertical line, write down the Hamiltonian for the system and hence the equations of motion. Prove or disprove that (i) Hamiltonian H represents the constant of motion (ii) H represents total energy of the system.

Solution: This example is solved in Chapter 1. (please refer to Example 24). The Lagrangian of the system is given by

$$L = \frac{1}{2}m_1(\dot{r}^2 + r^2\dot{\theta}^2) + \frac{1}{2}m_2\dot{r}^2 + m_2g(l - r). \qquad ...(132)$$

We see that the co-ordinate θ is cyclic in the Lagrangian L and hence the corresponding generalized momentum is conserved.

$$\Rightarrow \qquad p_\theta = \frac{\partial L}{\partial\dot{\theta}} = m_1r^2\dot{\theta} = \text{const.}$$

Similarly, we find

$$p_r = \frac{\partial L}{\partial\dot{r}} = (m_1 + m_2)\,\dot{r}.$$

Now the Hamiltonian function is defined as

$$H = \dot{r}\,\frac{\partial L}{\partial \dot{r}} + \dot{\theta}\,\frac{\partial L}{\partial \dot{\theta}} - L,$$

$$H = \frac{1}{2}(m_1 + m_2)\,\dot{r}^2 + \frac{1}{2}\,m_1 r^2 \dot{\theta}^2 - m_2 g(l - r).$$

Eliminating $\dot{r}$ and $\dot{\theta}$ we obtain

$$H = \frac{p_r^2}{2(m_1 + m_2)} + \frac{p_\theta^2}{2m_1 r^2} - m_2 g(l - r). \qquad \text{...(133)}$$

The Hamilton canonical equations of motion yield

$$\dot{p}_r = -\frac{\partial H}{\partial r} = \frac{p_\theta^2}{m_1 r^3} - m_2 g, \quad \dot{p}_\theta = -\frac{\partial H}{\partial \theta} = 0. \qquad \text{...(134)}$$

$$\dot{r} = \frac{\partial H}{\partial p_r} = \frac{p_r}{(m_1 + m_2)}, \quad \dot{\theta} = \frac{\partial H}{\partial p_\theta} = \frac{p_\theta}{m_1 r^2}. \qquad \text{...(135)}$$

From equations (134) and (135), we find the equation of motion in the form

$$(m_1 + m_2)\,\ddot{r} - \frac{p_\theta^2}{m_1 r^3} + m_2 g = 0. \qquad \text{...(136)}$$

(i) It is evident from equation (132) that the Lagrangian L does not involve time t explicitly, hence the Hamiltonian H will represent a constant of motion. The proof is left as an exercise for the readers.

(ii) Further, the kinetic energy of the system is a homogeneous quadratic function of generalized velocities. Hence by the Theorem 8, we infer that the Hamiltonian H will represent the total energy of the system.

(iii) To prove that the total energy is conserved, multiply the equation of motion (136) by $\dot{r}$, to get

$$(m_1 + m_2)\,\ddot{r}\dot{r} - \frac{p_\theta^2 \dot{r}}{m_1 r^3} + m_2 g\dot{r} = 0.$$

This we write as

$$\frac{d}{dt}\left[(m_1 + m_2)\,\frac{\dot{r}^2}{2}\right] + \frac{d}{dt}\left(\frac{p_\theta^2}{2m_1 r^2}\right) + \frac{d}{dt}(m_2 g r) = 0.$$

Integrating and then eliminating $\dot{r}$ we get

$$\frac{p_r^2}{2(m_1 + m_2)} + \frac{p_\theta^2}{2m_1 r^2} + m_2 g r = \text{const.} \qquad \text{...(137)}$$

Infact this is Hamiltonian H, proving that the first integral of equation of motion is the total energy and is conserved.

Example 27: A particle of mass m is moving on a xy-plane which is rotating about z-axis with angular velocity ω. The Lagrangian is given by

$$L = \frac{1}{2} m\left[(\dot{x} - \omega y)^2 + (\dot{y} + \omega x)^2 \right] - V(x, y).$$

Show that the Hamiltonian H is given by

$$H = \frac{1}{2m}\left(p_x^2 + p_y^2 \right) + p_x \omega y - p_y \omega x + V.$$

Find the equations of motion and hence prove or disprove that

   (i) H represents a constant of motion and

   (ii) H represents the total energy.

Solution: The Lagrangian of the system is derived in Example 56 in the Chapter 1. The generalized momenta p_x and p_y are given by

$$p_x = \frac{\partial L}{\partial \dot{x}} = m(\dot{x} - \omega y) \Rightarrow \dot{x} = \frac{p_x}{m} + \omega y \qquad \text{...(138)}$$

$$p_y = \frac{\partial L}{\partial \dot{y}} = m(\dot{y} + \omega x) \Rightarrow \dot{y} = \frac{p_y}{m} - \omega x \qquad \text{...(139)}$$

The Hamiltonian H is defined by $H = \dot{x}p_x + \dot{y}p_y - L$. This becomes

$$H = p_x \dot{x} + p_y \dot{y} - \frac{1}{2} m\left[(\dot{x} - \omega y)^2 + (\dot{y} + \omega x)^2 \right] + V(x, y).$$

Eliminating $\dot{x}$ and $\dot{y}$ from the above equation we get the Hamiltonian of the system in the form

$$H = \frac{1}{2m}\left(p_x^2 + p_y^2 \right) + \omega(p_x y - p_y x) + V. \qquad \text{...(140)}$$

The Hamilton's canonical equations of motions give

$$\dot{p}_x = -\frac{\partial H}{\partial x} = p_y \omega - \frac{\partial V}{\partial x}, \; \dot{p}_y = -\frac{\partial H}{\partial y} = -p_x \omega - \frac{\partial V}{\partial y},$$

$$\dot{x} = -\frac{\partial H}{\partial p_x} = \frac{p_x}{m} + \omega y, \; \dot{y} = \frac{\partial H}{\partial p_x} = \frac{p_y}{m} - \omega x. \qquad \text{...(141)}$$

Solving these equations we obtain the equations which describe the motion of the particle as

$$m(\ddot{x} - 2\omega \dot{y} - \omega^2 x) = -\frac{\partial V}{\partial x}, \; m(\ddot{y} + 2\omega \dot{x} - \omega^2 y) = -\frac{\partial V}{\partial y} \qquad \text{...(142)}$$

It is evident from the Lagrangian that it does not involve time t explicitly, hence from the Theorem 8 we conclude that the Hamiltonian H represents constant of motion. We see that the kinetic energy of the system is not homogeneous quadratic function of generalized velocities; hence by the Theorem 8 we conclude that H will not represent total energy of the system. This can be seen from the equation $E = T + V$. This gives

$$E = \frac{1}{2m}\left(p_x^2 + p_y^2\right) + V(x, y). \qquad \qquad ...(143)$$

We see from equations (140) and (143) that the Hamiltonian H does not represent total energy.

Example 28:   A particle of mass m is constrained to move on a circle with radius r which itself is rotating about its vertical diameter with a constant angular velocity ω. If θ is the inclination of the radius vector of the particle with the axis of rotation, then find the Hamiltonian H and show that whether it represents a (i) constant of motion and (ii) total energy of the system.

Solution: This example is solved in the Chapter 1 (refer Example 46). We recall here the Lagrangian function L for the determination of Hamiltonian H. It is given by

$$L = \frac{1}{2}\, mr^2\left(\dot{\theta}^2 + \omega^2 \sin^2 \theta\right) + mgr(1 - \cos \theta), \qquad ...(144)$$

Hamiltonian H of the system is obtained from the definition (23) in the form

$$H = \frac{p_\theta^2}{2mr^2} - \frac{1}{2}\, mr^2\omega^2 \sin^2 \theta - mgr(1- \cos \theta). \qquad ...(145)$$

The Hamilton's canonical equations of motion (27) yield the equation

$$\ddot{\theta} - \omega^2 \sin \theta \cos \theta - \frac{g}{r} \sin \theta = 0. \qquad ...(146)$$

We notice from the Lagrangian (144) that it does not contain the time t and the kinetic energy is not a homogeneous quadratic function of generalized velocities, hence from the Theorem 8 we immediately conclude that the Hamiltonian is a constant of motion but does not represent the total energy of the system. Mathematical proof is left as an exercise to the readers.

Remark: This is an example, where the system is conservative and scleronomic but kinetic energy is not a homogeneous quadratic function of generalized velocity due to the rotation of the system and hence Hamiltonian H does not represent total energy.

Example 29: A body of mass m is thrown up an inclined plane which is moving horizontally with constant velocity v. Find the Hamiltonian H and show that whether it represents a (i) constant of motion and (ii) total energy of the system.

Solution: This is another problem whose solution is given in Example 25 of the Chapter 1. The system is conservative and constraints involved are rheonomic. The Lagrangian is given by

$$L = \frac{1}{2} m\left(\dot{r}^2 + 2\dot{r}v \cos \theta\right) - mgr \sin \theta. \qquad \text{...(147)}$$

The constant term containing velocity is ignored from the Lagrangian.

The Hamiltonian H of the system is obtained from the definition (23) in the form

$$H = \frac{p_r^2}{2m} - v \cos \theta \, p_r + \frac{1}{2} mv^2 \cos^2 \theta + mgr \sin \theta. \qquad \text{...(148)}$$

The Hamilton's canonical equations of motion (27) give the equation of motion of the body as

$$m\ddot{r} + mg \sin \theta = 0. \qquad \text{...(149)}$$

This is an example of a rheonomic system but the Lagrangian function of the system is independent of t, hence the Hamiltonian H is a constant of motion. Also the kinetic energy is not homogeneous quadratic function of generalized velocities hence H does not represent the total energy of the system. A reader can prove this inference mathematically.

Example 30: A particle has Lagrangian

$$L = \frac{1}{2}\left[f(\theta) \, \dot{\theta}^2 + 2g(\theta) \, \dot{\theta}\omega + \omega^2 h(\theta)]\right] - V.$$

Show that $\frac{1}{2}\left[f(\theta) \, \dot{\theta}^2 - \omega^2 h(\theta)\right] + V = $ const. What does this constant represent?

Solution: We notice from the given Lagrangian function that the system has one degree of freedom and one generalized coordinate θ. Hence the corresponding Lagrange's equation of motion gives

$$\frac{1}{2} f'(\theta) \, \dot{\theta}^2 + f(\theta) \, \ddot{\theta} - \frac{1}{2} h'(\theta) \, \omega^2 + \frac{\partial V}{\partial \theta} = 0. \qquad \text{...(150)}$$

Now consider

$$\frac{d}{dt}\left[\frac{1}{2}(f(\theta) \, \dot{\theta}^2 - \omega^2 h(\theta)) + V\right] = \dot{\theta}\left[\frac{1}{2}(f'(\theta) \, \dot{\theta}^2 + 2f(\theta) \, \ddot{\theta} - \omega^2 h'(\theta)) + \frac{\partial V}{\partial \theta}\right].$$

It follows from the equation of motion (150) that

$$\frac{d}{dt}\left[f(\theta)\,\dot{\theta}^2 - \omega^2 h(\theta)) + V\right] = 0.$$

Integrating the equation we obtain

$$\frac{1}{2}\left[f(\theta)\,\dot{\theta}^2 - \omega^2 h(\theta)\right] + V = \text{const.} \qquad \text{...(151)}$$

The definition of the Hamiltonian function (23) yields the expression

$$H = \frac{1}{2f(\theta)}\left[p_\theta^2 - 2p_\theta g(\theta)\,\omega + \omega^2 g^2(\theta)\right] - \frac{1}{2}\,\omega^2 h(\theta) + V. \qquad \text{...(152)}$$

The Lagrangian L does not involve time t explicitly hence one can prove that the Hamiltonian H will be the constant of motion. However, the kinetic energy of the system is not the homogeneous quadratic function of generalized velocities; therefore, the Hamiltonian H will not represent the total energy of the system. The proof is left as an exercise for the readers. Further, the definition of generalized momentum gives

$$\dot{\theta} = \frac{1}{f(\theta)}[p_\theta - g(\theta)\,\omega].$$

Eliminating $\dot{\theta}$ from equation (151) we get equation (152). This shows that the constant of integration in equation (151) is the Hamiltonian H.

Example 31: A bead is sliding on a uniformly rotating wire. Find Hamiltonian and total energy and show that H is not a constant of motion and does not represent total energy.

Solution: Refer the Example 45 of Chapter 1, in which we have solved the problem by Lagrange's method. The Lagrangian of the system is given by

$$L = \frac{1}{2}\,m\left(\dot{r}^2 + \dot{r}^2\omega^2\right) - mgr\,\sin \omega t. \qquad \text{...(153)}$$

The Hamiltonian function H is obtained by simplifying equation (23) in the form

$$H = \frac{p_r^2}{2m} - \frac{1}{2}\,mr^2\omega^2 + mgr\,\sin \omega t. \qquad \text{...(154)}$$

The Hamilton's canonical equations of motion (27) yield the equation of motion of the bead as

$$\ddot{r} - r\omega^2 + g\,\sin \omega t = 0. \qquad \text{...(155)}$$

One can do simple and straight forward calculations to show that H is not a constant and does not represent total energy too. However, we see that this is an example of a rheonomic system and the Lagrangian L involves time t explicitly and hence by Theorem 8, the Hamiltonian H does not represent a constant of motion. Furthermore, the kinetic energy of the system is not a homogeneous quadratic function of generalized velocities, and also the potential energy involves time t explicitly, hence by Theorem 8; H does not represent a total energy too.

Example 32: If all the co-ordinates of a dynamical system of n degrees of freedom are cyclic then prove that the problem can be solved completely by integration.

Solution: The Hamiltonian function H for a dynamical system of n degrees of freedom is given by $H = H(p_j, q_j, t)$. It is given that all the co-ordinates are cyclic. This means that they are absent in the Hamiltonian. Thus we have $H = H(p_j, t), j = 1, 2, ..., n$.

Hence the Hamiltonian equations of motion are

$$\dot{p}_j = -\frac{\partial H}{\partial q_j} = 0 \Rightarrow p_j = \text{const.} \ \forall j \qquad \text{...(156)}$$

and

$$\dot{q}_j = \frac{\partial H}{\partial p_j} = f_j(p_1, p_2, \dots p_n, t). \ (\text{say}) \qquad \text{...(157)}$$

Integrating equations (157) we get $q_j = \int f_j(p_1, p_2, \dots p_n, t) \, dt + c_j$, where c_j are constants of integration. Thus the co-ordinates of a dynamical system are obtained just by integrating and the system can be located.

Example 33: Write Hamilton's equations of motion for a compound pendulum.

Solution: A compound pendulum is discussed by Lagrangian formulation in the Example 39 of the Chapter 1. The Lagrangian of the compound pendulum is given by

$$L = \frac{1}{2} I\dot{\theta}^2 + mgl \cos \theta, \qquad \text{...(158)}$$

where I is the moment of inertia. The only generalized co-ordinate is θ. The corresponding generalized momentum of the pendulum is given by

$$p_\theta = \frac{\partial L}{\partial \dot{\theta}} = I\dot{\theta} \Rightarrow \dot{\theta} = \frac{p_\theta}{I}. \qquad \text{...(159)}$$

Thus the Hamiltonian function of the pendulum is given by

$$H = p_\theta \dot{\theta} - \frac{1}{2} I\dot{\theta}^2 - mgl \cos \theta.$$

Eliminating $\dot{\theta}$ the Hamiltonian function becomes

$$H = \frac{1}{2I} p_\theta^2 - mgl \cos \theta. \qquad \text{...(160)}$$

The Hamilton's canonical equations give the required equation of compound pendulum in the form

$$\dot{\theta} + \frac{mgl}{I} \sin \theta = 0.$$

Example 34: A bead slides on a wire in the shape of a cycloid described by equations

$$x = a(\theta - \sin \theta), \ y = a(1 + \cos \theta), \ 0 \le \theta \le 2\pi.$$

Find the Hamiltonian H, hence the equations of motion. Also prove or disprove that (i) H represents a constant of motion, (ii) H represents a total energy.

Solution: A particle is constraint to move on the cycloid. From the equations of cycloids we find $\dot{x} = a\dot{\theta}(1 - \cos \theta)$, $\dot{y} = -a \sin \theta \dot{\theta}$. The kinetic energy $T = \frac{1}{2} m(\dot{x}^2 + \dot{y}^2)$ and the potential energy $V = mgy$ of the particle become

$$T = ma^2\dot{\theta}^2(1 + \cos \theta), \ V = mga(1 + \cos \theta).$$

Hence the Lagrangian of the particle becomes

$$L = ma^2\dot{\theta}^2(1 - \cos \theta) - mga(1 + \cos \theta). \qquad \text{...(161)}$$

Using the definition (23), the Hamiltonian H of the particle is obtained as

$$H = ma^2\dot{\theta}^2(1 - \cos \theta) + mga(1 + \cos \theta). \qquad \text{...(162)}$$

By usual procedure we eliminate $\dot{\theta}$ from equation (162) to get the required Hamiltonian H as

$$H = \frac{p_\theta^2}{4ma^2(1 - \cos \theta)} + mga(1 + \cos \theta). \qquad \text{...(163)}$$

The Hamilton's canonical equations of motion give

$$\dot{p}_\theta = -\frac{\partial H}{\partial \theta} = \frac{p_\theta^2}{4ma^2} \frac{\sin \theta}{(1 - \cos \theta)^2} + mga \sin \theta, \qquad \text{...(164)}$$

$$\dot{\theta} = \frac{\partial H}{\partial p_\theta} = \frac{p_\theta}{2ma^2(1 - \cos \theta)}. \qquad \text{...(165)}$$

From equations (164) and (165) we obtain the equation of motion of the particle

$$\ddot{\theta}(1 - \cos \theta) + \frac{p_\theta^2}{8m^2a^4} \frac{\sin \theta}{(1 - \cos \theta)^3} - \frac{g}{2a} \sin \theta = 0. \qquad \text{...(166)}$$

Eliminating p_θ from equation (166) we obtain the equation which describes the motion of the particle in the form

$$2ma^2(1 - \cos\theta)\,\ddot\theta + ma^2\sin\theta\dot\theta^2 - mga\sin\theta = 0. \qquad \ldots(167)$$

(i) H is a constant of motion: Differentiate equation (163) with respect to time t we get

$$\frac{dH}{dt} = \frac{2p_\theta\dot p_\theta}{4ma^2(1 - \cos\theta)} - \frac{p_\theta^2}{4ma^2}\frac{\sin\theta\dot\theta}{(1 - \cos\theta)^2} - mga\sin\theta\dot\theta.$$

Using equations (164) and (165) we readily get $\dfrac{dH}{dt} = 0$. This shows that the Hamiltonian H is a constant of motion.

(ii) H represents the total energy: We find from the expressions for kinetic energy and potential energy that

$$T + V = ma^2\dot\theta^2(1 - \cos\theta) + mga(1 + \cos\theta). \qquad \ldots(168)$$

Eliminating $\dot\theta$ from equation (168) we get equation (163) that gives the required expression for the Hamiltonian. This proves Hamiltonian represents total energy. Now multiply equation (167) by $\dot\theta$ we get

$$2ma^2(1 - \cos\theta)\,\ddot\theta + ma^2\sin\theta\dot\theta^3 - mga\sin\theta\dot\theta = 0.$$

This can be written as

$$\frac{d}{dt}[ma^2(1 - \cos\theta)\,\dot\theta^2 + mga(1 + \cos\theta)] = 0.$$

Integrating we get

$$H = T + V = ma^2\dot\theta^2(1 - \cos\theta) + mga(1 + \cos\theta) = \text{const.}$$

This shows that the Hamiltonian H represents the constant of total energy. The results can also be concluded from the Lagrangian (161) as it is independent of time t and kinetic energy is homogeneous quadratic function of generalized velocities.

Example 35: Obtain the Hamilton's equation of motion for a one dimensional harmonic oscillator. Show also that the path of the oscillator in the phase space is an ellipse with semi axes

$$a = \left(\frac{2E}{K}\right)^{\frac{1}{2}} \text{ and } b = (2mE)^{\frac{1}{2}}.$$

Solution: The one dimensional harmonic oscillator consists of a mass attached to one end of a spring and other end of the spring is fixed. If the spring is pressed and released then on account of the elastic property of the spring, the spring exerts a force $\overline{F}$ on the body in the opposite direction. This is called restoring force. It is found that this force is proportional to the displacement of the body from its equilibrium position.

$$\overline{F} \propto x \Rightarrow \overline{F} = -kx,$$

where k is the spring constant and negative sign indicates the force is opposite to the direction of displacement. Hence the potential energy of the particle is given by

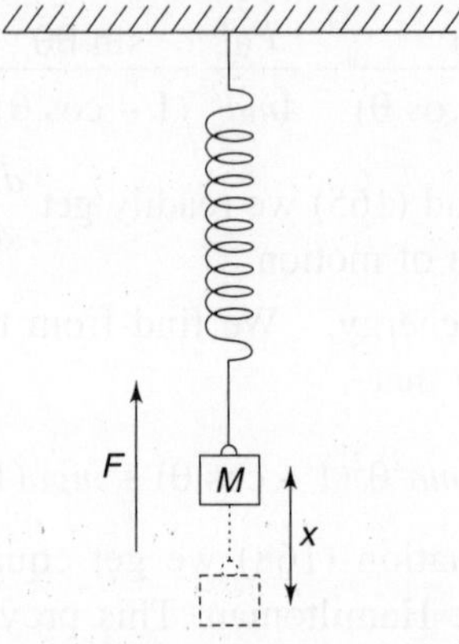

$$V = -\int F dx \Rightarrow V = \int kx\, dx + c,$$

$$V = \frac{kx^2}{2} + c$$

where c is the constant of integration. By choosing the horizontal plane passing through the position of equilibrium as the reference level, we have $V = 0$ at $x = 0$. This gives $c = 0$.

Hence potential energy of the particle is $V = \dfrac{1}{2} kx^2$. The kinetic energy of the one dimensional harmonic oscillator is $T = \dfrac{1}{2} m\dot{x}^2$. Hence the Lagrangian of the system becomes

$$L = \frac{1}{2} m\dot{x}^2 - \frac{1}{2} kx^2. \qquad \qquad ...(169)$$

The Lagrange's equation motion gives

$$\ddot{x} + \omega^2 x = 0, \ \omega^2 = \frac{k}{m}, \qquad \qquad ...(170)$$

ω is the frequency of oscillation. The Hamiltonian H of the oscillator is defined as

$$H = \dot{x} p_x - \frac{1}{2} m\dot{x}^2 + \frac{1}{2} kx^2,$$

where $p_x = \dfrac{\partial L}{\partial \dot{x}} = m\dot{x} \Rightarrow \dot{x} = \dfrac{p_x}{m}$. Substituting this in the above equation we get the Hamiltonian

$$H = \frac{p_x^2}{2m} + \frac{1}{2} kx^2. \qquad \ldots(171)$$

Solving the Hamilton's canonical equations of motion we readily get the equation (170) as the equation of motion. Obviously Hamiltonian H represents the total energy of the oscillator. From equation (171), we write

$$H = \frac{p_x^2}{2m} + \frac{1}{2} kx^2 = E.$$

This can be put in the form

$$\frac{p_x^2}{2mE} + \frac{x^2}{\left(\dfrac{2E}{k}\right)} = 1 \Rightarrow \frac{x^2}{\left(\sqrt{\dfrac{2E}{k}}\right)^2} + \frac{p_x^2}{(\sqrt{2mE})^2} = 1.$$

This is an ellipse in the phase space (x, p_x) with semi axes $a = \left(\dfrac{2E}{k}\right)^{\frac{1}{2}}$ and $b = (2mE)^{\frac{1}{2}}$.

Principle of Least Action

Action in mechanics

In Mechanics the time integral of twice the kinetic energy is called the action. Thus

$$A = \int_{t_0}^{t_1} 2T\,dt$$

is called the action.

i.e. The integral

$$A = \int_{t_0}^{t_1} \sum_j p_j \dot{q}_j\,dt \qquad \ldots(172)$$

is called action in Mechanics.

Principle of least action

There is another variational principle associated with the Hamiltonian formulation and is known as the principle of least action. It involves a new type of variation which we call the Δ-variation. In Δ-variation the co-ordinates of the end points remain fixed while the time is allowed to vary. The varied paths may terminate at different points, but still position co-ordinates are held fixed. Mathematically, we have

$$\delta I = \frac{\partial I}{\partial \alpha} d\alpha, \quad \text{and} \quad \Delta I = \frac{dI}{d\alpha} d\alpha. \qquad \ldots(173)$$

Thus for the family of paths represented by the equation $q_j = q_j(\alpha, t)$, $t = t(\alpha)$, we have

$$\Delta q_j = \frac{dq_j}{d\alpha}\, d\alpha = \left(\frac{\partial q_j}{\partial \alpha} + \dot{q}_j \frac{dt}{d\alpha}\right) d\alpha.$$

$$\Delta q_j = \left(\frac{\partial q_j}{\partial \alpha}\, d\alpha + \dot{q}_j \frac{dt}{d\alpha}\, d\alpha\right),$$

$$\Delta q_j = \delta q_j + \dot{q}_j \Delta t. \qquad\qquad ...(174)$$

This shows that the total variation is the sum of two variations.

WORKED EXAMPLES

Example 36: If $f = f(q_j, \dot{q}_j, t)$ then show that $\Delta f = \delta f + \Delta t \cdot \dfrac{df}{dt}$.

Solution: Consider a system of particles moving from one point to another. Let the family of paths between these two points be given by the equation $q_j = q_j(t, \alpha)$. In Δ variation time is not held fixed, it depends on the path. This implies that $t = t(\alpha)$. Since

$$f = f(q_j, \dot{q}_j, t)$$

Taking the Δ variation of f we obtain

$$\Delta f = \sum_j \left(\frac{\partial f}{\partial q_j}\, \Delta q_j + \frac{\partial f}{\partial \dot{q}_j}\, \Delta \dot{q}_j\right) + \frac{\partial f}{\partial t}\, \Delta t. \qquad ...(175)$$

From equation (174), we find replacing q_j by $\dot{q}_j$ the equation

$$\Delta \dot{q}_j = \delta \dot{q}_j + \ddot{q}_j \Delta t. \qquad\qquad ...(176)$$

Eliminating Δq_j and $\Delta \dot{q}_j$ from equation (175) by using equations (174) and (176), we get

$$\Delta f = \sum_j \left(\frac{\partial f}{\partial q_j}\, \delta q_j + \frac{\partial f}{\partial \dot{q}_j}\, \delta \dot{q}_j + \frac{\partial f}{\partial t}\, \delta t\right) + \sum_j \left(\frac{\partial f}{\partial q_j}\, \dot{q}_j + \frac{\partial f}{\partial \dot{q}_j}\, \ddot{q}_j + \frac{\partial f}{\partial t}\right) \Delta t.$$

Note here that the term δt is added because it is zero, since in δ variation time t is held fixed and consequently δ change in time t is zero. This equation is nothing but

$$\Delta f = \delta f + \Delta t \cdot \frac{df}{dt} \qquad\qquad ...(177)$$

Since f is arbitrary, we can write it as

$$\Delta = \delta + \Delta t \cdot \frac{d}{dt}.$$

...(178)

Theorem 11   For a conservative system for which the Hamiltonian H is conserved, the principle of least action states that

$$\Delta \int_{t_0}^{t_1} \sum_j p_j \dot{q}_j \, dt = 0.$$

Proof: Consider a conservative system for which the Hamiltonian H is conserved. Let AB be the actual path and CD be the varied path. In Δ-variation the end points of the two paths are not terminated at the same point. The end points A and B after Δt take the positions C and D such that the position co-ordinates of A, C and B, D are held fixed. Now we know the action is given by

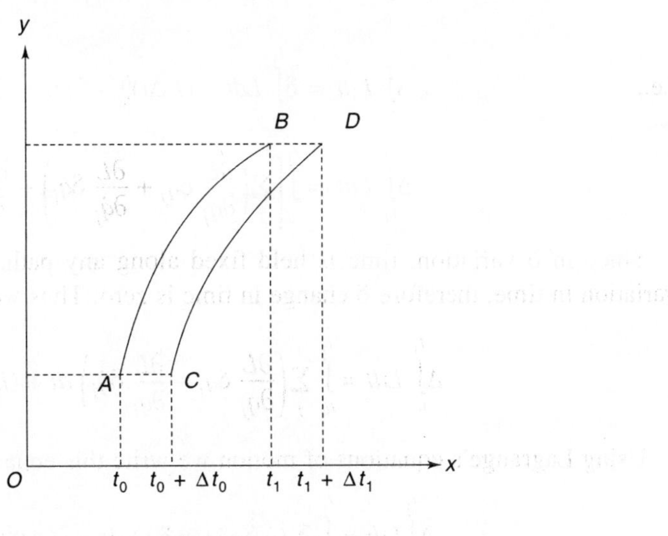

$$A = \int_{t_0}^{t_1} \sum_j p_j \dot{q}_j \, dt$$

$$A = \int_{t_0}^{t_1} (L + H) \, dt$$

$$A = \int_{t_0}^{t_1} L \, dt + H(t)\big|_{t_0}^{t_1}$$

$$A = \int_{t_0}^{t_1} L \, dt + H(t_1 - t_0).$$

Thus,
$$\Delta A = \Delta \int_{t_0}^{t_1} L\,dt + (H\Delta t)_{t_0}^{t_1}. \qquad \qquad ...(179)$$

Since the time limits are also subject to change in Δ variation therefore Δ can't be taken inside the integral.

Let
$$\int L\,dt = I \Rightarrow L = \dot{I}$$

Therefore, we have
$$\int_{t_0}^{t_1} L\,dt = I(t_1) - I(t_0).$$

This gives
$$\Delta \int_{t_0}^{t_1} L\,dt = \Delta I(t_1) - \Delta I(t_0).$$

Using the equation (177) we have
$$\Delta \int_{t_0}^{t_1} L\,dt = \delta I(t_1) - \delta I(t_0) + \dot{I}(t_1)\,\Delta t_1 - \dot{I}(t_0)\,\Delta t_0$$

i.e.,
$$\Delta \int_{t_0}^{t_1} L\,dt = \delta \int_{t_0}^{t_1} L\,dt + (L\Delta t)_{t_0}^{t_1}.$$

$$\Delta \int_{t_0}^{t_1} L\,dt = \int_{t_0}^{t_1}\left[\sum_j \left(\frac{\partial L}{\partial q_j}\,\delta q_j + \frac{\partial L}{\partial \dot{q}_j}\,\delta \dot{q}_j \right) + \frac{\partial L}{\partial t}\,\delta t \right] dt + (L\Delta t)_{t_0}^{t_1}.$$

Since in δ variation, time is held fixed along any path, hence there is no variation in time, therefore δ change in time is zero. Thus we have

$$\Delta \int_{t_0}^{t_1} L\,dt = \int_{t_0}^{t_1} \sum_j \left(\frac{\partial L}{\partial q_j}\,\delta q_j + \frac{\partial L}{\partial \dot{q}_j}\,\delta \dot{q}_j \right) dt + (L\Delta t)_{t_0}^{t_1}.$$

Using Lagrange's equations of motion we write this equation as

$$\Delta \int_{t_0}^{t_1} L\,dt = \int_{t_0}^{t_1} \sum_j (\dot{p}_j \delta q_j + p_j \delta \dot{q}_j)\,dt + (L\Delta t)_{t_0}^{t_1}.$$

Since
$$\delta \frac{dq_j}{dt} = \frac{d}{dt}\,\delta q_j.$$

Hence we have

$$\Delta \int_{t_0}^{t_1} L\,dt = \int_{t_0}^{t_1} \sum_j \left(\dot{p}_j \delta q_j + p_j \frac{d}{dt}\,\delta q_j \right) dt + (L\Delta t)_{t_0}^{t_1}.$$

$$\Rightarrow \qquad \Delta \int_{t_0}^{t_1} L\,dt = \int_{t_0}^{t_1} \frac{d}{dt}\left[\sum_j (p_j \delta q_j) \right] dt + (L\Delta t)_{t_0}^{t_1}.$$

Eliminating δ by using equation (178), the above integral becomes

$$\Delta\int_{t_0}^{t_1} L\,dt = \int_{t_0}^{t_1} d\left[\sum_j p_j\left(\Delta - \Delta t \frac{d}{dt}\right)q_j\right]dt + (L\Delta t)_{t_0}^{t_1}.$$

$$\Delta\int_{t_0}^{t_1} L\,dt = \left[\sum_j p_j\Delta q_j\right]_{t_0}^{t_1} - \left[\sum_j p_j\dot{q}_j\Delta t\right]_{t_0}^{t_1} + (L\Delta t)_{t_0}^{t_1}$$

Since in Δ variation, position co-ordinates at the end points are fixed.

$$\Rightarrow \qquad (\Delta q_j)_{t_0}^{t_1} = 0.$$

Consequently above equation reduces to

$$\Delta\int_{t_0}^{t_1} L\,dt = -\left[\sum_j (p_j\dot{q}_j - L)\,\Delta t\right]_{t_0}^{t_1},$$

$$\Delta\int_{t_0}^{t_1} L\,dt = -(H\Delta t)_{t_0}^{t_1}.$$

Substituting this in equation (179), we get $\Delta A = 0$. *i.e.*,

i.e.,
$$\Delta\int_{t_0}^{t_1} \sum_j p_j\dot{q}_j\,dt = 0.$$

This shows that the system moves in space such that Δ-variation of the line integral of twice the kinetic energy is zero. This proves the principle of least action.

Example 37: A system of two degrees of freedom is described by the Hamiltonian

$$H = q_1 p_1 - q_1 q_2 - q_2 p_2 - aq_1^2 + bq_2^2, \; a, b \text{ are const.}$$

Show that (i) $\dfrac{p_1 - aq_1}{q_2}$, (ii) $\dfrac{p_2 - bq_2}{q_1}$, (iii) $q_1 q_2$ (iv) H are constant of motion.

Solution: From the given Hamiltonian of a dynamical system we see that q_1, q_2 are the generalized co-ordinates. The Hamilton's canonical equations of motion become

$$\dot{p}_j = -\frac{\partial H}{\partial q_j} \Rightarrow \dot{p}_1 = 2aq_1 - p_1 \text{ and } \dot{p}_2 = p_2 - 2bq_2, \qquad ...(180)$$

and
$$\dot{q}_j = \frac{\partial H}{\partial p_j} \Rightarrow \dot{q}_1 = q_1 \text{ and } \dot{q}_2 = -q_2. \qquad ...(181)$$

(1) Consider

$$\frac{d}{dt}\left(\frac{p_1 - aq_1}{q_2}\right) = \frac{q_2(\dot{p}_1 - a\dot{q}_1) - (p_1 - aq_1)\,\dot{q}_2}{q_2^2}$$

Using equations (180) and (181), we obtain

$$\frac{d}{dt}\left(\frac{p_1 - aq_1}{q_1}\right) = 0 \Rightarrow \frac{p_1 - aq_1}{q_2} = \text{const.}$$

Similarly, we prove that

$$\frac{p_1 - aq_1}{q_2} = \text{const.}, \quad \frac{p_2 - bq_2}{q_1} = \text{const.}, \quad q_1q_2 = \text{const.}$$

Now to prove the Hamiltonian H is also constant, differentiating the Hamiltonian function with respect to t to get

$$\frac{dH}{dt} = \dot{q}_1 p_1 + q_1 \dot{p}_1 - \dot{q}_2 p_2 - q_2 \dot{p}_2 - 2aq_1\dot{q}_1 + 2bq_2\dot{q}_2.$$

Using equations (180) and (181), we see that $\dfrac{dH}{dt} = 0 \Rightarrow H = \text{const.}$

This shows that H is a constant of motion.

Example 38: For the Lagrangian $L = \dfrac{1}{2Z}\,\dot{q}^2 - \dfrac{1}{2}\left(X - \dfrac{Y^2}{Z}\right)q^2 - \dfrac{Y}{Z}\,q\dot{q},$

where X, Y, Z are time-dependent. Show that the Hamiltonian H is given by

$H = \dfrac{1}{2}\left(Zp^2 + 2Ypq + Xq^2\right)$ and the conjugate momentum $p = \left(\dfrac{\dot{q} - Yq}{Z}\right).$

Solution: We see from the given Lagrangian that q is the generalized co-ordinate and the functions X, Y, Z are time-dependent. Therefore the generalized momentum is given by

$$p = \frac{\partial L}{\partial \dot{q}} = \frac{\dot{q}}{Z} - \frac{Y}{Z}q \Rightarrow p = \frac{\dot{q} - Yq}{Z}. \qquad \text{...(182)}$$

The Hamiltonian function (23) becomes

$$H = p\dot{q} - \frac{1}{2Z}\dot{q}^2 + \frac{1}{2}\left(X - \frac{Y^2}{Z}\right)q^2 + \frac{Y}{Z}\,q\dot{q}. \qquad \text{...(183)}$$

Eliminating $\dot{q}$ from equations (182) and (183), we get

$$H = p(pZ + Yq) - \frac{1}{2Z}(pZ + Yq)^2 + \frac{1}{2}\left(X - \frac{Y^2}{Z}\right)q^2 + \frac{Y}{Z}\,q(pZ + Yq).$$

On simplifying, we get

$$H = \frac{1}{2}(p^2Z + 2Ypq + Xq^2). \qquad \text{...(184)}$$

The equations of motion are

$$\dot{p} = -\frac{\partial H}{\partial q} = -(Yp + Xq), \; \dot{q} = \frac{\partial H}{\partial p} = (Zp + Yp).$$

These equations can be written as

$$\begin{pmatrix} \dot{p} \\ \dot{q} \end{pmatrix} = \begin{pmatrix} -Y & -X \\ Z & Y \end{pmatrix} \begin{pmatrix} p \\ q \end{pmatrix}. \qquad \qquad ...(185)$$

Example 39: A Lagrangian for a particle of charge q moving in the electromagnetic field of force is given by $L = \frac{1}{2} mv^2 + q(\bar{v} \cdot \bar{A}) - q\phi$. Find the Hamiltonian H and the generalized momenta.

Solution: The given Lagrangian of a particle moving in the electromagnetic field can also be given by

$$L = \frac{1}{2} m(\dot{x}^2 + \dot{y}^2 + \dot{z}^2) + q(\dot{x}A_x + \dot{y}A_y + \dot{z}A_z) - q\phi. \qquad ...(186)$$

where ϕ is a scalar potential function of co-ordinates only. We see that x, y, z are the generalized co-ordinates. Hence the corresponding generalized momenta become

$$p_j = \frac{\partial L}{\partial \dot{q}_j} \Rightarrow p_x = m\dot{x} + qA_x, \; p_y = m\dot{y} + qA_y, \; p_z = m\dot{z} + qA_z. \qquad ...(187)$$

Solving these equations for velocity components we get

$$\dot{x} = \frac{1}{m}(p_x - qA_x), \; \dot{y} = \frac{1}{m}(p_y - qA_y), \; \dot{z} = \frac{1}{m}(p_z - qA_z). \qquad ...(188)$$

The Hamiltonian function (23) of the particle is given by

$$H = \dot{x}p_x + \dot{y}p_y + \dot{z}p_z - \frac{1}{2} m(\dot{x}^2 + \dot{y}^2 + \dot{z}^2) - q(\dot{x}A_x + \dot{y}A_y + \dot{z}A_z) + q\phi. \quad ...(189)$$

Eliminating $\dot{x}$, $\dot{y}$, $\dot{z}$ from equation (189) by using equation (188), we get

$$H = \frac{1}{2m}(p_x^2 + p_y^2 + p_z^2) - \frac{q}{m}(p_xA_x + p_yA_y + p_zA_z) +$$

$$+ \frac{1}{2m} q^2(A_x^2 + A_y^2 + A_z^2) + q\phi. \qquad \qquad ...(190)$$

This can be written in vector notions as

$$H = \frac{1}{2m}(\bar{p} - q\bar{A})^2 + q\phi. \qquad \qquad ...(191)$$

This is the required Hamiltonian of the particle moving in the electromagnetic field. The Hamilton's equation of motion $\dot{q}_j = \dfrac{\partial H}{\partial p_j}$ gives the same set of equations (188), while the equation $\dot{p}_j = -\dfrac{\partial H}{\partial q_j}$ gives

$$\dot{p}_x = -\frac{\partial H}{\partial x} = \frac{q}{m}\frac{\partial}{\partial x}\left(p_x A_x + p_y A_y + p_z A_z\right) - \frac{q^2}{2m}\frac{\partial}{\partial x}\left(A_x^2 + A_y^2 + A_z^2\right) - q\frac{\partial \phi}{\partial x}.$$

This can be written as

$$\dot{p}_x = q\frac{\partial}{\partial x}\left(\bar{v}\cdot\bar{A}\right) - q\frac{\partial \phi}{\partial x},$$

Similarly, other two components are given by

$$\dot{p}_y = q\frac{\partial}{\partial y}\left(\bar{v}\cdot\bar{A}\right) - q\frac{\partial \phi}{\partial y},$$

$$\dot{p}_z = q\frac{\partial}{\partial y}\left(\bar{v}\cdot\bar{A}\right) - q\frac{\partial \phi}{\partial z},$$

All these three equations can be put in to the single equation as

$$\dot{p} = -q\nabla\phi + q\nabla\left(\bar{v}\cdot\bar{A}\right). \qquad \text{...(192)}$$

Example 40: A Lagrangian for a particle of unit mass is given by

$$L = \frac{1}{2}\dot{x}^2 - \frac{1}{2}\omega^2 x^2 - \alpha x^3 + \beta x\dot{x}^2.$$

Find the Hamiltonian and hence show that H represents the constant of motion and also it represents total energy. Is the energy of the particle conserved? Justify.

Solution: We see that x is the only generalized co-ordinate. The generalized momentum is given by

$$p_x = \frac{\partial L}{\partial \dot{x}} = \dot{x}(1 + 2\beta x) \Rightarrow \dot{x} = \frac{p_x}{(1 + 2\beta x)}. \qquad \text{...(193)}$$

The Lagrangian equation of motion gives

$$\frac{d}{dt}\left[\dot{x}(1 + 2\beta x)\right] - \left(\beta\dot{x}^2 - \omega^2 x - 3\alpha x^2\right) = 0,$$

$$\Rightarrow \qquad \ddot{x}(1 + 2\beta x) + \beta\dot{x}^2 + \omega^2 x + 3\alpha x^2 = 0. \qquad \text{...(194)}$$

Hence the Hamiltonian function is obtained from the equation (23) as

$$H = \dot{x}p_x - \frac{1}{2}\dot{x}^2 + \frac{1}{2}\omega^2 x^2 + \alpha x^3 - \beta x \dot{x}^2.$$

Using equation (193) eliminating $\dot{x}$ from the above equation we obtain the Hamiltonian function as

$$H = \frac{p_x^2}{2(1 + 2\beta x)} + \frac{1}{2}\omega^2 x^2 + \alpha x^3. \qquad \qquad ...(195)$$

Now to show H represents a constant of motion, differentiation of equation (195) with respect to t gives

$$\frac{dH}{dt} = \frac{p_x \dot{p}_x}{1 + 2\beta x} - \frac{\beta \dot{x} p_x^2}{(1 + 2\beta x)^2} + \omega^2 x \dot{x} + 3\alpha x^2 \dot{x},$$

$$\frac{dH}{dt} = \frac{p_x}{1 + 2\beta x}\left(\dot{p}_x - \frac{\beta \dot{x} p_x}{1 + 2\beta x} + \omega^2 x + 3\alpha x^2\right).$$

Substituting the value of p_x and $\dot{p}_x$ from equation (193), we get

$$\frac{dH}{dt} = \frac{p_x}{(1 + 2\beta x)}\left[\ddot{x}(1 + 2\beta x) + \beta \dot{x}^2 + \omega^2 x + 3\alpha x^2\right].$$

Using the equation of motion (194) we readily get

$$\frac{dH}{dt} = 0 \Rightarrow H = \text{const.}$$

This shows that the Hamiltonian H represents the constant of motion. The total energy of the particle is obtain from the equation

$$E = \dot{x}\,\frac{\partial L}{\partial \dot{x}} - L,$$

$$E = \frac{1}{2}\dot{x}^2(1 + 2\beta x) + \frac{1}{2}\omega^2 x^2 + \alpha x^3. \qquad \qquad ...(196)$$

Using equation (193) we have

$$E = \frac{p_x^2}{2(1 + 2\beta x)} + \frac{1}{2}\omega^2 x^2 + \alpha x^3. \qquad \qquad ...(197)$$

This shows from equations (195) and (197) that

$$E = H = T + V.$$

Thus the Hamiltonian H represents the total energy of the particle. Consequently we have total energy is conserved. The given Lagrange's is independent of time t, hence by the Theorem 8 the constancy of Hamiltonian is confirmed. Also one can separate kinetic energy and potential energy form the Lagrangian function and we have the kinetic energy as the homogeneous quadratic function of generalized velocities. This confirms that H represents total energy of the particle.

Example 41: A particle of mass m moves on the surface of a cone $x^2 + y^2 = z^2 \tan^2 \theta$. Find the Hamiltonian and the equation of motion. Show that (i) H represents the constant of motion and (ii) H also represents the constant of total energy. (iii) Find also the first integral of equation of motion and show that it represents total energy.

Solution: The surface of a cone $x^2 + y^2 = z^2 \tan^2 \theta$ is described by the parametric equations

$$x = r \sin \theta \cos \phi, \ y = r \sin \theta \sin \phi, \ z = r \cos \theta,$$

where θ is a constant semi-vertical angle of the cone. The Lagrangian of the particle is determined (vide., the Example (58) of Chapter 1) and is given by

$$L = \frac{1}{2} m\left(\dot{r}^2 + r^2 \sin^2 \theta \dot{\phi}^2\right) - mgr \cos \theta. \qquad \text{...(198)}$$

We see that ϕ is cyclic co-ordinate in the Lagrangian. Hence the corresponding generalized momentum is conserved. *i.e.*,

$$p_\phi = \frac{\partial L}{\partial \dot{\phi}} = mr^2 \sin^2 \theta \dot{\phi} = \text{const.}$$

$$\Rightarrow \qquad \dot{\phi} = \frac{p_\phi}{mr^2 \sin^2 \theta}. \qquad \text{...(199)}$$

Also we find

$$p_r = \frac{\partial L}{\partial \dot{r}} = m\dot{r} \Rightarrow \dot{r} = \frac{P_r}{m}. \qquad \text{...(200)}$$

Using equation (23) the Hamiltonian function H is given by

$$H = \dot{r} p_r + \dot{\phi} p_\phi - \frac{1}{2} m\left(\dot{r}^2 + r^2 \sin^2 \theta \dot{\phi}^2\right) + mgr \cos \theta,$$

Eliminating $\dot{r}$ and $\dot{\phi}$ we obtain the Hamiltonian H as

$$H = \frac{p_r^2}{2m} + \frac{p_\phi^2}{2mr^2 \sin^2 \theta} + mgr \cos \theta. \qquad \text{...(201)}$$

From the Hamilton's canonical equations of motion we have

$$\dot{p}_r = -\frac{\partial H}{\partial r} = \frac{p_\phi^2}{mr^3 \sin^2 \theta} - mg \cos \theta. \qquad \text{...(202)}$$

From equations (200) and (202) we readily obtain the equation of motion of the particle as

$$\ddot{r} - \frac{p_\phi^2}{m^2 r^3 \sin^2 \theta} + g \cos \theta = 0. \qquad \ldots (203)$$

(i) **Hamiltonian H is a constant of motion:** Differentiation of the equation (201) with respect to t gives

$$\frac{dH}{dt} = \frac{p_r \dot{p}_r}{m} - \frac{p_\phi^2 \, \dot{r}}{mr^3 \sin^2 \theta} + mg\dot{r} \cos \theta.$$

Eliminating $\dot{p}_r$ on using equation (200), we obtain

$$\frac{dH}{dt} = p_r \left[\ddot{r} - \frac{p_\phi^2}{m^2 r^3 \sin^2 \theta} + g \cos \theta \right].$$

Due to equation of motion (203), we find

$$\frac{dH}{dt} = 0 \Rightarrow H = \text{const.}$$

(ii) **Hamitonian H represents a constant of total energy:** Consider therefore

$$E = T + V,$$

$$E = \frac{1}{2} m(\dot{r}^2 + r^2 \sin^2 \theta \dot{\phi}^2) + mgr \cos \theta.$$

Eliminating $\dot{r}$, $\dot{\phi}$ on using equations (199) and (200), we obtain

$$E = \frac{p_r^2}{2m} + \frac{p_\phi^2}{2mr^2 \sin^2 \theta} + mgr \cos \theta = H. \qquad \ldots (204)$$

This shows that the Hamiltonian H represents total energy. Now to find the first integral of motion, multiply equation (203) by $\dot{r}$ we get

$$\ddot{r}\dot{r} - \frac{p_\phi^2 \, \dot{r}}{m^2 r^3 \sin^2 \theta} + g \cos \theta \dot{r} = 0,$$

$$\Rightarrow \qquad \frac{d}{dt}\left(\frac{\dot{r}^2}{2}\right) + \frac{p_\phi^2}{m^2 \sin^2 \theta} \frac{d}{dt}\left(\frac{1}{r^2}\right) + g \cos \theta \frac{d}{dt}(r) = 0.$$

Integrating the above equation we get

$$\frac{\dot{r}^2}{2} + \frac{p_\phi^2}{2m^2 r^2 \sin^2 \theta} + gr \cos \theta = \text{const.} \qquad \ldots (205)$$

Putting the value of $\dot{r}$ from equation (200), we get

$$\frac{p_r^2}{2m} + \frac{p_\phi^2}{2m^2 r^2 \sin^2 \theta} + gr \cos \theta = \text{const.} \qquad \text{...(206)}$$

This shows from equations (204) and (206) that total energy is also constant. Thus H represents both a constant of motion and a constant of total energy.

Remark: As we have pointed out that the parametric equations of a surface are not unique. Hence one can represent cone $x^2 + y^2 = z^2 \tan^2 \theta$, where θ is the semi vertical angle of the cone, parametrically in the following form too

$$x = r \cos \phi, \ y = r \sin \phi, \ z = r \cot \theta.$$

However, in this case the Hamiltonian function of the particle can be obtained in the form

$$H = \frac{p_r^2 \sin^2 \theta}{2m} + \frac{p_\phi^2}{2mr^2} + mgr \cot \theta.$$

and the equation motion is

$$\ddot{r} - r \sin^2 \theta \dot{\phi}^2 + g \cos \theta \sin \theta = 0.$$

Similarly, one can also prove that the Hamiltonian H represents both the constant of motion and the constant of total energy.

Example 42: Find the Hamiltonian and the equation of motion for a particle of mass m constrained to move on the surface obtained by revolving the parabola $y^2 = x$ about x-axis.

Solution: The surface obtained by revolving the parabola $y^2 = x$ about x-axis is given by

$$y^2 + z^2 = x. \qquad \text{...(207)}$$

The parametric equations representing this surface are

$$x = u^2, \ y = u \sin v, \ z = u \cos v, \qquad \text{...(208)}$$

where from equation (208) we find

$$\dot{x} = 2u\dot{u}, \ \dot{y} = \dot{u} \sin v + u \cos v \dot{v}, \ \dot{z} = \dot{u} \cos v - u \sin v \dot{v}.$$

Hence the kinetic energy of the particle $T = \dfrac{1}{2} m(\dot{x}^2 + \dot{y}^2 + \dot{z}^2)$ becomes

$$T = \frac{1}{2} m\left[u^2 \dot{v}^2 + (1 + 4u^2)\, \dot{u}^2\right]. \qquad \text{...(209)}$$

The potential energy of the particle is given by

$$V = mgy = mgu \sin v \qquad \text{...(210)}$$

Hence the Lagrangian of the function becomes

$$L = \frac{1}{2} m\left[u^2\dot{v}^2 + \left(1 + 4u^2\right)\dot{u}^2\right] - mgu \sin v. \qquad \ldots(211)$$

The generalized momenta corresponding to the generalized co-ordinates u and v are given by

$$p_u = \frac{\partial L}{\partial \dot{u}} = m(1 + 4u^2)\,\dot{u}, \qquad \ldots(212)$$

and

$$p_v = \frac{\partial L}{\partial \dot{v}} = mu^2\dot{v}. \qquad \ldots(213)$$

Hence the Hamiltonian function of the particle becomes

$$H = \frac{p_u^2}{2m(1 + 4u^2)} + \frac{p_v^2}{2mu^2} + mgu \sin v. \qquad \ldots(214)$$

Hamilton's canonical equations of motion (27) become

$$\dot{p}_u = \frac{4up_u^2}{(1 + 4u^2)^2} + \frac{p_v^2}{mu^3} - mg \sin v \text{ and } \dot{u} = \frac{\partial H}{\partial p_u} = \frac{p_u}{m(1 + 4u^2)}.$$

Eliminating p_u between these equations we obtain

$$(1 + 4u^2)\,\ddot{u} + u(4\dot{u}^2 - \dot{v}^2) + g \sin v = 0 \qquad \ldots(215)$$

$$\dot{p}_v = -\frac{\partial H}{\partial v} = -mgu \cos v,$$

$$\dot{v} = \frac{\partial H}{\partial p_v} = \frac{p_v}{mu^2}.$$

Eliminating p_v between these equations we obtain

$$\ddot{v} + \frac{2}{u}\,\dot{u}\dot{v} + \frac{g}{u} \cos v = 0. \qquad \ldots(216)$$

Equations (215) and (216) are the required equations of motion for the particle moving on the surface obtained by revolving the parabola $y^2 = x$ about x-axis.

Example 43: A bead of mass m slides on a frictionless wire in the form of parabola under the influence of gravity, which rotates about the z-axis with constant angular velocity ω. Find the Hamiltonian H of the system. Is H represents total energy? Is $H = E$?

Solution: Let $P(x, y, z)$ be the position coordinates of a particle sliding along a wire in the form of a parabola which is rotating about z-axis with angular velocity ω. The equation of the parabola is given by

$$z = ar^2, \qquad \qquad \text{...(217)}$$

where

$$x = r \cos \omega t, \; y = r \sin \omega t. \qquad \qquad \text{...(218)}$$

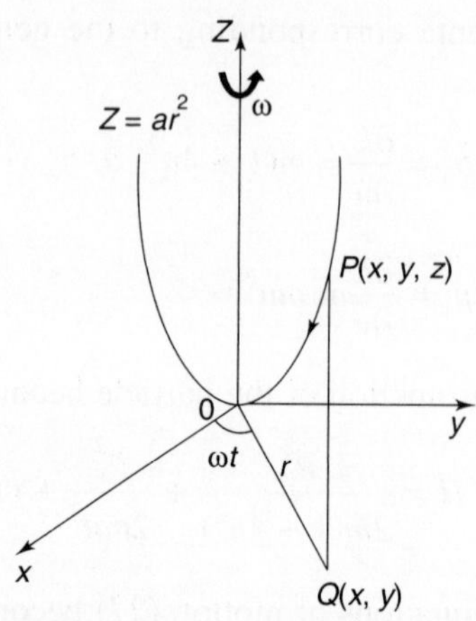

We see that x, y are related by the equation of constraint $y = x \tan \omega t$. Also z and r are related by another constraint (217). Hence the system has only one degree of freedom and hence one generalized coordinate r. The kinetic energy of the system $T = \frac{1}{2} m\left(\dot{x}^2 + \dot{y}^2 + \dot{z}^2\right)$ becomes $T = \frac{1}{2} m\left(\dot{r}^2 + r^2 w^2 + \dot{z}^2\right)$, where from equation (217) we have $\dot{z} = 2ar\dot{r}$. Hence kinetic energy becomes

$$T = \frac{1}{2} m\left[\dot{r}^2\left(1 + 4a^2r^2\right) + \omega^2 r^2\right]. \qquad \qquad \text{...(219)}$$

The potential energy of the system is given by $V = mgar^2$. Hence the Lagrangian of the system becomes

$$L = \frac{1}{2} m\left[\dot{r}^2\left(1 + 4a^2r^2\right) + \omega^2 r^2\right] - mgar^2. \qquad \qquad \text{...(220)}$$

$$p_r = \frac{\partial L}{\partial \dot{r}} = m\dot{r}(1 + 4a^2r^2) \Rightarrow \dot{r} = \frac{p_r}{m\left(1 + 4a^2r^2\right)}.$$

The equation of motion is obtained as

$$m\left(1 + 4a^2r^2\right)\ddot{r} + 4ma^2r\dot{r}^2 - m\omega^2 r + 2magr = 0. \qquad \qquad \text{...(221)}$$

Hence the Hamiltonian of the system $H = \dot{r}p_r - L$ yields

$$H = \frac{p_r^2}{2m\left(1 + 4a^2r^2\right)} - \frac{m\omega^2 r^2}{2} + mgar^2. \qquad \qquad \text{...(222)}$$

We also find the total energy of the system $T + V$ as

$$T + V = \frac{p_r^2}{2m(1 + 4a^2r^2)} + \frac{m\omega^2 r^2}{2} + mgar^2. \qquad \ldots(223)$$

We see that $H \neq T + V$. This conclusion can also be derived from the Theorem 8. We see that the system is conservative and rheonomic and the kinetic energy T is not a homogeneous quadratic function of generalized velocity, hence by Theorem 8 the Hamiltonian H does not represent total energy. However, the Lagrangian L does not involve time t, the Hamiltonian is conserved. This can easily be established by proving that $\dfrac{dH}{dt} = 0$.

Example 44: Suppose a mechanical system has the Lagrangian $L = \dfrac{1}{4}\dot{q}^2 - \dfrac{1}{9}q^2$. Then show that the path in the Hamiltonian phase space (q, p) is an ellipse.

Solution: A system has only one degree of freedom and hence one generalized coordinate q. Hence the Hamiltonian of the system is obtained as

$$H = p^2 + \frac{q^2}{9}. \qquad \ldots(224)$$

The Hamilton's canonical equations of motion yield the equation of motion in the form

$$\ddot{q} + \frac{4}{9}q = 0. \qquad \ldots(225)$$

The solution of this second order differential equation is given by

$$q = c_1 \cos\left(\frac{2}{3}t\right) + c_2 \sin\left(\frac{2}{3}t\right). \qquad \ldots(226)$$

It is obvious to see that Hamiltonian H is the total energy of the system. Hence from equation (224) we have

$$p^2 + \frac{q^2}{9} = E.$$

This can be written as

$$\frac{q^2}{(\sqrt{9E})^2} + \frac{p^2}{(\sqrt{E})^2} = 1. \qquad \ldots(227)$$

This shows that in the particle's phase space *i.e.*, in the (q, p) plane, the path of the system is an ellipse.

Exercise

1. The Lagrangian of an anharmonic oscillator of unit mass is

$$L = \frac{1}{2}\dot{x}^2 - \frac{1}{2}\omega^2 x^2 - \alpha x^3 + \beta x\dot{x},$$

   α, β are constants. Find the Hamiltonian and the equation of motion. Show also that (i) H is a constant of motion and (ii) $H \neq T + V$.

   Ans: $H = \frac{1}{2}(p_x - \beta x)^2 + \frac{1}{2}\omega^2 x^2 + \alpha x^3.$

   Equation of motion $\ddot{x} + \omega^2 x + 3\alpha x^2 = 0.$

2. Find the Hamiltonian and the equations of motion for a particle constrained to move on the surface obtained by revolving the line $x = z$ about z-axis. Does it represent the constant of motion and the constant of total energy?

   [**Hint:** Surface of revolution is a cone $x^2 + y^2 = z^2$,]

   $$H = \frac{p_r^2}{4m} + \frac{p_\phi^2}{2mr^2} + mgr.$$

   $$\ddot{r} - \frac{p_\phi^2}{2m^2 r^3} + \frac{g}{2} = 0, \ p_\phi = mr^2\dot{\phi} - a \text{ constant of motion.}$$

3. A sphere of radius 'a' and mass m rests on the top of a fixed trough sphere of radius 'b'. The first sphere is slightly displaced so that it rolls without slipping. Obtain the Hamiltonian of the system and hence the equation of motion. Also prove that H represents a constant of motion and also total energy.

   Ans: $H = \frac{7}{10}m(a + b)^2\,\dot{\phi}^2 + mg(a + b)\cos\phi.$

4. A particle is constrained to move on the plane curve $xy = c$, c is a constant, under gravity. Obtain the Hamiltonian H and the equations of motion. Prove that the Hamiltonian H represents the constant of motion and total energy.

   Ans: Refer Example (17) of Chapter 1 for the Lagrangian L given in equation (205). The Hamiltonian H becomes

   $$H = \frac{p_x^2}{2m\left(1 + \dfrac{c^2}{x^4}\right)} + \frac{mgc}{x}.$$

   The Hamilton's canonical equation of motion

$$m\ddot{x}\left(1 + \frac{c^2}{x^4}\right) - 2\,\frac{c^2 m}{x^5}\,\dot{x}^2 - \frac{mgc}{x^2} = 0.$$

5. A particle moves on the surface characterized by $x = r\cos\phi$, $y = r\sin\phi$, $z = r\cot\theta$. Find the Hamiltonian H and prove that it represents the constant of motion and also the constant of total energy.

Ans:
$$H = \frac{p_r^2 \sin^2\theta}{2m} + \frac{p_\phi^2}{2mr^2} + mgr\cot\theta.$$

The equation of motion is $\ddot{r} - r\sin^2\theta\,\dot{\phi}^2 + g\cos\theta\sin\theta = 0$.

6. Find the Hamiltonian and the Hamilton's canonical equations of motion for the Lagrangian given by

$$L(r,\dot{r},\theta,\dot{\phi}) = \frac{1}{2}m(\dot{r}^2 + r^2\dot{q}^2) + mgr\cos\theta - \frac{1}{2}k\,(r - r_0)^2, \text{ where } k, m, g,$$
r_0 are constants.

Ans:
$$H = \frac{p_r^2}{2m} + \frac{p_\theta^2}{2mr^2} - mgr\cos\theta + \frac{1}{2}k\,(r - r_0)^2$$

Equations of motion: $m\ddot{r} - mr\dot{\theta}^2 - mg\cos\theta + k(r - r_0) = 0,$

$$\ddot{\theta} + \frac{2}{r}\dot{r}\dot{\theta} + \frac{g}{r}\sin\theta = 0.$$

7. Describe by using Lagrange's procedure the motion of a particle of mass m moving near the surface of the Earth under the Earth's gravitational field.

Ans: $L = \dfrac{1}{2}m(\dot{x}^2 + \dot{y}^2 + \dot{z}^2) - mgz$, and equations of motion are

$$\ddot{x} = 0,\ \ddot{y} = 0,\ \ddot{z} = -g.$$

8. A particle of mass m is fired in to the air with initial velocity u at an angle θ with the ground. Neglecting all forces except gravity and the resistance of the air assumed proportional to the velocity, find the Lagrangian L and the equations of motion. Find also the Hamiltonian H and discuss its physical interpretation.

Ans: Lagrangian $L = \dfrac{1}{2}m(\dot{x}^2 + \dot{y}^2) - mgy$, $H = \dfrac{1}{2m}(p_x^2 + p_y^2) + mgy$,

Equations of motion: $m\ddot{x} + k\dot{x} = 0,\ m\ddot{y} + k\dot{y} + mg = 0$.

Physical interpretation of H : H represents total energy but it is not conserved though Lagrangian is independent of time t.

Two Body Central Force Motion

INTRODUCTION

In this chapter we shall discuss the motion of a particle under a particular type of force called the central force. The motion of this type is of particular interest, because this type of motion is closely related to the mechanics of nature. For example, the motion of planets around the Sun, the motion of satellites around the Earth, the motion of two charged particles with respect to each other.

Some Definitions

Central force

A force is said to be a central force if it is always directed towards a fixed point or away from the fixed point. It acts along the position vector $\bar{r}$ of the particle and is a function of distance r of the particle from the fixed point. The fixed point is another massive particle.

Mathematically, the central force is expressed as

$$\overline{F} = f(r)\,\hat{r}, \qquad\qquad ...(1)$$

where $\hat{r}$ is the unit vector along the direction of the position vector $\bar{r}$ and $f(r)$ is some function of the distance r. If this function $f(r) < 0$ *i.e.*, the force is negative, then the central force is called attractive and if $f(r) > 0$, the central force is called repulsive e.g. The force between two interacting bodies is a central force. Two bodies on each other exert such forces. The motion of the Earth around the Sun, the motion of the Moon around the Earth are some examples of central force motion.

The trajectory followed by the particle under central force is called central orbit. We have proved in the Chapter 1 that the central force is conservative; hence the potential energy of the particle is a function of position r only. *i.e.*, $V = V(r)$.

In the central force field, the position vector $\bar{r}$ of a particle is always parallel to the direction of force $\overline{F}$. Hence $\bar{r} \times \overline{F} = 0$. This implies $\dfrac{dL}{dt} = N = \bar{r} \times \overline{F} = 0$.

This shows that angular momentum of the particle in a central force field is conserved.

Areal velocity   It is the area swept out by the radius vector per unit time. To find the expression for the areal velocity, let C be the path followed by the particle under the action of a force directed towards a fixed point O. This fixed point O is called the center of force.

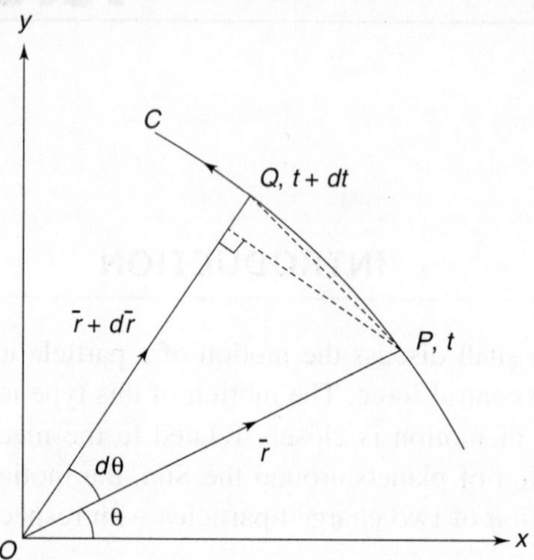

Let $\bar{r}$ be the position vector of the particle at some instant t and $\bar{r} + d\bar{r}$ be the position vector of the particle at time $t + dt$. If the radius vector rotates through an angle $d\theta$ in time dt, then the area swept out by the radius vector $\bar{r}$ in time dt is given by dA = *area of the sector OPQ*. Since in the limiting case as $dt \to 0$ we have Arc PQ = Chord PQ. Hence, in the limiting case, we have

$$dA = \text{area of } \triangle OPQ$$

$$= \frac{1}{2}\, OQ \cdot PR$$

$$= \frac{1}{2}(\bar{r} + d\bar{r}) \cdot \bar{r}\, d\theta.$$

Since $d\bar{r}$, $d\theta$ terms are very small, the higher order terms than first in $d\bar{r}$, $d\theta$ are neglected hence we have

$$dA = \frac{1}{2}\, r^2 d\theta.$$

Thus the area per unit time becomes

$$\dot{A} = \frac{1}{2}\, r^2 \dot{\theta}. \qquad \qquad \ldots(2)$$

The quantity $\dot{A} = \frac{1}{2}\, r^2 \dot{\theta}$ is called the areal velocity of the particle.

Apse   An apse is a point on a central orbit at which the radius vector drawn from the center of force to the orbit has a minimum or maximum value. The distance of an apse from the center of the force is called apsidal distance.

Perihelion   It is the point on the orbit nearest the Sun, when the Sun is at the center of attraction. This occurs when $\theta - \theta_0 = 0$, where θ is the inclination of the radius vector of the particle with the x-axis through the center of the force and θ_0 is the inclination of axis of the conic with the x-axis.

Aphelion   It is the point on a planetary orbit farthest from the center of the force. This occurs when $\theta - \theta_0 = 180°$.

Note: If the orbit of the planet is an ellipse, then the semi-major axis of the ellipse is half of the sum of the two apsidal distances.

Two body problems   We consider the motion of two particles under the influence of one on the other. In this case each one of which is a source of central force and the potential energy of the system is a function of their separation. We show below in the following theorem that the motion of two body problem can always be reduced to one body problem.

Theorem 1   Set up the Lagrangian for two bodies moving under central force about their center of mass and show that it can be reduced to an equivalent one body problem.

Proof: Consider the motion of two particles of mass m_1 and m_2 under the action of the mutual interaction between them. If the mass of the second particle is comparable to that of the mass of the first, it is no longer remains fixed during the interaction, rather the two particles move about their common center of mass. The force of attraction between the particles is conservative, hence the potential energy of the system is just function of their separation r. Therefore we have,

$$V = V(|r_1 - r_2|) = V(\bar{r}),$$

where r_1, r_2 are the position vectors of particles of masses m_1 and m_2 with respect to the fixed point O respectively. Let $\bar{R}$ be the position vector of the center of mass with respect to the fixed point O. Let also r_1' and r_2' be the positions of m_1 and m_2 with respect to the center of mass. The system of particles has six degrees of freedom and hence six generalized co-ordinates. Now the motion of the system of two particles m_1 and m_2 can be thought of as made up of motion of the center of mass and motion about the center of mass. Hence we choose six generalized co-ordinates as three components of the position vector of the center of mass $\bar{R}$ and three components of the difference vector $\bar{r} = |\bar{r}_1 - \bar{r}_2|$. Also we have from the figure.

$$\bar{r} = |\bar{r}_1 - \bar{r}_2| = |\bar{r}_1' - \bar{r}_2'|. \qquad \qquad \text{...(3)}$$

Naturally the Lagrangian function of the system will be of the form

$$L = T(\dot{\vec{R}}, \dot{\vec{r}}) - V(r), \qquad \qquad \text{...(4)}$$

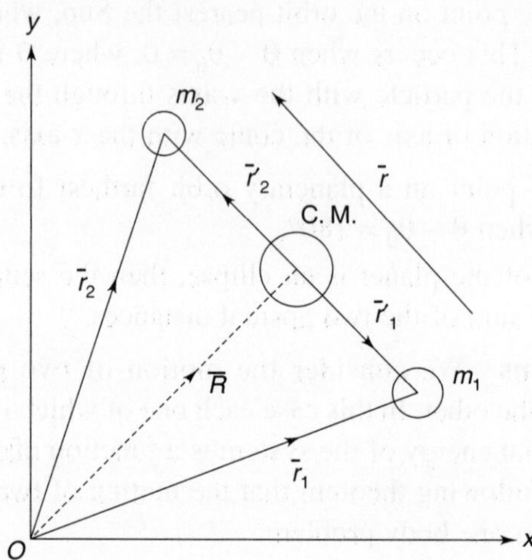

where T is the kinetic energy of the system of two masses and can be written as the sum of the kinetic energies of the motion of the center of mass and the kinetic energy of the particles about the center of mass. Thus we write

$$T = \frac{1}{2}(m_1 + m_2)\,\dot{\vec{R}}^2 + \frac{1}{2}m_1\dot{\vec{r}}_1'^2 + \frac{1}{2}m_2\dot{\vec{r}}_2'^2. \qquad \text{...(5)}$$

For equilibrium of the system, we have

$$m_1|\vec{r}_1'| = m_2|\vec{r}_2'|,$$

$$\Rightarrow \qquad m_1\vec{r}_1' = -m_2\vec{r}_2'.$$

The negative sign indicates that the direction of $\vec{r}_2'$ is opposite to that of $\vec{r}_1'$. This implies that $\vec{r}_1' = -\left(\dfrac{m_2}{m_1}\right)\vec{r}_2'$. From the equation (3) we write

$$\vec{r}_2' = \left(\frac{m_1}{m_1+m_2}\right)\vec{r}.$$

Consequently, from the equation (3) we have $\vec{r}_1' = -\left(\dfrac{m_1}{m_1 + m_2}\right)\vec{r}$. Substituting this in the equation (5) we obtain the expression for kinetic energy of the system as

$$T = \frac{1}{2}(m_1 + m_2)\,\dot{\vec{R}}^2 + \frac{1}{2}\left(\frac{m_1 m_2}{m_1 + m_2}\right)\dot{\vec{r}}^2. \qquad \text{...(6)}$$

Hence the Lagrangian of the system becomes

$$L = \frac{1}{2}(m_1 + m_2)\,\dot{\overline{R}}^2 + \frac{1}{2}\left(\frac{m_1 m_2}{m_1 + m_2}\right)\dot{\overline{r}}^2 - V(r). \qquad ...(7)$$

i.e.,
$$L = \frac{1}{2}\,M\dot{\overline{R}}^2 + \frac{1}{2}\,\mu\dot{\overline{r}}^2 - V(r), \qquad ...(8)$$

where the total mass of the system is $M = m_1 + m_2$, and $\mu = \dfrac{m_1 m_2}{m_1 + m_2}$ is the reduced mass of the system. We notice from the equation (8) that the three components of $\overline{R}$ do not appear in the Lagrangian L and consequently the three components of $\overline{R}$ are cyclic. This implies that $\dfrac{\partial L}{\partial R} = 0$. Also from equation (8), we have $\dfrac{\partial L}{\partial \dot{R}} = M\dot{\overline{R}}$. Thus from Lagrange's equation of motion for $\overline{R}$, we have

$$\dot{\overline{R}} = \text{const.} \qquad ...(9)$$

This shows that the centre of mass of the system is either moving with constant velocity or is at rest. Now the r-Lagrange's equation of motion gives

$$\mu\ddot{\overline{r}} = f(r), \qquad ...(10)$$

where
$$f(r) = -\frac{\partial V}{\partial r}.$$

Equation (10) is the equation of motion for the system. This equation does not contain $\overline{R}$ and $\dot{\overline{R}}$. Hence the equation of motion does not involve the co-ordinates of the center of mass. Consequently, we ignore the first term from the Lagrangian (8) and we are left with the term in the Lagrangian as

$$L = \frac{1}{2}\,\mu\dot{\overline{r}}^2 - V(r). \qquad ...(11)$$

This is the Lagrangian for the system of single particle of mass μ at a distance $\overline{r}$ from a fixed centre of force. Thus the problem of the central force motion of two bodies about their centre of mass can always be reduced to an equivalent one body problem with mass μ and position vector $\overline{r}$.

Theorem 2   Prove that central force motion is always motion in a plane.

Proof: In the central force field, the angular momentum is always constant during motion. The constancy of angular momentum implies that both radius vector and the linear momentum vector must lie in the plane perpendicular to the angular momentum. That is the motion of the particle is such that the radius vector of the particle is always perpendicular to the fixed angular momentum. Hence the motion of a particle in central field of force is confined to a plane.

Equation of motion and the first integral

Theorem 3 A particle of mass m is moving under the inverse square law of attractive force. Set up the Lagrangian and equation of motion. Obtain the first integral of motion. Show also that it is constant of total energy.

Proof: Consider a single particle of mass m moving in a central force field. The force is always directed along the radius vector $\bar{r}$, so that the potential energy V is the function of r only. The motion of the particle under central force is a motion in a plane, hence the system has two degree of freedom and hence two generalized coordinates. Let (r, θ) be the generalized coordinates of the particle, then its kinetic energy and potential energy are given by

$$T = \frac{1}{2}\, m(\dot{r}^2 + r^2\dot{\theta}^2),\ V = V(r).$$

Hence the Lagrangian L of the particle becomes

$$L = \frac{1}{2}\, m(\dot{r}^2 + r^2\dot{\theta}^2) - V(r). \qquad \text{...(12)}$$

We see that θ does not appear in L, hence θ is cyclic co-ordinate. This gives $\dfrac{\partial L}{\partial \theta} = 0$. Hence the Lagrange's equation of motion gives

$$\frac{d}{dt}\left(\frac{\partial L}{\partial \dot{\theta}}\right) = 0 \Rightarrow \frac{d}{dt}(mr^2\dot{\theta}) = 0.$$

or $$mr^2\dot{\theta} = l\text{-a constant.} \qquad \text{...(13)}$$

This proves that the angular momentum of the particle moving in conservative field of force is conserved. We may also write equation (13) as

$$\frac{1}{2}\, r^2\dot{\theta} = \text{const.} \qquad \text{...(14)}$$

This shows that areal velocity (area swept out radius vector per unit time) is constant. This is known as Kepler's second law of planetary motion. Now to find the equation of motion of the particle, we write from r-Lagrange's equation of motion

$$m\ddot{r} - mr\dot{\theta}^2 = -\frac{\partial V}{\partial r}$$

or $$m\ddot{r} - mr\dot{\theta}^2 = f(r), \qquad \text{...(15)}$$

where $$f(r) = -\frac{\partial V}{\partial r} \qquad \text{...(16)}$$

is the law of attractive force. Eliminating $\dot{\theta}$ between equations (13) and (15) we obtain the equation of motion of the particle moving in the inverse square law of attractive force in the form

$$m\ddot{r} - \frac{l^2}{mr^3} = f(r).\qquad\ldots(17)$$

Conservation of energy and first integral: We write equation of motion (17) in the form

$$m\ddot{r} = -\frac{\partial}{\partial r}\left(\frac{1}{2}\frac{l^2}{mr^2} + V\right).$$

Multiply both the sides of above equation by $\dot{r}$ to get

$$m\ddot{r}\dot{r} = -\frac{\partial}{\partial r}\left(\frac{1}{2}\frac{l^2}{mr^2} + V\right)\frac{dr}{dt}$$

$$\Rightarrow\qquad \frac{d}{dt}\left(\frac{1}{2}m\dot{r}^2\right) = -\frac{d}{dt}\left(\frac{1}{2}\frac{l^2}{mr^2} + V\right)$$

$$\Rightarrow\qquad \frac{d}{dt}\left(\frac{1}{2}m\dot{r}^2 + \frac{1}{2}\frac{l^2}{mr^2} + V\right) = 0.$$

Integrating we obtain the first integral of equation of motion as

$$\frac{1}{2}m\dot{r}^2 + \frac{1}{2}\frac{l^2}{mr^2} + V = \text{const.}\qquad\ldots(18)$$

Now to show the total energy is conserved, we have

$$E = \frac{1}{2}m(\dot{r}^2 + r^2\dot{\theta}^2) + V(r).\qquad\ldots(19)$$

Eliminating $\dot{\theta}$ between equations (13) and (19), we get

$$E = \frac{1}{2}m\dot{r}^2 + \frac{1}{2}\frac{l^2}{mr^2} + V(r).\qquad\ldots(20)$$

From equations (18) and (20), we see that

$$E = \frac{1}{2}m\dot{r}^2 + \frac{1}{2}\frac{l^2}{mr^2} + V(r) = \text{const.}\qquad\ldots(21)$$

This represents the conservation of energy directly from the equation of motion.

The Kepler's laws of planetary motion

Aristotle (340 B.C.) described that the Earth is at the centre and all planets including the Sun revolve around it. The Aristotle theory is known as geocentric theory. This theory was first challenged by *N.* Copernicus (1473-1543), and proposed a rival framework. According to him the Sun is fixed in space and all planets including the Earth revolve around the Sun in a circular orbit. His theory was known as "heliocentric theory", helio in Greek means the Sun. The next improvement was made by Kepler (1571-1630). Only on the basis of observations he described how planets move and discovered the exact laws behind the movements of the planets. However, he did not give any explanation for the planetary motion. He announced three laws of planetary motion 60 years before Newton's laws (1686) of motion were formulated. Latter these laws of planetary motion stand as an empirical background to Newton's dynamical theory. These laws are

Kepler's first law   The planet moves around the Sun in an elliptic orbit with the Sun at one of the foci of the orbit. *i.e.*, Orbit of the planet is an ellipse.

Kepler's second law   The radius vector from the Sun to the planet sweeps out equal area in equal interval of time. *i.e.*, Areal velocity is constant.

Kepler's third law   The square of time taken by the planet to complete one orbit is directly proportional to the cube of the semi-major axis of the orbit.

These laws describe how planet moves, where as Newton's law of gravitation supplied the reason why the planets move according to Kepler's laws.

Deduction of Kepler's laws

Theorem 4   Find the orbit described by the planet under the inverse square law of attractive force. Classify the orbit on the basis of total energy.

Proof: Consider a planet of mass m moving under the central field of force. The Lagrangian of the planet is defined by the equation (12). We see that θ is cyclic in L. This implies the corresponding generalized (angular) momentum is conserved.

$$\Rightarrow \qquad p_\theta = \frac{\partial L}{\partial \dot\theta} = mr^2\dot\theta = l\text{-a constant.} \qquad \qquad ...(22)$$

From energy conservation theorem the total energy is given by

$$E = \frac{1}{2}\,m\dot{r}^2 + \frac{1}{2}\frac{l^2}{mr^2} + V(r).$$

Solving this equation for $\dot{r}$ we get

$$\dot{r} = \sqrt{\frac{2}{m}\left[E - V - \frac{l^2}{2mr^2}\right]},$$

$$\Rightarrow \qquad dt = \frac{dr}{\sqrt{\dfrac{2}{m}\left[E - V - \dfrac{l^2}{2mr^2}\right]}}. \qquad \qquad ...(23)$$

This equation gives r as a function of time t. From equation (22), we have

$$l\,dt = mr^2 d\theta. \qquad \qquad ...(24)$$

Eliminating t between equations (23) and (24), we get

$$d\theta = \frac{l}{mr^2} \cdot \frac{dr}{\sqrt{\dfrac{2}{m}\left(E - V - \dfrac{l^2}{2mr^2}\right)}} \cdot$$

On integrating the above equation between the limits θ_0 to θ we get

$$\int_{\theta_0}^{\theta} d\theta = \int_{r_0}^{r} \frac{l}{mr^2}\left[\frac{2}{m}\left(E - V - \frac{l^2}{2mr^2}\right)\right]^{-\frac{1}{2}} dr$$

where
$$\theta = \theta_0 \text{ and } r = r_0 \text{ at } t = 0.$$

$$\Rightarrow \qquad \theta = \theta_0 + \int_{r_0}^{r} \frac{1}{r^2}\left[\frac{2mE}{l^2} - \frac{2mV}{l^2} - \frac{1}{r^2}\right]^{-\frac{1}{2}} dr$$

Change the variable r to u by putting $u = \dfrac{1}{r} \Rightarrow du = -\dfrac{1}{r^2}\,dr$. Hence the above integral becomes

$$\theta = \theta_0 - \int_{u_0}^{u}\left[\frac{2mE}{l^2} - \frac{2mV}{l^2} - u^2\right]^{-\frac{1}{2}} du. \qquad \qquad ...(25)$$

Equations (23) and (25) locate the position of the planet on its path at any instant t. Since the central force varies inversely square of the distance, we have

$$f(r) = -\frac{K}{r^2}, \quad \text{for} \quad K = -Gm_1 m_2.$$

This force is conservative, hence we have $V = -\dfrac{K}{r} = -Ku$. Hence equation (25) becomes

$$\theta = \theta_0 - \int\left[\frac{2mE}{l^2} + \frac{2mKu}{l^2} - u^2\right]^{-\frac{1}{2}} du, \qquad \qquad ...(26)$$

where the integration is taken as indefinite. The integral on the right hand side of equation (26) is of the standard form $\int \dfrac{dx}{\sqrt{c + bx + ax^2}}$ whose solution is given by $\dfrac{1}{\sqrt{-a}} \cos^{-1}\left[\dfrac{-(b + 2ax)}{\sqrt{b^2 - 4ac}} \right]$. Using this we obtain the solution of (26) in the form

$$\theta = \theta_0 - \cos^{-1}\left\{ \frac{-\dfrac{2mK}{l^2} + 2u}{\dfrac{2mK}{l^2}\sqrt{1 + \dfrac{2El^2}{mK^2}}} \right\}$$

$$\theta = \theta_0 - \cos^{-1}\left\{ \frac{\dfrac{l^2 u}{mK} - 1}{\sqrt{1 + \dfrac{2El^2}{mK^2}}} \right\}$$

$$\sqrt{1 + \frac{2El^2}{mK^2}} \cos(\theta - \theta_0) = \frac{l^2 u}{mK} - 1$$

or
$$u = \frac{1}{r} = \frac{mK}{l^2}\left[1 + \sqrt{1 + \frac{2El^2}{mK^2}} \cdot \cos(\theta - \theta_0) \right]. \qquad ...(27)$$

This is the equation of conic of the form $\dfrac{1}{r} = c[1 + e \cos(\theta - \theta_0)]$, where the eccentricity of the conic is given by

$$e = \sqrt{1 + \frac{2El^2}{mK^2}} \quad \text{and} \quad c = \frac{mK}{l^2}. \qquad ...(28)$$

Thus the equation of the path of the planet with respect to the Sun is always a conic. The nature of the conic depends upon the value of the eccentricity e.

1. If $e > 1 \Rightarrow \sqrt{1 + \dfrac{2El^2}{mK^2}} > 1 \Rightarrow E > 0 \Rightarrow$ the conic (28) is hyperbola.

2. If $e = 1 \Rightarrow \sqrt{1 + \dfrac{2El^2}{mK^2}} = 1 \Rightarrow E = 0 \Rightarrow$ the conic (28) is parabola.

3. If $e < 1 \Rightarrow \sqrt{1 + \dfrac{2El^2}{mK^2}} < 1 \Rightarrow E < 0 \Rightarrow$ the conic (28) is an ellipse.

4. If $e = 0 \Rightarrow \sqrt{1 + \dfrac{2El^2}{mK^2}} = 0 \Rightarrow E = -\dfrac{mK^2}{2l^2} \Rightarrow$ the conic (28) is a circle.

Theorem 5 Prove Kepler's second law of planetary motion.

Proof: In the case of a planet of mass m moving in the conservative central force, the Lagrangian L is given by the equation (12). From the conservation of angular momentum we have $mr^2\dot{\theta} = l$, - a constant. By rearranging the equation we write

$$\frac{1}{2} r^2 \dot{\theta} = \text{const.} \qquad \qquad ...(29)$$

This shows that areal velocity (area swept out radius vector per unit time) is constant. This is the Kepler's second law of planetary motion. This shows that "the conservation of angular momentum is equivalent to saying that the areal velocity is constant".

Theorem 6 Prove the Kepler's third law of planetary motion.

Proof: Let a planet of mass m be moving along an elliptical orbit. Let dA be an infinitesimal area swept out by the radius vector in infinite time. Then the total area of the orbit is given by

$$A = \int_0^T dA = \int_0^T \dot{A}\, dt, \quad \dot{A} = \frac{dA}{dt} \qquad \qquad ... (30)$$

where T denotes the periodic time of the planet and $\dot{A}$ is the area swept out by the radius vector per unit time and is given by $\dot{A} = \dfrac{1}{2} r^2 \dot{\theta}$. Further, the conservation of angular momentum of the planet gives $\dot{\theta} = \dfrac{l}{mr^2}\cdot$ Hence the total area of the orbit becomes

$$A = \int_0^T \frac{l}{2m}\, dt,$$

$$\Rightarrow \qquad A = \frac{l}{2m}\, T. \qquad \qquad ...(31)$$

Since the orbit of the planet is an ellipse hence the area of the ellipse is given by

$$A = \pi ab. \qquad \qquad ...(32)$$

Consequently, the equation (31) gives

$$T = \frac{2\pi mab}{l}, \qquad \qquad ...(33)$$

where a and b are the semi-major and semi-minor axes of the ellipse respectively. In case of ellipse the relation between them is $b = a\sqrt{1 - e^2}$, and the eccentricity of the ellipse is given by the equation (28). Using this equation we get the value of the semi-minor axis as

$$b = a\sqrt{-\frac{2El^2}{mK^2}}. \qquad \qquad \dots(34)$$

The orbit of the planet is an ellipse given in equation (27). The semi-major axis of an ellipse is half the sum of the two apsidal distances r_1 and r_2 which occur respectively at $\theta - \theta_0 = 0$ and $\theta - \theta_0 = 180°$. Thus $a = \frac{1}{2}(r_1 + r_2)$, where from equation (27) we obtain the values of r_1 and r_2 as $r_1 = \frac{1}{c(1 + e)}$, $r_2 = \frac{1}{c(1 - e)}$.

Hence $a = \frac{1}{c(1 - e^2)}$. Substituting the values of c and e from equation (28) in the above equation, we obtain

$$a = -\frac{K}{2E} \qquad \qquad \dots(35)$$

$$\Rightarrow \qquad \qquad E = -\frac{K}{2a}. \qquad \qquad \dots(36)$$

This shows that in the case of an elliptical orbit the total energy depends solely on the major axis of the ellipse. Substituting this value in equation (34) we get

$$b = (a)^{\frac{1}{2}}\sqrt{\frac{l^2}{mK}}.$$

Consequently from the equation (33) we obtain the square of the period of revolution

$$T^2 = \frac{4\pi^2 ma^3}{K}. \qquad \qquad \dots(37)$$

This shows that the square of the period of revolution of the planet around the Sun is directly proportional to the cube of semi-major axis of the ellipse. This is Kepler's third law of planetary motion. The law is same for all planets. It has been proved that the motion of a planet about the Sun is equivalent to one-body problem and m in equation (37) must be the reduced mass of the system and is given by

$$m = \frac{m_e M_0}{m_e + M_0},$$ ...(38)

where m_e = mass of the Earth, M_0 = mass of the Sun. Also the force of attraction between the Earth and the Sun is the inverse square law of force given by $f(r) = -\dfrac{K}{r^2}$, where $K = Gm_e M_0$. Under these conditions equation (37) becomes

$$T^2 = \frac{4\pi^2}{G(m_e + M_0)} a^3.$$ ...(39)

By neglecting the mass of the planet as compared the Sun we get

$$T^2 = \frac{4\pi^2}{GM_0} a^3.$$ ...(40)

This equation shows that the square of the period of revolution of the planet is independent of its mass.

Note: The planetary mass in equation (39) is not always completely negligible as compared to the Sun. Kepler's third law is rigorously true for the electron orbits in the Bohr atom.

WORKED EXAMPLES

Example 1:   Show that in case of elliptical orbit, under central force, the two apsidal distances are equal to $a(1 - e)$ and $a(1 + e)$.

Solution: Let a planet of mass m move under the action of central force and describes an elliptical orbit (27). The eccentricity of the conic is given in the equation (28). In the case of elliptical orbit the energy of the planet depends solely on the semi-major axis of the ellipse. *i.e.,*

$$E = -\frac{K}{2a}.$$ ...(41)

Substituting the value of E in equation (28) we get

$$\frac{l^2}{mK} = a(1 - e^2).$$

Consequently, the equation of the orbit of the planet (27) reduces to

$$r = \frac{a(1 - e^2)}{1 + e\cos(\theta - \theta_0)}.$$ ...(42)

The apsidal distances occur when $\theta - \theta_0 = 0$ and $\theta - \theta_0 = 180°$. Putting this in the equation of the conic (42) we readily get the required apsidal distances as

$$r_1 = a(1 - e), \quad r_2 = a(1 + e).$$

Example 2: A particle of mass m moves with velocity v under the inverse square law of attractive force, prove that its velocity v when the particle describes (1) an elliptic orbit is $v^2 = \dfrac{K}{m}\left(\dfrac{2}{r} - \dfrac{1}{a}\right)$ and (2) a parabolic orbit is $v^2 = \dfrac{2K}{mr}$.

Solution: In the case of an elliptical orbit, the total energy depends only on the semi-major axis of the ellipse and is given in the equation (41). From the total energy of the particle we have $T = E - V$. The potential energy of the particle moving under inverse square law of attractive force is given by $V = -\dfrac{K}{r}$. Using equation (41) the kinetic energy of the particle becomes

$$\frac{1}{2}mv^2 = -\frac{K}{2a} + \frac{K}{r}. \qquad \qquad \text{...(43)}$$

From which we obtain the velocity of the particle as

$$v^2 = \frac{K}{m}\left(\frac{2}{r} - \frac{1}{a}\right). \qquad \qquad \text{...(44)}$$

Now if the orbit is a parabola, then the eccentricity of the orbit is $e = 1$. This implies that the total energy of the particle E is zero. Hence from equation (43) we have

$$v^2 = \frac{2K}{mr}. \qquad \qquad \text{...(45)}$$

Note: Since ellipse becomes parabola if $a \to \infty$. Therefore equation (45) can also be obtained from equation (44) by taking its limit as $a \to \infty$.

Escape Velocity To launch a spaceship to the Moon or to more distant parts of the universe, we have to make sure that it leaves the confines of the Earth. However, we must remember that as we go farther and farther from the Earth there is a significant decrease in force of attraction. To find the speed that will carry a spaceship beyond the confines of the Earth's gravity, we start with the total energy of the particle which is given by $E = T + V$, where the potential energy of the particle moving under inverse square law of attractive force is given by $V = -\dfrac{K}{r}$. Thus we have

$$\frac{1}{2}mv^2 = E + \frac{K}{r}. \qquad \qquad \text{...(46)}$$

If v_e is the escape velocity, then the initial kinetic energy of the particle is $\dfrac{1}{2}mv_e^2$. This kinetic energy must at least be equal to the energy required for the particle to escape the gravitational pull of the Earth. Therefore, for velocity to be positive at far distance say $r \to \infty$, we must have $E \geq 0$. Thus for minimum escape velocity from the Earth's gravitational pull at the surface of the Earth, we have $E = 0$ and $r = R_e$. Substituting this in equation (46) we get

$$\frac{1}{2}mv_e^2 = \frac{K}{R_e},$$

where R_e is the radius of the Earth.

$$\Rightarrow \qquad \frac{K}{m} = \frac{1}{2}R_e v_e^2. \qquad \qquad ...(47)$$

Substituting this value in the equation (44) we obtain the expression for the velocity of the particle at any point of its orbit as

$$v = v_e \sqrt{R_e\left(\frac{1}{r} - \frac{1}{2a}\right)}. \qquad \qquad ...(48)$$

Note: The minimum velocity for escape from the Earth surface is equal to 11.2 km/sec.

Derivation of Newton's Law of Gravitation from Kepler's Laws of Planetary Motion

We have proved Kepler's laws of planetary motion in the previous part. These laws describe how planets move with respect to the sun, but Kepler did not find the cause responsible for such well defined movement. Newton supplied the reason why planets move according to the Kepler's laws and the cause is the law of force of gravitational attraction. We will derive this law below. The derivation will reveal that the real empirical backgroud for the development of Newton's law of gravitation was Kepler's laws of planetary motion and was not just a brain wave initiated by a falling apple.

Theorem 7 Derive Newton's law of gravitation from Kepler's laws of planetary motion.

Proof: According to the Kepler's first law of motion, the orbit of the planet is a conic, and is given by the equation

$$\frac{l}{r} = 1 + e\cos\theta, \qquad \qquad ...(49)$$

where e is the eccentricity of the orbit, and l the semi-latus rectum of the conic. In case of elliptic orbit, the length of the semi-latus rectum is given by

$$l = a(1 - e^2), \qquad \qquad ...(50)$$

where 'a' is the semi-major axis of the ellipse. Hence the equation of the orbit (49) becomes

$$r = \frac{a(1 - e^2)}{1 + e\cos\theta}. \qquad \qquad ...(51)$$

The Kepler's second law of planetary motion gives that the areal velocity is constant. *i.e.*,

$$\frac{1}{2} r^2 \dot\theta = \text{const.}$$

$$\Rightarrow \qquad r^2 \dot\theta = h. \qquad\qquad …(52)$$

The force on the planet at any instant is given by the Newton's second law of motion.

$$F = m\ddot{r}. \qquad\qquad …(53)$$

Further the radial component and transverse component of acceleration are respectively given by $\ddot{r} - r\dot\theta^2$ and $r\ddot\theta + 2\dot{r}\dot\theta$. Thus the force on the planet can be written as

$$\overline{F} = m\ddot{r} = m(\ddot{r} - r\dot\theta^2)\,\hat{r} + m(r\ddot\theta + 2\dot{r}\dot\theta)\,\hat\theta, \qquad …(54)$$

where $\hat{r}$ and $\hat\theta$ are unit vectors along the radial and the transverse directions respectively. Differentiating equation (51) and (52) with respect to time t we get

$$\dot{r} = \frac{re \sin\theta\dot\theta}{1 + e\cos\theta}, \qquad\qquad …(55)$$

and

$$2\dot{r}\dot\theta + r\ddot\theta = 0. \qquad\qquad …(56)$$

On using equations (49) and (52) in equation (55), we get

$$\dot{r} = \frac{he \sin\theta}{l}. \qquad\qquad …(57)$$

Equation (56) shows that the transverse component of acceleration is zero. Hence the force on the planet is only along radial direction and is given by

$$\overline{F} = m(\ddot{r} - r\dot\theta^2)\,\hat{r}. \qquad\qquad … (58)$$

Differentiating equation (57) with respect to the time t we get

$$\ddot{r} = \frac{he}{l}\cos\theta\dot\theta.$$

$$\Rightarrow \qquad \ddot{r} = \frac{h^2 e\cos\theta}{lr^2}. \qquad\qquad …(59)$$

Using equations (52) and (59) in equation (58), we get

$$\overline{F} = m\left(\frac{h^2 e\cos\theta}{lr^2} - r\frac{h^2}{r^4}\right)\hat{r}$$

$$\overline{F} = m\frac{h^2}{lr^2}\left(e\cos\theta - \frac{l}{r}\right)\hat{r}$$

$$\overline{F} = -\frac{h^2 m}{l r^2} \hat{r},$$

$$\overline{F} = -\frac{h^2 m}{a(1 - e^2) r^2} \hat{r}. \qquad ...(60)$$

Since for any planet m, a, e and h^2 are positive constant and $e < 1$, then the equation (60) shows that the force is not only central but also attractive in nature. Further, the area of the orbit is given by

$$A = \int_0^T dA = \int_0^T \dot{A} dt, \quad \dot{A} = \frac{dA}{dt}$$

$$\Rightarrow \qquad A = \int_0^T \frac{1}{2} r^2 \dot{\theta} dt$$

$$\Rightarrow \qquad A = \frac{h}{2} T. \qquad ...(61)$$

But the area of the ellipse is given by $A = \pi ab \Rightarrow \dfrac{h}{2} T = \pi ab$.

$$\Rightarrow \qquad h = \frac{2\pi ab}{T}.$$

However, we have the relation between the semi-major and semi minor axis of the ellipse as $b^2 = a^2(1 - e^2)$. Hence

$$h = r^2 \dot{\theta} = \frac{2\pi a^2 \sqrt{1 - e^2}}{T}. \qquad ...(62)$$

Now using Kepler's third law of planetary motion, we have

$$T^2 = K_0^2 a^3, \qquad ...(63)$$

where K_0^2 is the constant and is the same for all planets, and is given by (refer Theorem 6, equation (40))

$$K_0^2 = \frac{4\pi^2}{GM_0}. \qquad ...(64)$$

Therefore,

$$h^2 = \frac{4\pi^2 a(1 - e^2)}{K_0^2}. \qquad ...(65)$$

Putting this value in equation (60), we get

$$\overline{F} = -\frac{4m\pi^2}{K_0^2 r^2} \hat{r}. \qquad ...(66)$$

This shows that the force of attraction between any planet and the Sun is proportional to the mass of the planet. Now by Newton's third law of motion, the planet must be attracting the Sun with the same but opposite force. This is possible only if the factor $\dfrac{1}{K_0^2}$ is proportional to the mass of the Sun M_0. This leads finally to the equation

$$\overline{F} = -\frac{GmM_0}{r^2}\,\hat{r},$$

where G is the universal constant. This is the Newton's law of gravitation.

Differential Equation of the orbit

Theorem 8 Obtain the differential equation of a central orbit in the form

$$\frac{l^2 u^2}{m}\left(u + \frac{d^2 u}{d\theta^2}\right) = -f\left(\frac{1}{u}\right),$$

where f is a law of force and $u = \dfrac{1}{r}$ and l is the constant of angular momentum, m is the mass of the particle.

Solution: In the case of a particle moving in the central field force, the equation which describes the motion of the particle of mass m, is given by

$$m\ddot{r} - \frac{l^2}{mr^3} = f(r), \qquad\qquad \ldots(67)$$

where the law of force is given by $f(r) = -\dfrac{K}{r^2}.$ Change the variable r to u by putting

$$r = \frac{1}{u} \Rightarrow \dot{r} = -\frac{1}{u^2}\frac{du}{dt}$$

$$= -\frac{1}{u^2}\frac{du}{d\theta}\frac{d\theta}{dt}. \qquad\qquad \ldots(68)$$

In central motion, angular momentum of the particle is conserved.

$$\Rightarrow \qquad\qquad mr^2\dot{\theta} = l.$$

Eliminating $\dot{\theta}$ from the equation (68), we get

$$\dot{r} = -\frac{l}{m}\frac{du}{d\theta}. \qquad\qquad \ldots(69)$$

Differentiating with respect to time t again we get

$$\ddot{r} = -\frac{l}{m}\frac{d^2 u}{d\theta^2}\dot{\theta},$$

$$\ddot{r} = -\frac{l^2 u^2}{m^2} \frac{d^2 u}{d\theta^2}.$$

Substituting this in equation (67) we obtain

$$\frac{l^2 u^2}{m}\left(\frac{d^2 u}{d\theta^2} + u\right) = -f\left(\frac{1}{u}\right). \qquad \qquad \text{...(70)}$$

This equation determines the orbit when the law of force is given and conversely, when the orbit of the planet is known then we can determine law of force. Equation (70) establishes the interdependence of the Kepler's laws of planetary motion and the Newton's law of gravitation.

WORKED EXAMPLES

Example 3: A particle of mass m moves under inverse square law of force. Determine the orbit and classify the same.

Solution: The differential equation of the orbit of the particle of mass m moving under inverse square law of attractive force given by the equation (70). We write this equation as

$$\frac{d^2 u}{d\theta^2} + u = \frac{mK}{l^2}. \qquad \qquad \text{...(71)}$$

The general solution of the equation is given by

$$u = \frac{mK}{l^2} + b\,\cos(\theta - \theta_0), \qquad \qquad \text{...(72)}$$

$$u = \frac{1}{r} = \frac{mK}{l^2}\left[1 + \frac{l^2 b}{mK}\cos(\theta - \theta_0)\right]. \qquad \qquad \text{...(73)}$$

Comparing this equation with the standard equation of the ellipse (27), we have the eccentricity of the orbit as

$$e = \frac{l^2 b}{mK} \text{ and } c = \frac{mK}{l^2}.$$

This orbit of the planet is an ellipse if $e < 1$, hyperbola if $e > 1$, parabola if $e = 1$, and a circle if $e = 0$.

Example 4: A particle describes a circular orbit under the influence of an attractive central force directed to a point on the circle. Show that the force varies as the inverse fifth power of the distance.

Solution: A particle describes a circle under the action of a force which is directed towards a fixed point on the circle. Let the centre of the force be the origin. The polar equation of the circle of radius '*a*' passing through the origin and centre lies on the initial line through the centre of the force is given by

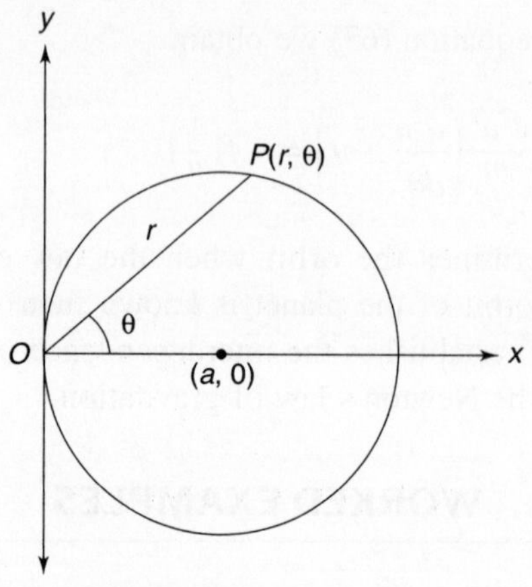

$$r = 2a \cos \theta. \qquad \ldots(74)$$

From equation (74), we have

$$u = \frac{1}{r} = \frac{\sec \theta}{2a}.$$

$\Rightarrow$

$$\frac{du}{d\theta} = \frac{\sec \theta \tan \theta}{2a},$$

$\Rightarrow$

$$\frac{d^2 u}{d\theta^2} = \frac{\sec^3 \theta + \sec \theta \tan^2 \theta}{2a}. \qquad \ldots(75)$$

Thus the equation of the orbit of the particle (70) becomes

$$f\left(\frac{1}{u}\right) = -\frac{l^2 u^2}{m}\left(\frac{\sec^3 \theta + \sec \theta \tan \theta}{2a} + \frac{\sec \theta}{2a}\right),$$

$$f\left(\frac{1}{u}\right) = -\frac{l^2 u^2}{2am}\left(\sec^3 \theta\right).$$

On using equation (74), we get

$$f\left(\frac{1}{u}\right) = -\frac{8l^2 a^2 u^5}{m},$$

$$f(r) = -\frac{8l^2 a^2}{m}\left(\frac{1}{r^5}\right),$$

$\Rightarrow$

$$f(r) \propto \frac{1}{r^5}.$$

This shows that the force varies as the inverse fifth power of the distance.

Example 5:   Find the central force under the action of which a particle will follow

$$r = a(1 + \cos \theta) \text{ orbit.}$$

Solution: A particle describes the orbit given by $r = 2a \cos^2\left(\dfrac{\theta}{2}\right)$. Changing the variable r to another variable u by substituting $u = \dfrac{1}{r}$ we get

$$u = \frac{1}{r} = \frac{\sec^2\left(\dfrac{\theta}{2}\right)}{2a}.$$

$$\frac{du}{d\theta} = \frac{1}{2a} \sec^2\left(\frac{\theta}{2}\right) \tan\left(\frac{\theta}{2}\right),$$

$$\frac{d^2u}{d\theta^2} = \frac{1}{4a}\left[2\sec^2\left(\frac{\theta}{2}\right)\tan^2\left(\frac{\theta}{2}\right) + \sec^4\left(\frac{\theta}{2}\right)\right]$$

$$\frac{d^2u}{d\theta^2} = \frac{1}{4a}\frac{\left(1 + 2\sin^2\left(\dfrac{\theta}{2}\right)\right)}{\cos^4\left(\dfrac{\theta}{2}\right)}.$$

The differential equation of the orbit (70) of the particle becomes

$$f\left(\frac{1}{u}\right) = \frac{l^2}{m} \, 3a\left(\frac{\sec^2\left(\dfrac{\theta}{2}\right)}{2a}\right)^4$$

Using equation of the orbit we find

$$f(r) = \frac{3al^2}{m}\left(\frac{1}{r^4}\right),$$

$$\Rightarrow \qquad\qquad f(r) \propto \left(\frac{1}{r^4}\right).$$

Example 6:   A particle moves on a curve $r^n = a^n \cos n\theta$ under the influence of a central force. Find the law of force.

Solution: Let $f(r)$ be the central force under the action of which a particle describes the orbit given by

$$r^n = a^n \cos n\theta. \qquad\qquad \text{...(76)}$$

Put $\qquad\qquad u = \dfrac{1}{r} = \dfrac{1}{a}(\sec n\theta)^{1/n} \qquad\qquad \text{...(77)}$

From which on differentiating with respect to θ we obtain

$$\frac{du}{d\theta} = \frac{1}{a}(\sec n\theta)^{1/n} \tan n\theta. \qquad \ldots(78)$$

Further, differentiating equation (78) with respect to θ we get

$$\frac{d^2u}{d\theta^2} = \frac{1}{a}\left[(\sec n\theta)^{1/n}\left(n \sec^2 n\theta + \tan^2 n\theta\right)\right]. \qquad \ldots(79)$$

Hence the differential equation of the path of the particle (70) becomes

$$\frac{1}{a}\left[(\sec n\theta)^{1/n}\left(n \sec^2 n\theta + \tan^2 n\theta\right)\right] + \frac{1}{a}(\sec n\theta)^{1/n} = \frac{ma^2}{l^2}(\sec n\theta)^{-2/n} f\!\left(\frac{1}{u}\right),$$

$$\left(n \sec^2 n\theta + \tan^2 n\theta\right) + 1 = \frac{ma^3}{l^2}(\sec n\theta)^{-3/n} f\!\left(\frac{1}{u}\right),$$

$$\frac{(n+1)}{\cos^2 n\theta} = \frac{ma^3}{l^2}(\sec n\theta)^{-3/n} f\!\left(\frac{1}{u}\right),$$

$$\Rightarrow \qquad f(r) = (n+1)\frac{l^2 a^{2n}}{m} r^{-(2n+3)},$$

$$\Rightarrow \qquad f(r) = \propto r^{-(2n+3)}.$$

Example 7:   A particle of mass m moves in an elliptical orbit under the action of an inverse square central force. Show that the velocity of the particle is extremum at the perihelion and aphelion points. Calculate the ratio of maximum and minimum speeds of the particle in its orbit and hence show that the eccentricity of the ellipse is given by $e = \dfrac{\alpha - 1}{\alpha + 1}$, where α is the ratio of the maximum velocity to the minimum velocity of the particle.

Solution: Given that a particle moves under the action of an inverse square law of attractive force. In this case the orbit described by the particle is an ellipse given by the equation (27). From this equation we write

$$r + er\cos(\theta - \theta_0) = \frac{l^2}{mk}. \qquad \ldots(80)$$

In the central force motion, the angular momentum of the particle is always conserved.

$$\Rightarrow \qquad mr^2\dot{\theta} = l. \qquad \ldots(81)$$

The kinetic energy of the particle is given by $T = \dfrac{1}{2}mv^2 = \dfrac{1}{2}m(\dot{r}^2 + r^2\dot{\theta}^2)$, from which the velocity of the particle is given by

$$v^2 = \dot{r}^2 + r^2\dot{\theta}^2. \qquad \ldots(82)$$

Differentiating equation (80) with respect to t we obtain

$$\dot{r} + e\dot{r}\cos(\theta - \theta_0) - re\sin(\theta - \theta_0)\,\dot{\theta} = 0$$

$\Rightarrow \qquad \dot{r} = \dfrac{r\dot{\theta}e\sin(\theta - \theta_0)}{1 + e\cos(\theta - \theta_0)}$

$\Rightarrow \qquad \dot{r}^2 = \dfrac{r^2\dot{\theta}^2 e^2 \sin^2(\theta - \theta_0)}{\left[1 + e\cos(\theta - \theta_0)\right]^2}$

Thus the velocity of the particle from equation (82) becomes

$$v^2 = r^2\dot{\theta}^2 \frac{\left[1 + e^2 + 2e\cos(\theta - \theta_0)\right]}{\left[1 + e\cos(\theta - \theta_0)\right]^2}$$

$\Rightarrow \qquad v = r\dot{\theta}\sqrt{\dfrac{1 + e^2 + 2e\cos(\theta - \theta_0)}{\left[1 + e\cos(\theta - \theta_0)\right]^2}}\, .$ $\hspace{2cm}$ ...(83)

Eliminating $\dot{\theta}$ on using (81) we get

$$v = \frac{l}{mr}\sqrt{\frac{1 + e^2 + 2e\cos(\theta - \theta_0)}{\left[1 + e\cos(\theta - \theta_0)\right]^2}}, \hspace{2cm} ...(84)$$

Now the velocity v to be maximum, we see that r should be minimum. The minimum value of r is obtained from the equation of the conic (27) by putting $\theta - \theta_0 = 0$. Thus from equation (27), we have

$$r_{\min} = \frac{l^2}{mK(1 + e)}. \hspace{2cm} ...(85)$$

Similarly, the velocity of the particle will be minimum when r is maximum. The maximum value of r is obtained from equation (27) on putting $\theta - \theta_0 = 180°$. Thus we have

$$r_{\max} = \frac{l^2}{mK(1 - e)}. \hspace{2cm} ...(86)$$

We see that the velocity of the particle is maximum at the perihelion point, while it is minimum at aphelion point. Using equations (85) and (86) in the equation (84) we obtain the maximum and minimum velocities as

$$v_{\max} = \frac{K}{l}(1 + e), \; v_{\min} = \frac{K}{l}(1 - e). \hspace{2cm} ...(87)$$

Thus the ratio of $v_{\max}$ to $v_{\min}$ becomes

$$\frac{v_{max}}{v_{min}} = \frac{1+e}{1-e} \qquad \ldots(88)$$

Solving this equation we get

$$e = \frac{v_{max} - v_{min}}{v_{max} + v_{min}},$$

$$e = \frac{\alpha - 1}{\alpha + 1}, \quad \alpha = \frac{v_{max}}{v_{min}}.$$

Example 8: Prove that the product of the minimum and maximum speeds of a particle moving in an elliptical orbit is $\left(\frac{2\pi a}{T}\right)^2$, where T is the periodic time of the particle.

Solution: When a particle describes an elliptical orbit; the maximum and the minimum values of its velocity occur at the perihelion and the aphelion points and are given by the equations (refer Example 7)

$$v_{max} = \frac{K}{l}(1+e), \quad v_{min} = \frac{K}{l}(1-e). \qquad \ldots(89)$$

The product of maximum velocity and minimum velocity gives

$$v_{max} \cdot v_{min} = \left(\frac{K}{l}\right)^2 (1-e^2). \qquad \ldots(90)$$

The semi major axis of the ellipse is equal to the one half of the sum of the two-apsidal distances. Therefore, we find

$$a = \frac{r_{max} + r_{min}}{2} = \frac{l^2}{2mK}\left(\frac{1}{1+e} + \frac{1}{1-e}\right),$$

$$a = \frac{l^2}{mK(1-e^2)},$$

$$\left(\frac{l^2}{1-e^2}\right) = amK.$$

Substituting this value in equation (90), we get

$$v_{max} \cdot v_{min} = \left(\frac{K}{ma}\right). \qquad \ldots(91)$$

However, from Kepler's third law of planetary motion we have

$$T^2 = 4\pi^2 a^3 \left(\frac{m}{K}\right). \qquad \ldots(92)$$

$$\Rightarrow \qquad \frac{4\pi^2}{T^2}a^2 = \frac{K}{ma}. \qquad\qquad ...(93)$$

Hence from equations (91) and (93) yield the required result

$$v_{max} \cdot v_{min} = \left(\frac{2\pi a}{T}\right)^2.$$

Example 9:   Starting from the energy conservation theorem, show that the eccentricity of the orbit of the particle moving under inverse square law of attractive force is given by

$$e = \sqrt{\frac{1 + 2EK^2}{mK^2}}.$$

Solution: The total energy of a particle moving under inverse square law of attractive force is given by

$$E = \frac{1}{2}m\dot{r}^2 + \frac{l^2}{2mr^2} + V(r), \qquad\qquad ...(94)$$

where $V = -\dfrac{K}{r}$. Thus the total energy becomes

$$E = \frac{1}{2}m\dot{r}^2 + \frac{l^2}{2mr^2} - \frac{K}{r}. \qquad\qquad ...(95)$$

Now to find the value of the energy, we consider the turning point at which r is extremum (minimum or maximum). At the turning point $\dot{r} = 0$. Hence equation (95) becomes

$$E = \frac{l^2}{2mr_{min}^2} - \frac{K}{r_{min}}, \qquad\qquad ...(96)$$

where r_{min} is given by the equation (85), substituting this in the equation (95) we readily obtain

$$E = \frac{mK^2}{2l^2}(e^2 - 1). \qquad\qquad ...(97)$$

Solving this equation for the eccentricity e of the orbit we obtain

$$e = \sqrt{\frac{1 + 2El^2}{mK^2}}.$$

Example 10:   Use Hamilton's procedure to find the differential equation for planetary motion under inverse square law of force.

Solution: Thus the Lagrangian of the motion is given by

$$L = \frac{1}{2} m \left(\dot{r}^2 + r^2 \dot{\theta}^2 \right) + \frac{K}{r}. \qquad \ldots(98)$$

We see that θ is cyclic in L, hence the generalized momentum corresponding to the cyclic co-ordinate is conserved. *i.e.*,

$$p_\theta = \frac{\partial L}{\partial \dot{\theta}} = mr^2 \dot{\theta} = l. \qquad \ldots(99)$$

Similarly, we find $\qquad p_r = \frac{\partial L}{\partial \dot{r}} = m\dot{r}. \qquad \ldots(100)$

The Hamiltonian of motion is given by

$$H = p_r \dot{r} + p_\theta \dot{\theta} - \frac{1}{2} m \left(\dot{r}^2 + r^2 \dot{\theta}^2 \right) - \frac{K}{r}.$$

Eliminating $\dot{r}$ and $\dot{\theta}$ by using equations (99) and (100), we get

$$H = \frac{1}{2m} \left(p_r^2 + \frac{p_\theta^2}{r^2} \right) - \frac{K}{r}. \qquad \ldots(101)$$

This is the Hamiltonian of motion. The Hamilton equations of motion corresponding to the generalized co-ordinates are

$$\dot{p}_r = \frac{\partial H}{\partial r} = \left(\frac{p_\theta^2}{mr^3} - \frac{K}{r^2} \right), \; \dot{p}_\theta = -\frac{\partial H}{\partial \theta} = 0. \qquad \ldots(102)$$

From equations (100) and (102), we have

$$\dot{p}_r = m\ddot{r} = \frac{p_\theta^2}{mr^3} - \frac{K}{r^2}, \qquad \ldots(103)$$

where p_θ is the constant of angular momentum. Replacing the variable r to u by putting $r = \frac{1}{u}$ we get

$$\dot{r} = -\frac{1}{u^2} \frac{du}{d\theta} \dot{\theta}$$

$$\Rightarrow \qquad \dot{r} = -\frac{p_\theta}{m} \frac{du}{d\theta},$$

$$\Rightarrow \qquad \ddot{r} = -\frac{p_\theta^2 u^2}{m^2}\frac{d^2 u}{d\theta^2}.$$

Hence equation (103) becomes

$$\Rightarrow \qquad -\frac{p_\theta^2 u^2}{m}\frac{d^2 u}{d\theta^2} = \frac{p_\theta^2 u^3}{m} - Ku^2,$$

$$\frac{d^2 u}{d\theta^2} + u = \frac{K}{p_\theta^2}\, m. \qquad\qquad \ldots(104)$$

This is the differential equation for planetary motion.

Example 11: A particle of mass m moves in a central force field defined by $\overline{F} = -\dfrac{K}{r^3}$, show that if E is the total energy supplied to the particle then its speed is given by

$$v = \sqrt{\frac{K}{mr^2} + \frac{2E}{m}}\,.$$

Solution: Let a particle of mass m be moving under the action of force given by

$$\overline{F} = -\frac{K}{r^3}\,. \qquad\qquad \ldots(105)$$

This force is conservative; hence the potential energy of the particle is given by

$$V = K\int_\infty^r \frac{1}{r^3}\, dr,$$

where integrating is performed between the limits from the standard position ∞ (this is chosen where the potential energy is zero) to the present position r of the particle. This gives $V(r) = -\dfrac{K}{2r^2}$. Similarly, the kinetic energy of the particle is given by $T = \dfrac{1}{2}\, m\dot{r}^2$. The total energy of the particle becomes

$$E = \frac{1}{2}\, m\dot{r}^2 - \frac{K}{2r^2},$$

$$\Rightarrow \qquad \dot{r}^2 = \frac{2E}{m} + \frac{K}{2r^2},$$

$$\Rightarrow \qquad \dot{r} = \sqrt{\frac{2E}{m} + \frac{K}{2r^2}}\,.$$

Example 12:   A particle moves in a circular orbit in a force field $f(r) = -\dfrac{K}{r^2}$. Suddenly K becomes $\dfrac{K}{2}$ without change in velocity of the particle. Show that the orbit becomes parabola.

Solution: The particle describes a circular orbit of radius r under the force $f(r) = -\dfrac{K}{r^2}$. This implies that the eccentricity of the orbit $e = 0$ and the semi-major axis a is equal to the semi minor axis b and is just the radius of the circle r. *i.e.*, $a = b = r$. In the case of elliptical orbit, the total energy of the particle is given by $E = \dfrac{K}{2a}$. However, the orbit is circular, hence $E = \dfrac{K}{2r}$. The potential energy of the particle becomes $V = -\dfrac{K}{r}$. The kinetic energy of the particle becomes $T = \dfrac{K}{2r}$. The motion of the particle is such that K becomes $\dfrac{K}{2}$ without change in velocity. Hence kinetic energy remains the same, but the potential energy changes to $V = -\dfrac{K}{2r}$. Therefore, total energy of the particle becomes

$$E = T + V = \frac{K}{2r} - \frac{K}{2r},$$

$$\Rightarrow \qquad E = 0.$$

This proves that the orbit of the particle is a parabola.

The virial theorem   Clausius derived the general form of the Virial theorem in connection with the search for an accurate equation of state for imperfect gases. The theorem is useful in the kinetic theory of gases and it can be used to derive Boyle's law for perfect gases. Here we shall consider only a special case of the general form of this theorem to derive an important property of the central force field. We first define the virial of the system.

Virial of the system   The virial of the system is defined as the average over a long period of time of $-\dfrac{1}{2}$ the sum over the particles in the system of the scalar product of the total force acting on the particle and its radius vector.

Theorem 9 (Virial)   "The total kinetic energy of the system averaged over a long period of time equals the virial of the system". *i.e.*,

$$\overline{T} = -\frac{1}{2} \overline{\sum_i \overline{F}_i \cdot \overline{r}_i}.$$

Proof: Consider a system of particles of masses m_i with position vectors $\overline{r}_i$ and applied forces $\overline{F}_i$ (conservative and non-conservative). Then the fundamental equation of motion of the ith particle is given by

$$\overline{F}_i = \dot{p}_i. \qquad\qquad \text{...(106)}$$

Define a quantity G such that

$$G = \sum_i \bar{p}_i \bar{r}_i, \qquad \qquad \ldots(107)$$

where the summation is taken over all particles of the system. Thus the total time derivative of this equation gives

$$\frac{dG}{dt} = \sum_i \dot{\bar{r}}_i \cdot \bar{p}_i + \sum_i \dot{\bar{p}}_i \cdot \bar{r}_i, \qquad \ldots(108)$$

where p_i is the momentum of the ith particle and is equal to $m_i \dot{\bar{r}}_i$. Therefore, equation (108) becomes

$$\frac{dG}{dt} = \sum_i m_i \dot{\bar{r}}_i^{\,2} + \sum_i \bar{F}_i \cdot \bar{r}_i.$$

i.e.,
$$\frac{dG}{dt} = 2T + \sum_i \bar{F}_i \cdot \bar{r}_i, \qquad \ldots(109)$$

where T is the kinetic energy of the system. The time average of equation (109) over a time interval τ is obtained by integrating both sides with respect to t between the limits 0 to τ and dividing by τ. Thus we obtain

$$\frac{1}{\tau} \int_0^\tau \frac{dG}{dt}\, dt = \overline{\frac{dG}{dt}} = 2\bar{T} + \overline{\sum_i \bar{F}_i \bar{r}_i},$$

where $\bar{T}$ is the average kinetic energy of the system over time interval τ and is defined by $\bar{T} = \frac{1}{\tau} \int_0^\tau T\, dt$ and $\overline{\sum_i \bar{F}_i \bar{r}_i}$ is the average value of $\sum_i \bar{F}_i \bar{r}_i$ over the time interval τ. Hence the above equation yields

$$2\bar{T} + \overline{\sum_i \bar{F}_i \bar{r}_i} = \frac{1}{\tau}[G(\tau) - G(0)]. \qquad \ldots(110)$$

1. Now consider the motion is periodic, *i.e.*, all co-ordinates repeat after a certain time. If this certain time is chosen to be the time period of motion then the right hand side of equation (110) vanishes. Hence we have

$$2\bar{T} + \overline{\sum_i \bar{F}_i \bar{r}_i} = 0$$

$$\bar{T} = -\frac{1}{2} \overline{\sum_i \bar{F}_i \bar{r}_i}. \qquad \ldots(111)$$

   This is known as the Virial Theorem.

2. Assume that the motion is not periodic. Let us consider the motion such that the co-ordinates and velocities for all particles in the system remain finite, so that there is an upper bound to G. Then by choosing

τ sufficiently large such that the right hand side of equation (110) can be made as small as we please. Thus in this case also, equation (110) reduces to equation (111) and that is the required Virial theorem.

Note: If the forces acting on the particles are derivable from potential, then the theorem becomes

$$\overline{T} = \frac{1}{2} \overline{\sum_i \nabla_i V \cdot \overline{r}_i}.$$

For a single particle moving under a central force the Virial theorem reduces to

$$\overline{T} = \frac{1}{2} \overline{\frac{\partial V}{\partial r}} r.$$

WORKED EXAMPLES

Example 13: If the particles attract each other according to inverse square law of force, prove that $\overline{2T + V} = 0$, where T is the total kinetic energy of the particles and V the potential energy.

Solution: For a conservative system of a single particle moving under a central force, the Virial theorem states that

$$\overline{T} = \frac{1}{2} \overline{\frac{\partial V}{\partial r}} r, \qquad \qquad ...(112)$$

where the central force is given by $f(r) = -\dfrac{\partial V}{\partial r} = -\dfrac{K}{r^2}.$ Substituting this value in equation (112), we get

$$\overline{T} = \frac{1}{2} \overline{\left(\frac{K}{r^2}\right)} r,$$

$$\Rightarrow \qquad \overline{T} = \frac{1}{2} \overline{\left(\frac{K}{r}\right)},$$

$$\Rightarrow \qquad \overline{T} = \frac{1}{2} \overline{V},$$

$$\Rightarrow \qquad \overline{2T + V} = 0.$$

Note: In general, if the potential is the power law function of r *i.e.*, $V = ar^{n+1}$. Then

$$\frac{\partial V}{\partial r} = (n+1)\, ar^n,$$

$$\frac{\partial V}{\partial r} r = (n+1)\, ar^{n+1},$$

$$\frac{\partial V}{\partial r} r = (n + 1) V$$

$$\Rightarrow \qquad \overline{T} = \frac{1}{2} (n + 1) \overline{V}.$$

Example 14: In a system, the total forces acting on particles are constrained by conservative forces $\overline{F}'_i$ and frictional forces $\overline{f}_i$ proportional to the velocity. Show that the Virial theorem holds for such system in the form $\overline{T} = -\frac{1}{2} \sum_i \overline{F}'_i \overline{r}_i$, provided the motion reaches a steady state. (*i.e.*, the motion is not allowed to die down as a result of frictional forces).

Solution: Given that the total forces acting on the system of particles is the sum of conservative forces $\overline{F}'_i$ and frictional forces $\overline{f}_i$. Thus we have

$$\overline{F}_i = \overline{F}'_i + \overline{f}_i. \qquad \qquad \ldots (113)$$

It is also given that the frictional forces are proportional to the velocity. *i.e.*,

$$\overline{f}_i = k\dot{\overline{r}}_i. \qquad \qquad \ldots (114)$$

Hence the equation of motion of the ith particle becomes

$$\dot{\overline{p}}_i = \overline{F}'_i + k\dot{\overline{r}}_i. \qquad \qquad \ldots (115)$$

Define the quantity G as $G = \sum_i \overline{p}_i \overline{r}_i$, where summation is taken over all particles of the system. Thus the total time derivative of this quantity is

$$\frac{dG}{dt} = \sum_i \dot{\overline{r}}_i \cdot \overline{p}_i + \sum_i \dot{\overline{p}}_i \cdot \overline{r}_i, \qquad \qquad \ldots (116)$$

where $p_i = m_i \dot{\overline{r}}_i$. Therefore, equation (116) becomes

$$\frac{dG}{dt} = \sum_i m_i \dot{\overline{r}}_i^2 + \sum_i (\overline{F}'_i + k\dot{\overline{r}}_i) \, \overline{r}_i.$$

i.e.,
$$\frac{dG}{dt} = 2T + \sum_i \overline{F}'_i \overline{r}_i + k \sum_i \dot{\overline{r}}_i \overline{r}_i,$$

where T is the kinetic energy of the system. We write this equation as

$$\frac{d}{dt}\left[G - k \sum_i \frac{1}{2} \overline{r}_i^2 \right] = 2T + \sum_i \overline{F}'_i \overline{r}_i. \qquad \qquad \ldots (117)$$

The time average of equation (117) over a time interval τ is obtained by integrating both sides of equation (117) with respect to t from 0 to τ and dividing by τ. *i.e.*,

$$\frac{1}{\tau} \int_0^\tau \frac{d}{dt}\left[G - k \sum_i \frac{1}{2} \overline{r}_i^2 \right] dt = 2\overline{T} + \sum_i \overline{F}'_i \overline{r}_i,$$

where over head bar denotes the average value of the respective quantities. This can be written as

$$2\bar{T} + \overline{\sum_i \vec{F}'_i \vec{r}_i} = \frac{1}{\tau}[\phi(\tau) - \phi(0)], \qquad \ldots(118)$$

where

$$\phi = G - k\sum_i \frac{1}{2}\,\vec{r}_i^2.$$

Since the motion is not allowed to die and if it is periodic with a period τ then we make right hand side of equation (118) to zero by choosing τ sufficiently large. Hence the virial of the system gives

$$2\bar{T} + \overline{\sum_i \vec{F}_i \vec{r}_i} = 0.$$

Exercise

1. A particle moving in a central force field located at $r = 0$ describes the spiral $r = e^{-\theta}$. Prove that the magnitude of the force is inversely proportional to r^3.

2. A particle of mass m moves in an elliptical orbit. If the particle has its largest and smallest orbital speeds $v_{\max}$ and $v_{\min}$ respectively, then show that the length of the major axis of the orbit is given by $\dfrac{T}{\pi}\sqrt{v_{\max} \cdot v_{\min}}$, where T is the particle's period. Hint: refer the Example 8.

3. Show that in an elliptical orbit of a planet around the Sun, the total energy solely depends on the major axis of the ellipse. Hint: refer Theorem 7.

4. If a central force is $f = -\dfrac{k}{r^3}$, show that the orbit of the particle is

$$\frac{1}{r} = A\cos\gamma(\theta - \theta_0), \text{ where } \gamma = \sqrt{1 - \frac{mk}{l^2}} \text{ and } A \text{ is a constant.}$$

5. A particle moves in a field of force given by $f = -\dfrac{k}{r^2} - \dfrac{2\alpha}{r^3}$, where k, α are constants. Show that the orbit of the particle is an ellipse given by $\dfrac{p}{r} = 1 + e\cos\gamma(\theta - \theta_0)$, where $\gamma = \sqrt{1 - \dfrac{2m\alpha}{l^2}}$, $e = \dfrac{A\gamma^2 l^2}{mk}$, $p = \dfrac{\gamma^2 l^2}{mk}$.

6. A particle moves in a potential given by $V = \dfrac{k}{r} + \dfrac{\alpha}{r^2}$, where k, α are constants.

Show that the orbit of the particle is given by $\dfrac{p}{r} = e\cos\gamma(\theta - \theta_0) - 1$,

where $\gamma = \sqrt{1 + \dfrac{2m\alpha}{l^2}}$, $e = \dfrac{A\gamma^2 l^2}{mk}$, $p = \dfrac{\gamma^2 l^2}{mk}$.

The Kinematics of Rigid Body

5

INTRODUCTION

In this chapter we define a rigid body and describe how the number of degrees of freedom of a rigid body with N particles is determined. There is a difference in the motion of a particle and the motion of a rigid body. In case of a rigid body there involves two types of motion viz.; the translation and the rotation, where as in the case of a particle there involves only translation motion. A rigid body with one point fixed has only rotational motion and is studied when the orientation of the rigid body is specified. Various sets of variables have been used to describe the orientation of rigid body. However, we will discuss in this chapter how the Eulerian angles and the complex Cayley-Klein parameters can be used for the description of rigid body with one point fixed.

Geometrically, matrix represents rotation; to study the rotation motion of a rigid body; we will find in this chapter the matrix of transformation in terms of Eulerian angles and Cayley-Klein parameters and also establish the relation between them.

Rigid Body: A rigid body is regarded as a system of many (at least three) non-collinear particles whose positions relative to one another remain fixed during motion. That is distance between any two of them remains constant through out the motion. The internal forces holding the particles at fixed distances from one another are known as forces of constraint. These forces of constraint obey the Newton's third law of motion.

Example 1:   Explain how the generalized co-ordinates of a rigid body with N particles reduce to six for its description.

Solution: *Generalized co-ordinates of a rigid body:* A system of N particles free from constraints can have $3N$ degrees of freedom and hence $3N$ generalized co-ordinates. But the constraints involved in rigid body with N particles are holonomic and scleronomic and are given by the equations

$$\bar{r}_{ij} = a_{ij}, \; i \neq j = 1,2,...,N \qquad \qquad ...(1)$$

where $\bar{r}_{ij}$ denotes the distance between the ith and j^{th} particles, and a_{ij} is the distance between them and is constant. Equation (1) is symmetric in i and j and $i \neq j$ as the distance of the ith particle from itself is zero, the possible number of constraints is therefore $\dfrac{N(N-1)}{2}$. We notice that for $N > 7$, $\dfrac{N(N-1)}{2} > 3N$. Therefore the actual number of degrees of freedom cannot be obtained simply by subtracting the number of constraints from $3N$. This is simply because all constraints in equation (1) are not independent.

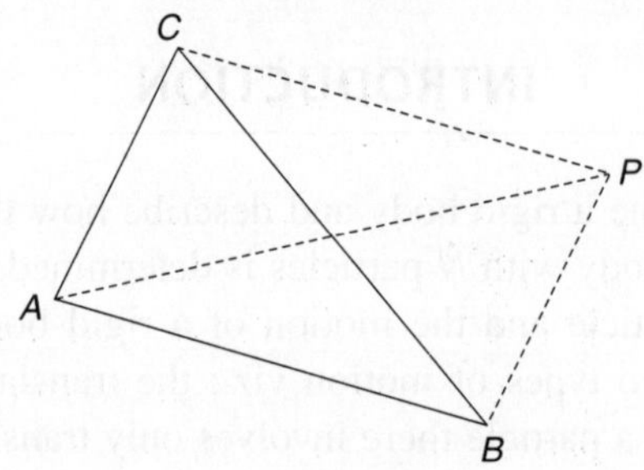

To show how the generalized co-ordinates of a rigid body with N particles reduce to six for its description, let a rigid body be regarded as a system of at least three non-collinear particles whose positions relative to one another remain fixed. Thus a system of three particles free from constraints has nine degrees of freedom but there involves three constraints that the distance between each of them is constant. These three constraints can be used to eliminate the dependent variables from the nine degrees of freedom. Hence the number of generalized co-ordinates reduces to only six. Thus the total number of degrees of freedom for three non-collinear particles A, B and C of a rigid body is equal to six. This is because each particle has three degrees of freedom and less three equations of constraints.

The position of each further particle say P requires three more co-ordinates for its description, but there will be three equations of constraints for this particle, because the distance of P from A, B and C is fixed. Thus three co-ordinates for P and less three equations of constraints for P gives zero degrees of freedom. This is true for every other particle of the system. Thus any other particle apart from A, B and C taken to specify the configuration of the rigid body will not add any degrees of freedom. Once the positions of three of the particles of the rigid body are determined the constraints fix the positions of all remaining particles.

Thus the configuration of the rigid body would be completely specified by only three particles *i.e.,* by six degrees of freedom, no matter how many particles it may contain.

Example 2:   Describe the motion of the rigid body.

Solution: A rigid body can have two types of motion viz., (i) a translational motion and (ii) a rotational motion. Thus a rigid body in motion can be completely specified if its position and orientation are known.

However, if one of the points of a rigid body is fixed, the translation motion of the body is absent and the body rotates about any line through the fixed point. Again, if we fix up a second point of the body, then the motion of the body is restricted to rotate about the line joining the two fixed points. Further, if we also fix the third point of the body non-collinear with other two, the position of the body is fixed and there is no motion of any kind. The co-ordinates of the third point alone will be able to locate the rigid body completely in space. It follows that the position of the rigid body is determined by any three non-collinear points of it that is by six degrees of freedom. Of the six generalized co-ordinates, three co-ordinates are used to describe translational motion and other three co-ordinates are used to describe rotational motion. Since a rigid body with one point fixed has no translational motion and hence it has three degrees of freedom and hence three generalized co-ordinates to describe the rotational motion.

Orthogonal Transformation: There are two types of orthogonal transformations viz.,

(1) finite orthogonal transformation and (ii) infinite orthogonal transformation.

Definitions:

Finite Orthogonal Transformation: Consider (x_1, x_2, x_3) and (x_1', x_2', x_3') be two co-ordinate systems. The general linear transformation between these two co-ordinate systems is defined by the following set of equations

$$x_1' = a_{11}x_1 + a_{12}x_2 + a_{13}x_3,$$

$$x_2' = a_{21}x_1 + a_{22}x_2 + a_{23}x_3, \qquad \qquad ...(2)$$

$$x_3' = a_{31}x_1 + a_{32}x_2 + a_{33}x_3.$$

where all a_{ij} are constants and usually referred as cosine of the angles between x_i' and x_j. These three equations can be combined in to a single equation as

$$x_i' = \sum_{j=1}^{3} a_{ij}x_j, \; i = 1, 2, 3. \qquad \qquad ...(3)$$

where $A = (a_{ij})$ is called the matrix of transformation. Now to call the transformation (3) as an orthogonal transformation, the elements of the matrix of transformation A have to satisfy certain conditions. To arrive to these conditions let a vector $\bar{r} = x_1 i + x_2 j + x_3 k$ defined in (x_1, x_2, x_3) co-ordinate system be transformed to (x_1', x_2', x_3') co-ordinate systems in the form $\bar{r} = x_1' i + x_2' j + x_3' k$.

Since the magnitude of the vector must be the same in both the co-ordinate system, we must have therefore

$$\sum_{i=1}^{3} x_i'^2 = \sum_{i=1}^{3} x_i^2 \qquad \qquad ...(4)$$

Using equation (3) in the equation (4) we obtain

$$\sum_{i=1}^{3}\left(\sum_{j=1}^{3} a_{ij}x_j\right)\left(\sum_{k=1}^{3} a_{ik}x_k\right) = \sum_{i=1}^{3} x_i^2,$$

$$\Rightarrow \qquad \sum_{j=1}^{3}\sum_{k=1}^{3}\left[\sum_{i=1}^{3} a_{ij}a_{ik}\right]x_jx_k = \sum_{i=1}^{3} x_i^2.$$

Equating the corresponding coefficients on both the sides of the above equation we get

$$\sum_{i=1}^{3} a_{ij}a_{ik} = \delta_{jk}, \qquad \qquad ...(5)$$

where δ_{jk} is the Kronecker delta symbol and is defined by

$$\delta_{jk} = 0 \ when \ j \neq k,$$

$$= 1 \ when \ j = k. \qquad \qquad ...(6)$$

Thus any transformation (3) satisfying the condition (5) is called as an orthogonal transformation.

Infinite Orthogonal Transformation

An infinitesimal rotation is an orthogonal transformation of co-ordinate axes in which the components of a vector are almost the same in both the sets of axes. The new coordinates differ from the old co-ordinates by an infinitesimal amounts.

Mathematically, an infinitesimal transformation is defined as

$$x_1' = x_1 + \varepsilon_{11}x_1 + \varepsilon_{12}x_2 + \varepsilon_{13}x_3,$$

$$x_2' = x_2 + \varepsilon_{21}x_1 + \varepsilon_{22}x_2 + \varepsilon_{23}x_3,$$

$$x_3' = x_3 + \varepsilon_{31}x_1 + \varepsilon_{32}x_2 + \varepsilon_{33}x_3, \qquad \qquad ...(7)$$

i.e., $\qquad \qquad x_i' = x_i + \varepsilon_{ij}x_j, \ i, j = 1, 2, 3, \qquad \qquad ...(8)$

where summation is defined over the repeated index j and ε_{ij} are the elements of the matrix of infinitesimal transformation and are infinitesimal. *i.e.,* second order terms in ε_{ij} can be neglected. Hence we write equations (8) as

$$x_i' = \delta_{ij}x_j + \varepsilon_{ij}x_j,$$

i.e., $\qquad \qquad x_i' = (\delta_{ij} + \varepsilon_{ij})x_j. \qquad \qquad ...(9)$

In matrix notations we write equation (9) as

$$X' = (I + \varepsilon) X, \qquad \qquad \text{...(10)}$$

where

$$X = \begin{pmatrix} x_1 \\ x_2 \\ x_3 \end{pmatrix}, \ X' = \begin{pmatrix} x_1' \\ x_2' \\ x_3' \end{pmatrix}, \ I = (\delta_{ij}), \ \varepsilon = (\varepsilon_{ij}).$$

The transformation (10) is called the infinite transformation and $I + \varepsilon$ is the matrix of infinitesimal transformation.

Some Properties of Orthogonal Transformation Matrix

Property 1: Finite rotation of a rigid body about a fixed point of the body is not commutative.

Solution: We now claim that two successive finite rotations of a rigid body about a fixed point of the body are not commutative.

Consider two successive linear transformations described by the matrices B and A corresponding to two successive displacements of the rigid body. Let the first transformation from x to x' be denoted by the matrix B and is defined by

$$x_k' = \sum_{j=1}^{3} b_{kj} x_j, \ k = 1, 2, 3, \qquad \qquad \text{...(11)}$$

where the matrix of transformation is $B = (b_{kj})$. Let the succeeding transformation from x' to x'' be defined by the matrix $A = (a_{ik})$ and is given by

$$x_i'' = \sum_{k=1}^{3} a_{ik} x_k', \ i = 1, 2, 3. \qquad \qquad \text{...(12)}$$

Now the transformation from x to x'' is obtained by combining the two equations (11) and (12) as

$$x_i'' = \sum_{j,k=1}^{3} a_{ik} b_{kj} x_j, \ i = 1, 2, 3.$$

This may also be written as

$$x_i'' = \sum_{j=1}^{3} c_{ij} x_j, \ i = 1, 2, 3. \qquad \qquad \text{...(13)}$$

where $C = (c_{ij})$ is the matrix of transformation from x to x'' and the elements of the matrix of transformation are defined by

$$c_{ij} = \sum_{k=1}^{3} a_{ik} b_{kj}. \qquad \qquad \text{...(14)}$$

These elements are obtained by multiplying the two matrices A and B. Thus the two successive linear transformations described by A and B is equivalent to a third linear transformation described by the matrix C, where $C = AB$. Since the matrix multiplication is not commutative in general, hence the two successive finite rotations of a rigid body about a fixed point of the body are not commutative.

Property 2: The product of two orthogonal transformations is again orthogonal transformation.

Solution: Consider two successive orthogonal linear transformations of a rigid body with one point fixed corresponding to two successive displacement of the rigid body and are described by the matrices B and A respectively. We know that the two successive orthogonal transformations is equivalent to a third linear transformation described by the matrix C, where $C = AB$, and its elements are defined by the equation (14), where $A = (a_{ik})$, $B = (b_{kj})$ are matrices of orthogonal transformations. Therefore, the elements of the matrices A and B satisfy the conditions

$$\sum_i a_{ij} a_{ik} = \delta_{jk}, \qquad\qquad \ldots(15)$$

$$\sum_i b_{ij} b_{ik} = \delta_{jk}. \qquad\qquad \ldots(16)$$

Consider now

$$\sum_i c_{ij} c_{ik} = \sum_i \left[\sum_m a_{im} b_{mj} \sum_l a_{il} b_{lk} \right]$$

$$= \sum_{i,m,l} a_{im} b_{mj} a_{il} b_{lk},$$

$$= \sum_{m,l} \left[\sum_i a_{im} a_{il} \right] b_{mj} b_{lk},$$

$$= \sum_{m,l} \delta_{ml} b_{mj} b_{lk},$$

$$= \sum_l b_{lj} b_{lk},$$

$$\sum_i c_{ij} c_{ik} = \delta_{jk}.$$

This proves that the product of two orthogonal transformations is again an orthogonal transformation.

Note: Though the matrix multiplication is not commutative in general, but it is associative when the product is defined. *i.e.*, $(AB)C = A(BC)$.

Property 3: In the case of an orthogonal transformation the inverse matrix is identified by the transpose of the matrix.

Solution: Consider an orthogonal transformation from x_i to x'_i described by the matrix A and is defined in equation (3) together with the condition (5). Let the matrix of inverse transformation from x'_i to x_i be described by the matrix $A^{-1} = (a'_{ij})$, a'_{ij} are the elements of the inverse matrix of transformation satisfying

$$\sum_i a'_{ij} a'_{ik} = \delta_{jk}. \qquad \qquad ...(17)$$

Also we have

$$AA^{-1} = I \Rightarrow \sum_i a_{ki}\, a'_{ij} = \delta_{kj}. \qquad \qquad ...(18)$$

Consider the double sum $\sum\limits_{k,i} a_{kl}\, a_{ki}\, a'_{ij}$ which can be evaluated either by summing over k first or over i first. Therefore, we evaluate the double sum first summing over k and then over i as

$$\sum_{k,i} a_{kl} a_{ki} a'_{ij} = \sum_i \left(\sum_k a_{kl} a_{ki} \right) a'_{ij}$$

$$= \sum_i \delta_{li} a'_{ij}$$

$$\sum_{k,i} a_{ki} a_{ki}\, a'_{ij} = a'_{lj}. \qquad \qquad ...(19)$$

Now evaluating the double sum over i first and then over k, we obtain

$$\sum_{k,i} a_{kl} a_{ki} a'_{ij} = \sum_k \left(\sum_i a_{ki} a'_{ij} \right) a_{kl}$$

$$= \sum_k \delta_{kj} a_{kl}$$

$$\sum_{k,i} a_{kl} a_{ki} a'_{ij} = a_{jl} \qquad \qquad ...(20)$$

From equations (19) and (20), we have

$$a'_{ij} = a_{ji}, \ \forall\ i, j,$$

$$\Rightarrow \qquad (a'_{ij}) = (a_{ji}),$$

$$\Rightarrow \qquad A^{-1} = A'. \qquad \qquad ...(21)$$

This proves that the inverse matrix of an orthogonal transformation identifies the transpose matrix.

Property 4: Show that the determinant of an orthogonal matrix is equal to ± 1.

Solution: Let A be the matrix of an orthogonal transformation and A^{-1} be its inverse matrix. Then we have $AA^{-1} = I$. However, we know that in the case of an orthogonal matrix, its inverse is identified by its transpose *i.e.,* $A^{-1} = A'$. It follows that

$$AA' = I.$$

Taking the determinant on both the sides of above equation we get

$$|AA'| = 1$$

$$\Rightarrow \qquad |A|\,|A'| = 1,$$

$$\Rightarrow \qquad |A|^2 = 1 \text{ as } |A| = |A'|,$$

$$\Rightarrow \qquad |A| = \pm\, 1.$$

Property 5: Infinitesimal rotation of a rigid body with one point fixed is commutative. Also find the inverse matrix of infinitesimal rotation.

Solution: Let $I + \varepsilon_1$ and $I + \varepsilon_2$ be two matrices of successive infinitesimal transformations. Consider the product

$$(I + \varepsilon_1)\,(I + \varepsilon_2) = I.I + I\,\varepsilon_2 + \varepsilon_1 I + \varepsilon_1\varepsilon_2$$

$$= I + \varepsilon_1 + \varepsilon_2.$$

Similarly, by changing the order of multiplication we find

$$(I + \varepsilon_2)\,(I + \varepsilon_1) = I.I + I\varepsilon_1 + \varepsilon_2 I + \varepsilon_2\varepsilon_1$$

$$= I + \varepsilon_1 + \varepsilon_2.$$

We see from above equations that

$$(I + \varepsilon_1)\,(I + \varepsilon_2) = (I + \varepsilon_2)\,(I + \varepsilon_1).$$

This shows that the product of the matrices of two successive infinitesimal transformations is commutative. To find the inverse matrix of an infinitesimal transformation, consider the product

$$(I + \varepsilon)\,(I - \varepsilon) = I - I\varepsilon + \varepsilon I - \varepsilon\varepsilon$$

$$(I + \varepsilon)\,(I - \varepsilon) = I.$$

This shows that the inverse matrix of an infinitesimal transformation is $I - \varepsilon$. *i.e.,*

$$(I + \varepsilon)^{-1} = (I - \varepsilon).$$

For orthogonal transformation, the transpose matrix identifies the inverse matrix. Hence we have

$$(I + \varepsilon)^{-1} = (I + \tilde{\varepsilon}) = (I - \varepsilon),$$

where $\tilde{\varepsilon}$ is the transpose of ε. Consequently we have $\tilde{\varepsilon} = -\,\varepsilon$. This shows that the matrix of infinite transformation is anti-symmetric.

Example 3: A constant vector X is given by $\overline{X} = i + 4j + 2\sqrt{3}k$ with respect to a particular co-ordinate system. Find the form of the vector with respect to co-ordinate system obtained from the first by rotating it about the x-axis through an angle $\dfrac{\pi}{3}$ in the anti-clockwise direction. Determine its magnitude and compare with $|\overline{X}|$.

Solution: The matrix of rotation about x-axis through an angle $\dfrac{\pi}{3}$ in the anti-clockwise direction is given by

$$A\left(\frac{\pi}{3}\right) = \begin{pmatrix} 1 & 0 & 0 \\ 0 & \cos\dfrac{\pi}{3} & \sin\dfrac{\pi}{3} \\ 0 & -\sin\dfrac{\pi}{3} & \cos\dfrac{\pi}{3} \end{pmatrix} = \begin{pmatrix} 1 & 0 & 0 \\ 0 & \dfrac{1}{2} & \dfrac{\sqrt{3}}{2} \\ 0 & -\dfrac{\sqrt{3}}{2} & \dfrac{1}{2} \end{pmatrix}.$$

The new vector with respect to the new co-ordinate axes obtained from the first by rotating through an angle $\dfrac{\pi}{3}$ about x-axis is given by

$$\overline{X}' = \begin{pmatrix} 1 & 0 & 0 \\ 0 & \dfrac{1}{2} & \dfrac{\sqrt{3}}{2} \\ 0 & -\dfrac{\sqrt{3}}{2} & \dfrac{1}{2} \end{pmatrix} \begin{pmatrix} 1 \\ 4 \\ 2\sqrt{3} \end{pmatrix} = \begin{pmatrix} 1 \\ 5 \\ -\sqrt{3} \end{pmatrix}.$$

$$\overline{X}' = i + 5j - \sqrt{3}k.$$

The magnitude of this new vector is $|\overline{X}'| = \sqrt{29}$. This shows that $|\overline{X}| = |\overline{X}'|$. Hence the magnitude of a vector is invariant under rotation.

Euler's Theorem 1 Show that the general displacement of a rigid body with one point fixed is a rotation about some axis passing through the fixed point.

Proof: Consider a rigid body with one point fixed and be taken as the origin of the body set of axes. Then the displacement of the rigid body involves no translation of the body axes, the only change is in its orientation. Hence the body set of axes at any time t can always be obtained by a single rotation of the space set of axes.

Thus any vector lying along the axis of rotation must have the same components in both the initial and final axes. Further, the orthogonality condition implies that the magnitude of a vector parallel to the axis of rotation is unaffected. It means that the vector R has same components in both the systems. This gives

$$R' = AR = R. \qquad \qquad \dots(22)$$

This is the special case of more general equation

$$R' = AR = \lambda R, \qquad \ldots(23)$$

where λ is called as the eigen (characteristic) value. Equation (23) can be written as

$$(A - \lambda I)\, R = 0. \qquad \ldots(24)$$

The equation (24) can have solution only when

$$|A - \lambda I| = 0. \qquad \ldots(25)$$

This equation is known as characteristic equation, and the roots of the characteristic equation are known as characteristic values. Since the matrix of rotation A is orthogonal, and then we have

$$A^{-1} = \tilde{A}, \qquad \ldots(26)$$

where $\tilde{A}$ is the transpose of A. This orthogonal matrix satisfies the equation

$$|A| = |\tilde{A}| = 1 \qquad \ldots(27)$$

It is just enough to prove that eigen value $\lambda = 1$. Consider the expression

$$(A - I)\tilde{A} = A\tilde{A} - I\tilde{A}$$

$$= AA^{-1} - \tilde{A}$$

$$(A - I)\,\tilde{A} = I - \tilde{A}. \qquad \ldots(28)$$

Taking the determinant of equation (28), we get

$$|(A - I)\,\tilde{A}| = |I - \tilde{A}|$$

$$|(A - I)|\,|A| = |I - \tilde{A}|$$

$$\Rightarrow \qquad |A - I| = -\,|A - I| \qquad \ldots(29)$$

$$\Rightarrow \qquad |A - I| = 0. \qquad \ldots(30)$$

Comparing equations (25) and (30), we get $\lambda = 1$. This proves the theorem.

Eulerian Angles

We have seen that a rigid body with one point fixed has three degrees of freedom and hence three generalized co-ordinates for its description. To describe the orientation of a rigid body about a fixed point of the body we use a matrix of rotation, whose elements are called the direction cosines, which are not linearly independent, therefore they are not suitable as generalized co-ordinates. So we cannot use them in the description of Lagrangian of the rigid body. Therefore three new independent parameters are necessary for the description of a rigid body with one point fixed. A number of such sets of parameters have been used in the literature but the most common and found to be useful is the set of Eulerian angles.

Euler has designed three independent parameters called as Eulerian angles, to describe the orientation of a rigid body with one point fixed. These angles can be used to write Lagrangian and hence the Lagrange's equations of motion of the rigid body. We shall define the Eulerian angles and show how these angles can be used for the description of the orientation of the rigid body.

Theorem 2   Define Eulerian angles. Obtain the matrix of transformation from space co-ordinates to body co-ordinates in terms of Eulerian angles. Prove further that this matrix is orthogonal and hence deduce the matrix of inverse transformation from the body set of axes to space set of axes.

Proof: Eulerian angles ϕ, θ, ψ are the three successive angles of rotation about a specified axes performed in specific sequence. These angles can be used as generalized co-ordinates to fix the orientation of a rigid body with one point fixed. Thus the orientation of a rotating body with one point fixed can be completely specified by three independent Eulerian angles. To discuss the rotation of the rigid body, let one of the points of the body be fixed. This implies that there is no translational motion but the body rotates about an axis passing through the fixed point. Consider two co-ordinate systems, one of which is (x, y, z) fixed in space (called an inertial frame) and the other (x', y', z') fixed in the body called the body set of axes (also known as non-inertial frame). It has been observed that the configuration of the rigid body is completely specified by locating the body set of axes relative to the co-ordinate axes fixed in space. This is achieved by finding the matrix of transformation from the space set of axes to the body set of axes. Therefore we shall carry out the transformation from space set of axes to body set of axes such that (x, y, z) coincides with (x', y', z'). This is achieved by three successive rotations about specified axes and is done in the following way.

The sequence of rotation starts by rotating the initial system of axes (x, y, z) through an angle ϕ anti-clockwise direction about z-axis. Let the resulting co-ordinate system be labeled as (x_1, y_1, z_1) axes as shown in the figure. In this case xy-plane becomes $x_1 y_1$ plane. The rotation is affected by the following transformation equations.

$$x_1 = x \cos \phi + y \sin \phi,$$

$$y_1 = -x \sin \phi + y \cos \phi,$$

$$z_1 = z. \qquad \qquad \dots(31)$$

These equations can be written in matrix form as

$$\begin{pmatrix} x_1 \\ y_1 \\ z_1 \end{pmatrix} = D \begin{pmatrix} x \\ y \\ z \end{pmatrix} \text{ or } X_1 = DX$$

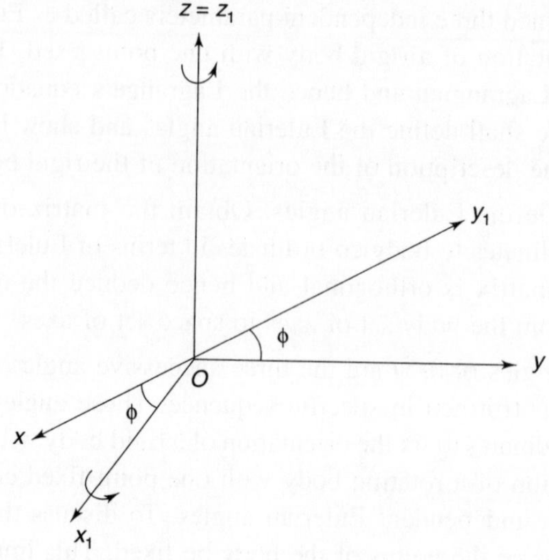

where

$$D = \begin{pmatrix} \cos\phi & \sin\phi & 0 \\ -\sin\phi & \cos\phi & 0 \\ 0 & 0 & 1 \end{pmatrix} \qquad ...(32)$$

is the matrix of transformation.

The second rotation is performed about the new x_1 axis. The axes x_1, y_1, z_1 are rotated about x_1 axis counter clockwise direction by an angle θ. Let the resultant set of co-ordinate axes be relabeled as x_2, y_2, z_2. Here x_2 axis being the line of intersection of xy-plane and x_1y_1 plane is called the line of nodes. The transformation equations from x_1, y_1, z_1 to new set of axes x_2, y_2, z_2 can be represented by the following set of equations:

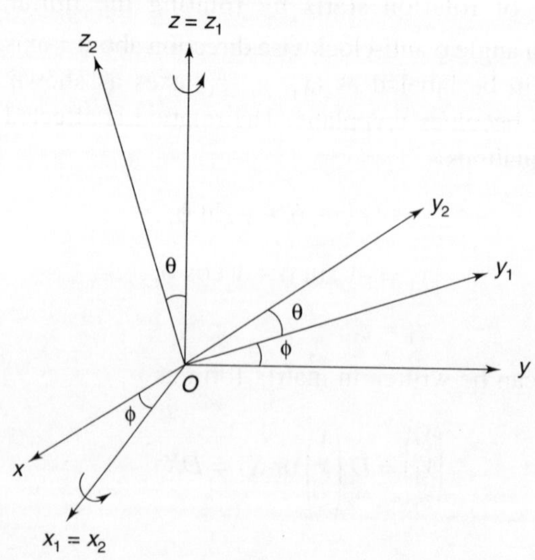

$$x_2 = x_1,$$

$$y_2 = y_1 \cos\theta + z_1 \sin\theta,$$

$$z_1 = -y_1 \sin\theta + z_1 \cos\theta, \qquad \qquad ...(33)$$

i.e.,
$$\begin{pmatrix} x_2 \\ y_2 \\ z_2 \end{pmatrix} = C \begin{pmatrix} x_1 \\ y_1 \\ z_1 \end{pmatrix} \text{ or } X_2 = CX_1$$

where

$$C = \begin{pmatrix} 1 & 0 & 0 \\ 0 & \cos\theta & \sin\theta \\ 0 & -\sin\theta & \cos\theta \end{pmatrix} \qquad \qquad ...(34)$$

is the matrix of transformation.

Finally, the third rotation is performed about z_2 axis. The x_2, y_2, z_2 axes are rotated counter clockwise direction by an angle ψ about z_2 axis to produce the third and the final set of axes x_3, y_3, z_3, which coincide, with body set of axes x', y', z'. This completes the transformation from space set of axes to body set of axes. This transformation is represented by the set of equations

$$x_3 = x' = x_2 \cos\psi + y_2 \sin\psi,$$

$$y_3 = y' = -x_2 \sin\psi + y_2 \cos\psi,$$

$$z_3 = z' = z_2 \qquad \qquad ...(35)$$

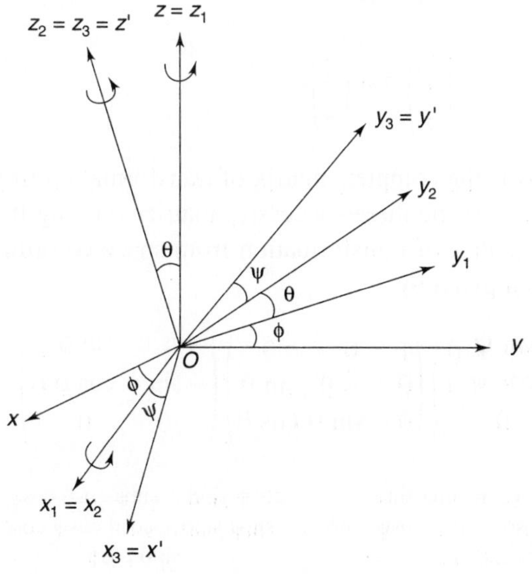

or

$$\begin{pmatrix} x' \\ y' \\ z' \end{pmatrix} = B \begin{pmatrix} x_2 \\ y_2 \\ z_2 \end{pmatrix}$$

or

$$X' = BX_2,$$

where

$$B = \begin{pmatrix} \cos\psi & \sin\psi & 0 \\ -\sin\psi & \cos\psi & 0 \\ 0 & 0 & 1 \end{pmatrix} \qquad \ldots(36)$$

is the matrix of transformation.

Thus the space set of axes x, y, z coincides with body set of axes through three successive rotations ϕ, θ, ψ, which are described by matrices D, C and B. The angles ϕ, θ, ψ are called Eulerian angles. The Eulerian angles completely specify the orientation of the x', y', z' system relative to the x, y, z system. Now we can obtain the complete matrix of transformation from x, y, z to x', y', z' by writing the matrix as the triple product of the separate rotations.

$$X' = BX_2$$

$$= B\,(CX_1)$$

$$= BC\,(X_1)$$

$$X' = BCDX$$

$$X' = AX,$$

$$\begin{pmatrix} x' \\ y' \\ z' \end{pmatrix} = A \begin{pmatrix} x \\ y \\ z \end{pmatrix}, \qquad \ldots(37)$$

where $A = BCD$ is the complete matrix of transformation from x, y, z to x', y', z' and is the product of the three successive matrices. Using the equations (32), (34) and (36) the matrix of transformation from space co-ordinates to the body co-ordinates is then given by

$$A = \begin{pmatrix} \cos\psi & \sin\psi & 0 \\ -\sin\psi & \cos\psi & 0 \\ 0 & 0 & 1 \end{pmatrix} \begin{pmatrix} 1 & 0 & 0 \\ 0 & \cos\theta & \sin\theta \\ 0 & -\sin\theta & \cos\theta \end{pmatrix} \begin{pmatrix} \cos\phi & \sin\phi & 0 \\ -\sin\phi & \cos\phi & 0 \\ 0 & 0 & 1 \end{pmatrix}$$

$$A = \begin{pmatrix} \cos\psi\cos\phi - \cos\theta\sin\psi\sin\phi, & \cos\psi\sin\phi + \sin\psi\cos\theta\cos\phi, & \sin\psi\sin\theta \\ -\sin\psi\cos\phi - \cos\theta\cos\psi\sin\phi, & -\sin\psi\sin\phi + \cos\theta\cos\psi\cos\phi, & \cos\psi\sin\theta \\ \sin\theta\sin\phi, & -\sin\theta\cos\phi & \cos\theta \end{pmatrix}$$

$$\ldots(38)$$

This is the required matrix of transformation. This matrix in Eulerian angles determines the orientation of the rigid body. We now show that this matrix A is orthogonal. Let the matrix A be represented by $A = (a_{ij})$. The condition for orthogonal transformation is defined in equation (5). We take $j = k = 1$, and consider the case

$$a_{11}^2 + a_{21}^2 + a_{31}^2 = (\cos \psi \cos \phi - \cos \theta \sin \psi \sin \phi)^2 +$$
$$+ (- \sin \psi \cos \phi - \cos \theta \cos \psi \sin \phi)^2 + (\sin \theta \sin \phi)^2,$$
$$a_{11}^2 + a_{21}^2 + a_{31}^2 = 1.$$

Similarly, we can show for all value of $j = k$. Now for $j \neq k$, $j, k = 1, 2, 3$ we consider the case

$$a_{11}a_{12} + a_{21}a_{22} + a_{31}a_{32} = (\cos \psi \cos \phi - \cos \theta \sin \psi \sin \phi) \cdot$$
$$\cdot (\cos \psi \sin \phi + \cos \theta \sin \psi \cos \phi) +$$
$$+ (- \sin \psi \cos \phi - \cos \theta \cos \psi \sin \phi) \cdot$$
$$\cdot (- \sin \psi \sin \phi + \cos \theta \cos \psi \cos \phi) +$$
$$+ (\sin \theta \sin \phi)(- \sin \theta \cos \phi).$$

Simple calculations show that

$$a_{11} a_{12} + a_{21} a_{22} + a_{31} a_{32} = 0.$$

Similarly, we can show for all $j \neq k$. that $\sum_{i} a_{ij}a_{jk} = 0$. Hence the matrix A is orthogonal. In the case of orthogonal matrix the inverse of A is the equal to the transpose of A. Thus the inverse of the matrix of transformation from body set of axes to the space set of axes is

$$A^{-1} = \begin{pmatrix} \cos\psi \cos\phi - \cos\theta \sin\psi \sin\phi, & - \cos\phi \sin\psi - \cos\theta \cos\psi \sin\phi, & \sin\theta \sin\phi \\ \sin\phi \cos\psi + \cos\theta \cos\phi \sin\psi, & - \sin\psi \sin\phi + \cos\theta \cos\psi \cos\phi, & - \cos\phi \sin\theta \\ \sin\theta \sin\psi, & \sin\theta \cos\psi & \cos\theta \end{pmatrix} \qquad \ldots(39)$$

WORKED EXAMPLES

Example 4: If the matrix of transformation from space set of axes to body set of axes is equivalent to a rotation through an angle χ about some axis through the origin then show that

$$\cos\left(\frac{\chi}{2}\right) = \cos\left(\frac{\phi + \psi}{2}\right) \cos\left(\frac{\theta}{2}\right).$$

Solution: The matrix of complete rotation from space set of axes to body set of axes in terms of Eulerian angles ϕ, θ, ψ is given in the equation (38). It is given that this matrix of rotation is equivalent to the matrix of rotation of co-ordinate axes through an angle χ about some axis with the same origin. Equivalently, it means that it is always possible by means of some similar transformation, to transform the matrix A defined in the equation (38) to the matrix B obtained by rotating the co-ordinate axes through an angle χ about some axis with the same origin. This matrix of rotation is given by either

$$B = \begin{pmatrix} \cos \chi & \sin \chi & 0 \\ -\sin \chi & \cos \chi & 0 \\ 0 & 0 & 1 \end{pmatrix} \text{ or } B = \begin{pmatrix} 1 & 0 & 0 \\ 0 & \cos \chi & \sin \chi \\ 0 & -\sin \chi & \cos \chi \end{pmatrix}, \qquad \ldots(40)$$

depending upon the axis of rotation.

It is well known that under similar transformation trace of the matrix is invariant. Therefore, we have Trace of A = Trace of B.

$\Rightarrow \qquad 2 \cos \chi + 1 = \cos \psi \cos \phi - \cos \theta \sin \psi \sin \phi + \cos \psi \cos \phi \cos \theta -$
$$- \sin \psi \sin \phi + \cos \theta,$$

$$2 \cos \chi + 1 = (\cos \psi \cos \phi - \sin \psi \sin \phi) + \cos \theta +$$
$$+ \cos \theta (\cos \psi \cos \phi - \sin \psi \sin \phi),$$
$$= [1 + \cos \theta] \cos (\psi + \phi) + \cos \theta.$$

$$2 \cos \chi + 1 = 2 \cos^2\left(\frac{\theta}{2}\right)\left[2 \cos^2\left(\frac{\psi + \phi}{2}\right) - 1\right] + \cos^2\left(\frac{\theta}{2}\right) - \sin^2\left(\frac{\theta}{2}\right),$$

$$2 \cos \chi + 1 = 4 \cos^2\left(\frac{\theta}{2}\right) \cos^2\left(\frac{\phi + \psi}{2}\right) - 1,$$

$$2 (\cos \chi + 1) = 4 \cos^2\left(\frac{\theta}{2}\right) \cos^2\left(\frac{\phi + \psi}{2}\right),$$

$$\cos^2\left(\frac{\chi}{2}\right) = \cos^2\left(\frac{\theta}{2}\right) \cos^2\left(\frac{\phi + \psi}{2}\right),$$

$$\cos\left(\frac{\chi}{2}\right) = \cos\left(\frac{\theta}{2}\right) \cos\left(\frac{\phi + \psi}{2}\right).$$

Example 5: Eulerian angles $\phi = \dfrac{\pi}{4}$, $\theta = \dfrac{\pi}{2}$ and $\psi = \dfrac{\pi}{4}$ bring space frame S in to coincidence with body frame S' with common origin. Find the transformation relating co-ordinates of S and S'. Show that if the transformation is equivalent to a rotation through an angle χ about some axis through origin then $\cos\left(\dfrac{\chi}{2}\right) = \dfrac{1}{2}$.

Solution: The matrix of complete transformation from space set of axes S to body set of axes S' is given in the equation (38), where in this case it is given that

$$\phi = \frac{\pi}{4}, \; \theta = \frac{\pi}{2} \text{ and } \psi = \frac{\pi}{4}.$$

Hence the matrix of transformation (38) becomes

$$A = \begin{vmatrix} \dfrac{1}{2} & \dfrac{1}{2} & \dfrac{1}{\sqrt{2}} \\[2mm] -\dfrac{1}{2} & \dfrac{-1}{2} & \dfrac{1}{\sqrt{2}} \\[2mm] \dfrac{1}{\sqrt{2}} & \dfrac{1}{\sqrt{2}} & 0 \end{vmatrix}. \qquad \ldots(41)$$

The matrix of rotation through an angle χ about some axis through the same origin is given by the equation (40). Given that the transformations defined by two matrices (38) and (40) are equivalent. This implies that the trace of the matrix A is equal to the trace of the matrix B. This gives

$$2 \cos \chi + 1 = 0,$$

$$2 \cos^2\left(\frac{\chi}{2}\right) - 1 = -\frac{1}{2},$$

$$\cos\left(\frac{\chi}{2}\right) = \frac{1}{2}.$$

Moments of Inertia and Products of Inertia

Moment of Inertia: A uniformly rotating body possesses the tendency to oppose any change in its state of rotation motion. This quantity is called the moment of inertia.

Theorem 3 Obtain the angular momentum of a rigid body about a fixed point of the body when the body rotates instantaneously with angular velocity ω in terms of inertia tensor.

Proof: Consider a rigid body composed of N particles having mass m_i and position vectors r_i with respect to the fixed point of the body. Since the translational motion is absent and the body rotates about an axis passes through the fixed point. Let ω be the instantaneous angular velocity of the body. If v_i is the linear velocity of the ith particle, then it is given by

$$v_i = \omega \times r_i. \qquad \ldots(42)$$

If L is the total angular momentum then it is equal to the sum of the angular momenta of an individual particle. Therefore we have

$$L = \sum_i I_i = \sum_{i=1}^{N} r_i \times m_i v_i$$

$$L = \sum_{i=1}^{N} r_i \times m_i \, (\omega \times r_i),$$

$$L = \sum_{i=1}^{N} m_i \, [r_i \times (\omega \times r_i)].$$

Using the vector identity

$$\bar{a} \times (\bar{b} \times \bar{c}) = (\bar{a}.\bar{c}) \, \bar{b} - (\bar{a}.\bar{b}) \, \bar{c},$$

we obtain

$$L = \sum_{i=1}^{N} m_i \, r_i^2 \, \omega - \sum_i m_i (r_i \cdot \omega) \, r_i. \qquad \ldots(43)$$

Let the components of the position vector r_i, the angular velocity ω and the angular momentum vector L be denoted by

$$r_i = i x_i + j y_i + k z_i,$$

$$\omega = i \omega_x + j \omega_y + k \omega_z,$$

$$L = i L_x + j L_y + k L_z. \qquad \ldots(44)$$

Using equations (44) we write the equation (43) as

$$i L_x + j L_y + k L_z = \sum_i m_i \left[(i\omega_x + j\omega_y + k\omega_z) r_i^2 - (x_i\omega_x + y_i\omega_y + z_i\omega_z)(ix_i + jy_i + kz_i) \right].$$

$$i L_x + j L_y + k L_z = i \left[\omega_x \sum_i m_i (r_i^2 - x_i^2) - \omega_y \sum_i m_i x_i y_i - \omega_z \sum_i m_i x_i z_i \right] +$$

$$+ j \left[- \omega_x \sum_i m_i x_i y_i + \omega_y \sum_i m_i (r_i^2 - y_i^2) - \omega_z \sum_i m_i y_i z_i \right] +$$

$$+ k \left[- \omega_x \sum_i m_i x_i z_i - \omega_y \sum_i m_i y_i z_i + \omega_z \sum_i m_i (r_i^2 - z_i^2) \right].$$

Equating the corresponding coefficients on both the sides of the equation we get

$$L_x = \omega_x \sum_i m_i (r_i^2 - x_i^2) - \omega_y \sum_i m_i x_i y_i - \omega_z \sum_i m_i x_i z_i, \qquad \ldots(45a)$$

$$L_y = - \omega_x \sum_i m_i x_i y_i + \omega_y \sum_i m_i (r_i^2 - y_i^2) - \omega_z \sum_i m_i y_i z_i, \qquad \ldots(45b)$$

$$L_z = -\omega_x \sum_i m_i x_i z_i - \omega_y \sum_i m_i y_i z_i + \omega_z \sum_i m_i (r_i^2 - z_i^2). \qquad \ldots(45c)$$

We write the components of angular momentum as

$$L_x = I_{xx}\omega_x + I_{xy}\omega_y + I_{xz}\omega_z,$$

$$L_y = I_{yx}\omega_x + I_{yy}\omega_y + I_{yz}\omega_z,$$

$$L_z = I_{zx}\omega_x + I_{zy}\omega_y + I_{zz}\omega_z, \qquad \ldots(46)$$

where the coefficients of ω_x, ω_y, ω_z are respectively given by

$$I_{xx} = \sum_i m_i (r_i^2 - x_i^2) = \sum_i m_i(y_i^2 + z_i^2),$$

$$I_{yy} = \sum_i m_i (r_i^2 - y_i^2) = \sum_i m_i(x_i^2 + z_i^2),$$

$$I_{zz} = \sum_i m_i (r_i^2 - z_i^2) = \sum_i m_i(x_i^2 - z_i^2), \qquad \ldots(47)$$

and are called the moment of inertia about x, y and z axes respectively. Equations (47) show that the moment of inertia is the sum over the particles in the system of the product of masses and the square of its perpendicular distance from the axis of rotation. Also the quantities I_{xy}, I_{xz} and I_{yz} are called the product of inertia and are defined by

$$I_{xy} = -\sum_i m_i r_i y_i = I_{yx},$$

$$I_{xz} = -\sum_i m_i x_i z_i = I_{zx},$$

$$I_{xy} = -\sum_i m_i y_i z_i = I_{zy}. \qquad \ldots(48)$$

Now the equation (46) can be written in the matrix form as

$$\begin{pmatrix} L_x \\ L_y \\ L_z \end{pmatrix} = \begin{pmatrix} I_{xx} & I_{xy} & I_{xz} \\ I_{yx} & I_{yy} & I_{yz} \\ I_{zx} & I_{zy} & I_{zz} \end{pmatrix} \begin{pmatrix} \omega_x \\ \omega_y \\ \omega_z \end{pmatrix} \qquad \ldots(49)$$

or $$L = \bar{I}\omega, \qquad \ldots(50)$$

where $\bar{I}$ is called the moment of inertia tensor or inertia tensor. Note that moment of inertia tensor is symmetric and hence it has only six independent components. The moment of inertia tensor depends only on the mass distribution in the body. It is given by

$$\ddot{I} = \begin{pmatrix} I_{xx} & I_{xy} & I_{xz} \\ I_{yx} & I_{yy} & I_{yz} \\ I_{zx} & I_{zy} & I_{zz} \end{pmatrix}. \qquad \qquad \text{...(51)}$$

The diagonal elements are called the moments of inertia of the body about the given point and the given set of body axes. The off diagonal components of moment of inertia tensor are called the product of inertia of the body about the given point and the given set of body axes. Note that it is always possible to find a set of axes with respect to which all the products of inertia tensor vanish leaving diagonal terms and the axis is called **Principal axis** of the body. Thus with respect to the principal axis of the body, the products of inertia vanish.

Kinetic Energy of a Rigid Body with one Point Fixed

Theorem 4 Find the kinetic energy of a rigid body rotating about a fixed point of the body when the moments of inertia and products of inertia of the body relative to the set of axes through fixed point are known.

Proof: Consider a rigid body composed of N particles having mass m_i and rotating with instantaneous angular velocity ω. If one of the points of the rigid body is fixed then the translational motion is absent and the body rotates about an axis passes through the fixed point. If v_i is the linear velocity of the ith particle and position vector r_i with respect to the fixed point, then we have the relation $v_i = \omega \times r_i$. The kinetic energy of the body is given by

$$T = \frac{1}{2} \sum_i m_i v_i^2. \qquad \qquad \text{...(52)}$$

Eliminating v_i from equation (52), we get

$$T = \frac{1}{2} \sum_i m_i v_i (\omega \times r_i).$$

On using the vector identity $\bar{a}.(\bar{b} \times \bar{c}) = \bar{b}.(\bar{c} \times \bar{a})$ we simplify the expression for the kinetic energy as

$$T = \frac{1}{2} \sum_i m_i \omega. (r_i \times v_i),$$

$$= \frac{1}{2} \omega \sum_i (r_i \times m_i v_i),$$

$$= \frac{1}{2} \omega \sum_i r_i \times p_i,$$

$$T = \frac{1}{2} \omega.L, \qquad \qquad \text{...(53)}$$

where $L = \sum_i r_i \times p_i$ is the total angular momentum. Using equation (50) we eliminate the angular momentum of the rigid body to get

$$T = \frac{1}{2}\, \omega.\ddot{I}.\omega. \qquad\qquad ...(54)$$

Solving the equation (49) we obtain the expression for angular momentum as

$$iL_x + jL_y + kL_z = i\,(I_{xx}\omega_x + I_{xy}\omega_y + I_{xz}\omega_z) + j(I_{yx}\omega_x + I_{yy}\omega_y + I_{yz}\omega_z) +$$
$$+ k(I_{zx}\omega_x + I_{zy}\omega_y + I_{zz}\omega_z).$$

Hence the equation (53) becomes

$$T = \frac{1}{2}\,(i\omega_x + j\omega_y + k\omega_z)\left[i\,(I_{xx}\omega_x + I_{xy}\omega_y + I_{xz}\omega_z) +\right.$$
$$\left. + j(I_{yx}\omega_x + I_{yy}\omega_y + I_{yz}\omega_z) + k(I_{zx}\omega_x + I_{zy}\omega_y + I_{zz}\omega_z)\right].$$

This equation reduces to the equation

$$T = \frac{1}{2}(I_{xx}\omega_x^2 + I_{yy}\omega_y^2 + I_{zz}\omega_z^2) + 2I_{xy}\omega_x\,\omega_y + 2I_{yz}\omega_y\,\omega_z + 2I_{zx}\omega_z\omega_x). \qquad ...(55)$$

If the body rotates about z-axis with angular velocity ω, we have $\omega_z = \omega$, $\omega_x = \omega_y = 0$. In this case equation (55) becomes

$$T = \frac{1}{2}\,I\cdot\omega^2, \qquad\qquad ...(56)$$

where I is the moment of inertia of the body about z-axis.

Kinetic Energy of a Rotating Rigid Body

Theorem 5 Obtain the kinetic energy of a rotating rigid body.

Proof: Consider a rigid body composed of N particles having masses m_i. The kinetic energy T at any instant, with respect to any fixed outside inertial observer is given by

$$T = \frac{1}{2}\sum_i m_i v_i^2, \qquad\qquad ...(57)$$

where v_i is the instantaneous inertial velocity of any i^{th} particle of a body and is given by

$$v_i = \bar{u} + \bar{\omega} \times \bar{r}_i, \qquad\qquad ...(58)$$

where $\bar{u}$ is the homogeneous translational velocity and is same for the whole body and $\bar{\omega}$ is the instantaneous angular velocity, which is also same for the whole body. Thus using equation (58) in the equation (57), we get

$$T = \frac{1}{2} \sum_{i=1}^{N} m_i (\overline{u} + \overline{\omega} \times \overline{r}_i)^2 .$$

$$T = \frac{1}{2} u^2 \sum_{i=1}^{N} m_i + \sum_{i=1}^{N} m_i \overline{r}_i (\overline{u} \times \overline{\omega}) + \frac{1}{2} \sum_{i=1}^{N} m_i (\overline{\omega} \times \overline{r}_i)(\overline{\omega} \times \overline{r}_i).$$

The total mass of the body is the sum of the masses of the particles of the body. It is given by $M = \sum_{i=1}^{N} m_i$ similarly, if $R_{c.m.}$ is the position vector of the centre of mass of the body then it is defined by

$$R_{c.m.} = \frac{\sum_{i=1}^{N} m_i \overline{r}_i}{M}.$$

Hence the expression for the kinetic energy becomes

$$T = \frac{1}{2} M u^2 + M R_{c.m.} \cdot (\overline{u} \times \overline{\omega}) + \frac{1}{2} \sum_{i=1}^{N} m_i \overline{\omega} \left[\overline{r}_i \times (\overline{\omega} \times \overline{r}_i) \right],$$

$$T = \frac{1}{2} M u^2 + M R_{c.m.} \cdot (\overline{u} \times \overline{\omega}) + \frac{1}{2} \sum_{i=1}^{N} m_i \left[\omega^2 r_i^2 - (\overline{r}_i \cdot \overline{\omega})^2 \right]. \qquad ...(59)$$

For $\overline{\omega} = \omega_1 i + \omega_2 j + \omega_3 k$, we have $\omega^2 = \omega_1^2 + \omega_2^2 + \omega_3^2$. This can also be written as

$$\omega^2 = \sum_{j,k=1}^{N} \omega_j \omega_k \delta_{jk} = \sum_{j=1}^{N} \omega_j^2 . \qquad ...(60)$$

Similarly, we write

$$(\overline{\omega} \cdot \overline{r}_i)^2 = (\omega_1 x_i + \omega_2 y_i + \omega_3 z_i)^2 ,$$

$$(\overline{\omega} \cdot \overline{r}_i)^2 = \omega_1^2 x_i^2 + \omega_2^2 y_i^2 + \omega_3^2 z_i^2 + 2\omega_1 \omega_2 x_i y_i + 2\omega_1 \omega_3 x_i z_i + 2\omega_2 \omega_3 y_i z_i .$$

This can also be written as

$$(\overline{\omega} \cdot \overline{r}_i)^2 = \sum_{j,k=1}^{N} \omega_j \omega_k (\overline{r}_i)_j (\overline{r}_i)_k , \qquad ...(61)$$

where $(\overline{r}_i)_j$ represents the jth component of the position vector $\overline{r}_i$. Using the notations given in the equations (60) and (61), we write equation (59) as

$$T = \frac{1}{2} M u^2 + M R_{c.m.} \cdot (\overline{u} \times \overline{\omega}) + \frac{1}{2} \sum_{i=1}^{N} m_i \left[\sum_{j,k=1}^{N} \omega_j \omega_k \delta_{jk} r_i^2 - \sum_{j,k=1}^{N} \omega_j \omega_k (\overline{r}_i)_j (\overline{r}_i)_k \right]$$

$$T = \frac{1}{2} Mu^2 + MR_{c.m.} \cdot (\overline{u} \times \overline{\omega}) + \frac{1}{2} \sum_{j,k=1}^{N} \left[\sum_{i=1}^{N} m_i \left[\delta_{jk} r_i^2 - (\overline{r}_i)_j (\overline{r}_i)_k \right] \right] \omega_j \omega_k.$$

or

$$T = \frac{1}{2} Mu^2 + MR_{c.m.} \cdot (\overline{u} \times \overline{\omega}) + \frac{1}{2} \sum_{j,k=1}^{N} I_{jk} \omega_j \omega_k, \qquad \qquad ...(62)$$

where $I_{jk} = \sum_{i=1}^{N} m_i \left[\delta_{jk} r_i^2 - (\overline{r}_i)_j (\overline{r}_i)_k \right]$ are the components of the moments of inertia tensor.

Let the equation (62) be written in the form $T = T_t + T_m + T_r$, where the first term T_t corresponds to the translational kinetic energy of the whole body at given instant. The second term T_m is a term having mixtures of rotational and translational velocities of individual particles and be termed as the mixed term. The third term T_r is purely rotational term, depending exclusively on the components of the instantaneous angular velocity and configuration of the whole body. If one of the points of a rigid body is fixed, then translational motion of the rigid body is zero, and the rotational kinetic energy of the rigid body with one point fixed is given by equation

$$T_r = \frac{1}{2} \sum_{j,k=1}^{N} I_{jk} \omega_j \omega_k. \qquad \qquad ...(63)$$

This is equivalent to the result obtained earlier given by

$$T = \frac{1}{2} \left(I_{xx} \omega_x^2 + I_{yy} \omega_y^2 + I_{zz} \omega_z^2 + 2I_{xy} \omega_x \omega_y + 2I_{yz} \omega_y \omega_z + 2I_{zx} \omega_z \omega_x \right) \qquad ...(64)$$

If the body rotates about z-axis with angular velocity ω, then we have $\omega_x = \omega_y = 0$ and $\omega_z = \omega$. In this case equation (63) becomes $T = \frac{1}{2} I.\omega^2$, where I is the moment of inertia of the body about z axis. Thus from equation (62) the total kinetic energy expression of the body with one point fixed rotating about z-axis with angular velocity ω becomes

$$T = \frac{1}{2} Mu^2 + MR_{c.m.} \cdot (\overline{u} \times \overline{w}) + \frac{1}{2} I\omega^2. \qquad \qquad ...(65)$$

Example 6: Prove the statement that "*a* change in time dt of the components of a vector $\overline{r}$ as seen by an observer in the body system of axes will differ from the corresponding change as seen by an observer in the space system.

Solution: consider two co-ordinate systems S and S', where S' is rotating uniformly with angular velocity ω, with respect to the frame S with the same origin O. Let S be the space set of axes and S' the body set of axes. Let i, j, k be the unit vectors associated with the co-ordinate axes of S frame and i', j', k' be the unit vectors associated with x', y', z' axes of the S' frame. Consider the

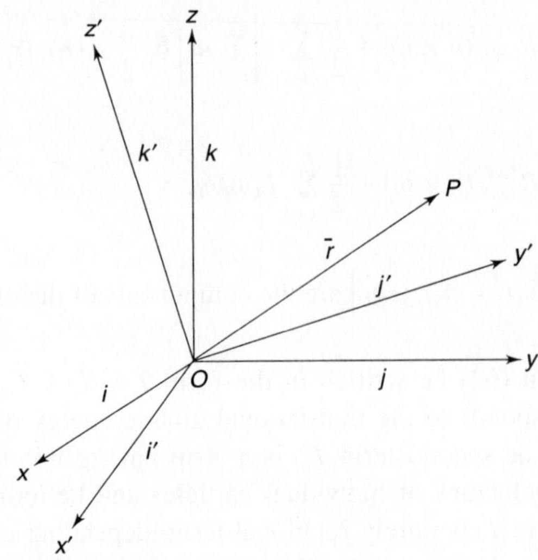

position vector $\bar{r}$ of a particle in a rigid body with respect to the body set of axes. It is represented by

$$\bar{r} = i'x' + j'y' + k'z'. \qquad \qquad ...(65)$$

Clearly such a vector appears constant when measured in the body set of axes. However, to an observer fixed in space set of axes, the components of the vector will vary in time. Let the components of the vector with respect to the space set of axes be given by

$$\bar{r} = ix + jy + kz. \qquad \qquad ...(66)$$

The time derivatives of $\bar{r}$ however will be different in the two systems. For the space (fixed) system S, we have

$$\left(\frac{d\bar{r}}{dt}\right)_{\text{fix}} = i\,\frac{dx}{dt} + j\,\frac{dy}{dt} + k\,\frac{dz}{dt}. \qquad \qquad ...(67)$$

Similarly, the time derivative of the vector $\bar{r}$ defined in S' with respect to the body set of axes is given by

$$\left(\frac{d\bar{r}}{dt}\right)_{\text{body}} = i'\,\frac{dx'}{dt} + j'\,\frac{dy'}{dt} + k'\,\frac{dz'}{dt}. \qquad \qquad ...(68)$$

However, as body rotates, the unit vectors of the body set of axes will be seen changing relative to the observer in the space set of axes. Hence we find the time derivative of the vector $\bar{r}$ in S' with respect to the fixed co-ordinate system as

$$\left(\frac{d\bar{r}}{dt}\right)_{fix} = i'\frac{dx'}{dt} + j'\frac{dy'}{dt} + k'\frac{dz}{dt} + x'\frac{di'}{dt} + y'\frac{dj'}{dt} + z'\frac{dk'}{dt}. \qquad \text{...(69)}$$

The first three terms on the right hand side of equation (69) are the time derivative of the vector in the rotating system, while the remaining three terms arise as a result of rotation of the system. Hence we write the equation (69) as

$$\left(\frac{d\bar{r}}{dt}\right)_{fix} = \left(\frac{d\bar{r}}{dt}\right)_{body} + x'\frac{di'}{dt} + y'\frac{dj'}{dt} + z'\frac{dk'}{dt}. \qquad \text{...(70)}$$

The linear velocity of a particle having the position vector $\bar{r}$ and rotating with angular ω is given by

$$\bar{v} = \frac{d\bar{r}}{dt} = \omega \times \bar{r}. \qquad \text{...(71)}$$

This formula can be applied to the unit vectors as a special case. Thus we write

$$\frac{di'}{dt} = \omega \times i', \quad \frac{dj'}{dt} = \omega \times j', \quad \frac{dk'}{dt} = \omega \times k'. \qquad \text{...(72)}$$

Hence equation (70) reduces to the equation

$$\left(\frac{d\bar{r}}{dt}\right)_{fix} = \left(\frac{d\bar{r}}{dt}\right)_{body} + x'\left(i'\omega_x + j'\omega_y + k'\omega_z\right) \times i' + y'\left(i'\omega_x + j'\omega_x + k'\omega_z\right) \times j' +$$
$$+ z'\left(i'\omega_x + j'\omega_y + k'\omega_z\right) \times k'.$$

$$\left(\frac{d\bar{r}}{dt}\right)_{fix} = \left(\frac{d\bar{r}}{dt}\right)_{body} + i'\left(z'\omega_y - y'\omega_z\right) - j'\left(z'\omega_x - x'\omega_z\right) + k'\left(y'\omega_x - x'\omega_y\right). \qquad \text{...(73)}$$

This is equivalent to

$$\left(\frac{d\bar{r}}{dt}\right)_{fix} = \left(\frac{d\bar{r}}{dt}\right)_{body} + \begin{vmatrix} i' & j' & k' \\ \omega_x & \omega_y & \omega_z \\ x' & y' & z' \end{vmatrix}.$$

$$\left(\frac{d\bar{r}}{dt}\right)_{fix} = \left(\frac{d\bar{r}}{dt}\right)_{body} + \omega \times \bar{r}. \qquad \text{...(74)}$$

This is the required relation between the two time derivatives of a vector with respect to two frames of references.

Note: The formula (74) can also be represented as

$$\left(\frac{d\bar{r}}{dt}\right)_{space} = \left(\frac{d\bar{r}}{dt}\right)_{rot} + \omega \times \bar{r}. \qquad \text{...(75)}$$

Euler's Equations of Motion

A set of equations governing the rotation of a rigid body referred to its own axis are known as Euler's equations of motion of a rigid body with one point fixed.

Theorem 6   Obtain the Euler's equations for motion of a rigid body when one point of the body remains fixed.

Proof: Consider a rigid of which one point is fixed. Hence translational motion of the body is absent and the body rotates about an axis passes through the fixed point. The rotation of the body takes place under the action of torque acting on it. Thus the equation of the rotational motion of the body in a fixed frame is given by

$$\text{Torque = rate of change of angular momentum.}$$

$$N = \left(\frac{dL}{dt}\right)_{\text{fix}}.$$

Using the equation (74) we write the equation of motion of the rigid body as

$$N = \left(\frac{dL}{dt}\right)_{\text{fix}} = \left(\frac{dL}{dt}\right)_{\text{body}} + \omega \times L, \qquad \ldots(76)$$

where ω is the angular velocity of the body and L is the angular momentum and is given by $L = \bar{I}\omega$, where $\bar{I}$ is the moment of inertia tensor and is constant with respect to the body frame of reference. We choose the principal axis of the body with respect to which the off diagonal elements of the moment of inertia tensor vanish and only the diagonal elements remain in the expression for $\bar{I}$. If I_1, I_2, I_3 are the principal moments of inertia then we have

$$\bar{I} = \begin{pmatrix} I_1 & 0 & 0 \\ 0 & I_2 & 0 \\ 0 & 0 & I_3 \end{pmatrix}.$$

In this case the expression for angular momentum is given by

$$L = I_1\omega_x i + I_2\omega_y j + I_3\omega_z k.$$

Differentiating this with respect to time t in the body frame we get

$$\left(\frac{dL}{dt}\right)_{\text{body}} = I_1\dot{\omega}_x i + I_2\dot{\omega}_y j + I_3\dot{\omega}_z k.$$

Also we find the value of

$$\omega \times L = \begin{vmatrix} i & j & k \\ \omega_x & \omega_y & \omega_z \\ I_1\omega_x & I_2\omega_y & I_3\omega_z \end{vmatrix}$$

$\Rightarrow \quad \omega \times L = i(I_3 - I_2)\,\omega_y\omega_z - j(I_3 - I_1)\,\omega_x\omega_z + k(I_2 - I_1)\,\omega_y\omega_x.$

Hence the equation of rotational motion of the rigid body becomes

$$N = \left(\frac{dL}{dt}\right)_{body} + \omega \times L = i\left[I_1\dot{\omega}_x + (I_3 - I_2)\,\omega_y\omega_z\right] + j\left[I_2\dot{\omega}_y + (I_1 - I_3)\,\omega_x\omega_z\right] +$$
$$+ k\left[I_3\dot{\omega}_z + (I_2 - I_1)\,\omega_x\omega_y\right].$$

If torque is expressed in term of its components as $N = iN_x + jN_x + kN_z$, then on equating the corresponding components on both sides of the above equations we obtain

$$N_x = I_1\dot{\omega}_x + (I_3 - I_2)\omega_y\omega_z,$$
$$N_y = I_2\dot{\omega}_y + (I_1 - I_3)\omega_x\omega_z,$$
$$N_z = I_3\dot{\omega}_z + (I_2 - I_1)\omega_x\omega_y. \qquad \ldots(77)$$

These are the required Euler's equations of motion of the rigid body with one point fixed.

Note: In the case of torque free motion of a rigid body, equations (77) reduce to

$$I_1\dot{\omega}_x = (I_2 - I_3)\omega_y\omega_z,$$
$$I_2\dot{\omega}_y = (I_3 - I_1)\omega_x\omega_z$$
$$I_3\dot{\omega}_z = (I_1 - I_2)\omega_x\omega_y. \qquad \ldots(78)$$

WORKED EXAMPLES

Example 7: If the rigid body with one point fixed rotates about the principal axis of the body, then show that (i) kinetic energy of the body and (ii) the magnitude of the angular momentum are constant throughout the motion.

Solution: We recall here the expression for the kinetic energy of a rigid body with one point fixed derived in equation (55) for our ready reference

$$T = \frac{1}{2}\left(I_{xx}\omega_x^2 + I_{yy}\omega_y^2 + I_{zz}\omega_z^2 + 2I_{xy}\omega_x\omega_y + 2I_{yz}\omega_y\omega_z + 2I_{zx}\omega_z\omega_x\right),$$

where I_{xx}, I_{yy}, I_{zz} and I_{xy}, I_{yz}, I_{zx} are the moments of inertia and product of inertia about the x, y and z axes respectively. If the body rotates about the principal axis of the body then the products of inertia tensors are zero and the moments of inertia tensors are constants. In this case the kinetic energy becomes

$$T = \frac{1}{2}\left(I_1\omega_x^2 + I_2\omega_y^2 + I_3\omega_z^2\right), \qquad \ldots(79)$$

where $I_1 = I_{xx}$, $I_2, = I_{yy}$, $I_3 = I_{zz}$ are constant moment of inertia. In the absence of torque the Euler's equations of motion of the rigid body are given by

$$I_1 \dot{\omega}_x + (I_3 - I_2)\, \omega_y \omega_z = 0,$$

$$I_2 \dot{\omega}_y + (I_1 - I_3)\, \omega_x \omega_z = 0,$$

$$I_3 \dot{\omega}_z + (I_2 - I_1)\, \omega_x \omega_y = 0. \qquad \qquad \ldots(80)$$

(i) Multiply each equation in (80) by ω_x, ω_y, ω_z respectively and adding we get

$$I_1 \omega_x \dot{\omega}_x + I_2 \omega_y \dot{\omega}_y + I_3 \omega_z \dot{\omega}_z = 0. \qquad \qquad \ldots(81)$$

This is nothing but

$$\frac{1}{2}\frac{d}{dt}(I_1 \omega_x^2 + I_2 \omega_y^2 + I_3 \omega_z^2) = 0$$

$$\Rightarrow \qquad \frac{dT}{dt} = 0,$$

$$\Rightarrow \qquad T = \frac{1}{2}(I_1 \omega_x^2 + I_2 \omega_y^2 + I_3 \omega_z^2) = \text{const.}$$

(ii) We now claim that the magnitude of the angular momentum is constant.

The moment of inertia tensor with respect to the principal axis of the body, is given by

$$\bar{I} = \begin{pmatrix} I_1 & 0 & 0 \\ 0 & I_2 & 0 \\ 0 & 0 & I_3 \end{pmatrix}$$

where I_1, I_2, I_3 are constant with respect to the principal axis of the body. Hence the expression for angular momentum becomes

$$L = I_1 \omega_x i + I_2 \omega_y j + I_3 \omega_z k,$$

$$\Rightarrow \qquad L^2 = L.L = I_1^2 \omega_x^2 + I_2^2 \omega_y^2 + I_3^2 \omega_z^2. \qquad \qquad \ldots(82)$$

Multiply each equation of (80) by $I_1 \omega_x$, $I_2 \omega_y$, $I_3 \omega_z$ respectively and adding we get

$$I_1^2 \omega_x \dot{\omega}_x + I_2^2 \omega_y \dot{\omega}_y + I_3^2 \omega_z \dot{\omega}_z = 0.$$

Equivalently we write this as

$$\frac{1}{2}\frac{d}{dt}(I_1^2 \omega_x^2 + I_2^2 \omega_y^2 + I_3^2 \omega_z^2) = 0,$$

$$\Rightarrow \qquad \frac{d}{dt}(L^2) = 0,$$

$$\Rightarrow \qquad L^2 = \text{const.}$$

This shows that the magnitude of the angular momentum of the rigid body is constant.

Components of Angular Velocity Vector along Body Set of Axes

Example 8: Show that the components of angular velocity vector along the body set of axes are given by

$$\omega_{x'} = \dot{\phi}\, \sin\theta\, \sin\psi + \dot{\theta}\, \cos\psi,$$

$$\omega_{y'} = \dot{\phi}\, \sin\theta\, \cos\psi - \dot{\theta}\, \sin\psi,$$

$$\omega_{z'} = \dot{\phi} + \cos\theta + \dot{\psi}.$$

Solution: Let (x, y, z) and (x', y', z') be the space (fixed) set of axes and body (rotating) set of axes respectively. Let a rigid body with one point fixed rotate instantaneously with angular velocity ω. We shall obtain the components of ω along the body set of axes. If ϕ, θ, ψ are the Eulerian angles, then their time derivatives $\dot{\phi}$, $\dot{\theta}$, $\dot{\psi}$ represent the angular speeds about the space z-axis, the line of nodes and the body z-axis respectively and not along the space set of axes or the body set of axes. We denote these three angular speeds by ω_ϕ, ω_θ, ω_ψ which are the three components of the angular velocity ω. Let $\overline{\omega} = (\omega_{x'}, \omega_{y'}, \omega_{z'})$ be the components of ω with respect to the body set of axes, x', y', z'.

If $\dot{\phi}\,(\omega_\phi)$ represents the angular speed about space z-axis then its components along the body set of axes are found by applying matrix of orthogonal transformations C through an angle θ about new x-axis and the matrix B through an angle ψ about new z-axis, as two orthogonal transformations are required to reach to body axes. If $(\omega_\phi)_{x'}$, $(\omega_\phi)_{y'}$, $(\omega_\phi)_{z'}$ are the components of $\dot{\phi}(\omega_\phi)$ along body set of axes, then we have

$$\begin{pmatrix} (\omega_\phi)_{x'} \\ (\omega_\phi)_{y'} \\ (\omega_\phi)_{z'} \end{pmatrix} = BC \begin{pmatrix} 0 \\ 0 \\ \dot{\phi} \end{pmatrix},$$

$$= \begin{pmatrix} \cos\psi & \sin\psi & 0 \\ -\sin\psi & \cos\psi & 0 \\ 0 & 0 & 1 \end{pmatrix} \begin{pmatrix} 1 & 0 & 0 \\ 0 & \cos\theta & \sin\theta \\ 0 & -\sin\theta & \cos\theta \end{pmatrix} \begin{pmatrix} 0 \\ 0 \\ \dot{\phi} \end{pmatrix},$$

$$\begin{pmatrix} (\omega_\phi)_{x'} \\ (\omega_\phi)_{y'} \\ (\omega_\phi)_{z'} \end{pmatrix} = \begin{pmatrix} \cos\psi, & \sin\psi\cos\theta, & \sin\psi\sin\theta \\ -\sin\psi, & \cos\psi\cos\theta, & \sin\theta\cos\psi \\ 0, & -\sin\theta, & \cos\theta \end{pmatrix} \begin{pmatrix} 0 \\ 0 \\ \dot{\phi} \end{pmatrix},$$

$$\Rightarrow \qquad (\omega_\phi)_{x'} = \dot{\phi} \sin \psi \sin\theta,$$

$$(\omega_\phi)_{y'} = \dot{\phi} \cos\psi \sin\theta,$$

$$(\omega_\phi)_{z'} = \dot{\phi} \cos \theta. \qquad \qquad \ldots(83)$$

Similarly, if $\omega_\theta(\dot{\theta})$ represents the angular speed along the line of nodes (new x'-axis), and $(\omega_\theta)_{x'}$, $(\omega_\theta)_{y'}$, $(\omega_\theta)_{z'}$, are the components of ω_θ $(\dot{\theta})$ about the body set of axes x', y', z', then to find these components, we apply the matrix of orthogonal transformation B-through an angle ψ about new z-axis to reach to the body axes; after θ rotation has been performed. Thus we have

$$\begin{pmatrix} (\omega_\theta)_{x'} \\ (\omega_\theta)_{y'} \\ (\omega_\theta)_{z'} \end{pmatrix} = B \begin{pmatrix} \dot{\theta} \\ 0 \\ 0 \end{pmatrix},$$

$$= \begin{pmatrix} \cos\psi & \sin\psi & 0 \\ -\sin\psi & \cos\psi & 0 \\ 0 & 0 & 1 \end{pmatrix} \begin{pmatrix} \dot{\theta} \\ 0 \\ 0 \end{pmatrix},$$

$$\Rightarrow \qquad (\omega_\theta)_{x'} = \dot{\theta} \cos\psi,$$

$$(\omega_\theta)_{y'} = - \dot{\theta} \sin\psi,$$

$$(\omega_\theta)_{z'} = 0. \qquad \qquad \ldots(84)$$

Now ω_ψ $(\dot{\psi})$ is already parallel to z' axis, no transformation is necessary. Hence, if $(\omega_\psi)_{x'}$, $(\omega_\psi)_{y'}$, $(\omega_\psi)_{z'}$ are components of ω_ψ with respect to the body set of axes, then they are given by

$$(\omega_\psi)_{x'} = 0,$$

$$(\omega_\psi)_{y'} = 0,$$

$$(\omega_\psi)_{z'} = \dot{\psi}. \qquad \qquad \ldots(85)$$

Thus the components of angular velocity ω, $(\omega_{x'}$, $\omega_{y'}$, $\omega_{z'}$, about body set of axes are defined by

$$\omega_{x'} = (\omega_\phi)_{x'} + (\omega_\theta)_{x'} + (\omega_\psi)_{x'},$$

$$\omega_{y'} = (\omega_\phi)_{y'} + (\omega_\theta)_{y'} + (\omega_\psi)_{y'},$$

$$\omega_{z'} = (\omega_\phi)_{z'} + (\omega_\theta)_{z'} + (\omega_\psi)_{z'}.$$

Using equations (83), (84) and (85), we readily obtain the components of angular velocity ω about body set of axes in the form

$$\omega_{x'} = \dot{\phi}\,\sin\theta\,\sin\psi + \dot{\theta}\,\cos\psi,$$

$$\omega_{y'} = \dot{\phi}\,\sin\theta\,\cos\psi - \dot{\theta}\,\sin\psi,$$

$$\omega_{z'} = \dot{\phi}\,\cos\theta + \dot{\psi}. \qquad\qquad ...(86)$$

Example 9: Show that the components of angular velocity vector along the space set of axes are given by

$$\omega_x = \dot{\theta}\,\cos\phi + \dot{\psi}\,\sin\theta\,\sin\phi,$$

$$\omega_y = \dot{\theta}\,\sin\phi - \dot{\psi}\,\sin\theta\,\cos\phi,$$

$$\omega_{z'} = \dot{\psi}\,\cos\theta + \dot{\phi}.$$

Solution: Let (x, y, z) and (x', y', z') be the space (fixed) set of axes and body (rotating) set of axes respectively. Let a rigid body one point fixed rotate instantaneously with angular velocity ω. We shall obtain the components of along the space set of axes. If ϕ, θ, ψ are the Eulerian angles, then their time derivatives $\dot{\phi}$, $\dot{\theta}$, $\dot{\psi}$ represent the angular speeds about the space z-axis, the line of nodes and the body z-axis respectively and not along the space set of axes or the body set of axes. We denote these three angular speeds by ω_ϕ, ω_θ, ω_ψ which are the three components of the angular velocity ω. Let $\overline{\omega} = (\omega_x, \omega_y, \omega_z)$ be the components of ω with respect to the space set of axes x, y, z.

If a rigid body is in rotation about an arbitrary axis, the components ω_x, ω_y, ω_z can be expressed in terms of Eulerian angles and the components $\dot{\phi}$, $\dot{\theta}$, $\dot{\psi}$. In order to go from body set of axes to space set of axes all the three rotational operations are to be performed in reverse order ψ-back, θ-back, and ϕ-back. Therefore, the components of ω_ψ i.e., $\dot{\psi}$ along space set of axes will be obtained by multiplying it with the inverse matrix A^{-1}, where A^{-1} is defined in the equation (39). Using this equation we readily obtain

$$(\omega_\psi)_x = \dot{\psi}\,\sin\phi\,\sin\theta,$$

$$(\omega_\psi)_y = -\,\dot{\psi}\,\cos\phi\,\sin\theta,$$

$$(\omega_\psi)_z = \dot{\psi}\,\cos\theta. \qquad\qquad ...(87)$$

Next ω_θ *i.e.*, $\dot{\theta}$ is along the line of nodes and the rotation ϕ about z' axis reversed. Therefore, the components of ω_θ, *i.e.*, $\dot{\theta}$ with respect to space set of axes are obtained by applying the inverse matrix of transformation of the product CD.

$$\begin{pmatrix} (\omega_\theta)_x \\ (\omega_\theta)_y \\ (\omega_\theta)_z \end{pmatrix} = (CD)^{-1} \begin{pmatrix} \dot{\theta} \\ 0 \\ 0 \end{pmatrix},$$

$$= \left[\begin{pmatrix} 1 & 0 & 0 \\ 0 & \cos\theta & \sin\theta \\ 0 & -\sin\theta & \cos\theta \end{pmatrix} \begin{pmatrix} \cos\phi & \sin\phi & 0 \\ -\sin\phi & \cos\phi & 0 \\ 0 & 0 & 1 \end{pmatrix} \right]^{-1} \begin{pmatrix} \dot{\theta} \\ 0 \\ 0 \end{pmatrix}$$

$$\begin{pmatrix} (\omega_\theta)_x \\ (\omega_\theta)_y \\ (\omega_\theta)_z \end{pmatrix} = \begin{pmatrix} \cos\phi & \sin\phi & 0 \\ -\sin\phi\cos\theta & \cos\phi\cos\theta & \sin\theta \\ \sin\theta\sin\phi & -\sin\theta\cos\phi & \cos\theta \end{pmatrix}^{-1} \begin{pmatrix} \dot{\theta} \\ 0 \\ 0 \end{pmatrix},$$

$$= \begin{pmatrix} \cos\phi & -\sin\phi\cos\theta & \sin\theta\sin\phi \\ \sin\phi & \cos\phi\cos\theta & -\sin\theta\cos\phi \\ 0 & \sin\theta & \cos\theta \end{pmatrix} \begin{pmatrix} \dot{\theta} \\ 0 \\ 0 \end{pmatrix}$$

$$\Rightarrow \quad (\omega_\theta)_x = \dot{\theta}\cos\phi,$$

$$(\omega_\theta)_y = \dot{\theta}\sin\phi,$$

$$(\omega_\theta)_z = 0. \qquad \qquad \dots(88)$$

Finally, the components of ω_ϕ *i.e.*, $\dot{\phi}$ with respect to the space set of axes are obtained by applying the inverse matrix of transformation D

$$\begin{pmatrix} (\omega_\phi)_x \\ (\omega_\phi)_y \\ (\omega_\phi)_z \end{pmatrix} = \begin{pmatrix} \cos\phi & \sin\phi & 0 \\ -\sin\phi & \cos\phi & 0 \\ 0 & 0 & 1 \end{pmatrix}^{-1} \begin{pmatrix} 0 \\ 0 \\ \dot{\phi} \end{pmatrix},$$

$$= \begin{pmatrix} \cos\phi & -\sin\phi & 0 \\ \sin\phi & \cos\phi & 0 \\ 0 & 0 & 1 \end{pmatrix} \begin{pmatrix} 0 \\ 0 \\ \dot{\phi} \end{pmatrix}$$

$$(\omega_\phi)_x = 0,$$

$$(\omega_\phi)_y = 0,$$

$$(\omega_\phi)_z = \dot{\phi}. \qquad \qquad \dots(89)$$

The components of angular velocity ω about space set of axes are thus defined by

$$\omega_x = (\omega_\phi)_x + (\omega_\theta)_x + (\omega_\psi)_x,$$

$$\omega_y = (\omega_\phi)_y + (\omega_\theta)_y + (\omega_\psi)_y,$$

$$\omega_z = (\omega_\phi)_z + (\omega_\theta)_z + (\omega_\psi)_z.$$

Using equations (87), (88), (89), we obtain

$$\omega_x = \dot{\theta} \cos \phi + \dot{\psi} \sin \theta \sin \phi,$$

$$\omega_y = \dot{\theta} \sin \phi - \dot{\psi} \sin \theta \cos \phi,$$

$$\omega_z = \dot{\psi} \cos \theta + \dot{\phi}. \qquad \qquad \dots(90)$$

These are the required components of angular velocity along the space set of axes.

Example 10:   Find the kinetic energy of rotation of a rigid body with respect to the principal axes in terms of Eulerian angles.

Solution: The kinetic energy of a rigid body with one point fixed with respect to the principal axis is obtained in the equation (79). It is given by

$$T = \frac{1}{2} (I_1 \omega_x^2 + I_2 \omega_y^2 + I_3 \omega_z^2), \qquad \qquad \dots(91)$$

where ω_x, ω_y, ω_z are the components of angular velocity vector with respect to the body set of axes and I_1, I_2, I_3 are the principal moments of inertia along the body set of axes taken as principal axes of the body. However, the components of angular velocity vector with respect to body set of axes are derived in the Example 8 (refer equation (86). These equations are given by

$$\omega_x = \dot{\phi} \sin \theta \sin \psi + \dot{\theta} \cos \psi,$$

$$\omega_y = \dot{\phi} \sin \theta \cos \psi - \dot{\theta} \sin \psi,$$

$$\omega_z = \dot{\phi} \cos \theta + \dot{\psi}. \qquad \qquad \dots(92)$$

Using equation (92) in the equation (91) we obtain the expression for kinetic energy of the rigid body with one point fixed in terms of Eulerian angles in the form

$$T = \frac{1}{2} \left[I_1 (\dot{\phi} \sin \theta \sin \psi + \dot{\theta} \cos \psi)^2 + \right.$$

$$\left. + I_2 (\dot{\phi} \sin \theta \cos \psi - \dot{\theta} \sin \psi)^2 + I_3 (\dot{\phi} \cos \theta + \dot{\psi})^2 \right]. \qquad \dots(93)$$

If the rigid body is symmetric about one of the principal axes say z-axis of the body, then we have $I_1 = I_2$ In this case kinetic energy of the rigid body becomes

$$T = \frac{1}{2} I_1(\dot{\phi}^2 \sin^2 \theta + \dot{\theta}^2) + \frac{1}{2} I_3(\dot{\phi} \cos \theta + \dot{\psi})^2. \qquad \ldots(94)$$

Example 11: Find Lagrangian and the equations of motion of a rigid body with one point fixed.

Solution: Consider a rigid body with one point fixed, then translational motion is absent and there is only rotational motion of the body about an axis passing through the fixed point. The Eulerian angles ϕ, θ, ψ completely determine the orientation of the body, hence ϕ, θ, ψ are the generalized co-ordinates. Hence the kinetic energy of the body is given by the equation (93) and the potential energy is $V = V(\phi, \theta, \psi)$. Hence the Lagrangian of the rigid body becomes

$$L = \frac{1}{2} [I_1(\dot{\phi} \sin \theta \sin\psi + \dot{\theta} \cos \psi)^2 +$$

$$+ I_2(\dot{\phi} \sin \theta \cos \psi - \dot{\theta} \sin \psi)^2 + I_3(\dot{\phi} \cos \theta + \dot{\psi})^2] - V. \quad \ldots(95)$$

We use the ψ – Lagrange's equation of motion for con-conservative system

$$\frac{d}{dt}\left(\frac{\partial T}{\partial \dot{\psi}}\right) - \frac{\partial T}{\partial \psi} = -\frac{\partial V}{\partial \psi}, \qquad \ldots(96)$$

where the generalized force $Q_j = -\dfrac{\partial V}{\partial q_j}$ is the torque on the rigid body. By the chain rule of differential calculus we have

$$\frac{\partial T}{\partial \dot{\psi}} = \frac{\partial T}{\partial \omega_x} \frac{\partial \omega_x}{\partial \dot{\psi}} + \frac{\partial T}{\partial \omega_y} \frac{\partial \omega_y}{\partial \dot{\psi}} + \frac{\partial T}{\partial \omega_z} \frac{\partial \omega_z}{\partial \dot{\psi}}, \qquad \ldots(97)$$

and

$$\frac{\partial T}{\partial \psi} = \frac{\partial T}{\partial \omega_x} \frac{\partial \omega_x}{\partial \psi} + \frac{\partial T}{\partial \omega_y} \frac{\partial \omega_y}{\partial \psi} + \frac{\partial T}{\partial \omega_z} \frac{\partial \omega_z}{\partial \psi}, \qquad \ldots(98)$$

Thus on using equations (79) and (92) we simplify the right hand sides of equations (97) and (98) and finally obtain

$$\frac{\partial T}{\partial \dot{\psi}} = I_3\omega_z \text{ and } \frac{\partial T}{\partial \psi} = (I_1 - I_2)\omega_x\omega_y.$$

Hence the ψ – Lagrange's equation of motion (96) becomes

$$I_3\dot{\omega}_z - (I_1 - I_2)\omega_x\omega_y = N_z, \qquad \ldots(99)$$

where $N_z = -\dfrac{\partial V}{\partial \psi}$. This is the z-equation of motion. The other two equations are obtained simply by permutation of the indices in equation (99) since identification of one of the principal axes, as the z-axis is arbitrary.

Example 12:   If a rectangular parallelepiped with its edges $2a$, $2a$, $2b$ rotates about its center of gravity under no force, prove that, its angular velocity about one principal axis is constant and about the other axis it is periodic.

Solution: The rigid body rotates under the action of no forces. In the absence of force the Euler's equation of motion are described in the set of equations (78). The rigid body is parallelepiped with its edges $2a$, $2a$, $2b$ therefore, the moments of inertia about the principal axes are given by

$$I_1 = I_2 = m\left(\frac{a^2 + b^2}{3}\right),\ I_3 = \frac{2}{3}\,ma^2, \qquad\qquad \text{...(100)}$$

where m is the mass of the parallelepiped. Substituting these values in the set of equations (78) we get

$$(a^2 + b^2)\,\dot{\omega}_x = (b^2 - a^2)\,\omega_y\omega_z,$$

$$(a^2 + b^2)\,\dot{\omega}_y = (a^2 - b^2)\,\omega_x\omega_z,$$

$$\dot{\omega}_z = 0 \Rightarrow \omega_z = n = \text{const.} \qquad\qquad \text{...(101)}$$

The last equation in (101) shows that the angular velocity about one principal axis is constant. Consequently, the other two equations give

$$(a^2 + b^2)\,\dot{\omega}_x = (b^2 - a^2)\,n\omega_y,$$

$$(a^2 + b^2)\,\dot{\omega}_y = (a^2 - b^2)n\omega_x. \qquad\qquad \text{...(102)}$$

Eliminating ω_y from equations (102) we obtain

$$\ddot{\omega}_x = -\left(n\,\frac{a^2 - b^2}{a^2 - b^2}\right)^2\omega_x. \qquad\qquad \text{...(103)}$$

This is a second order differential equation of simple harmonic motion showing that ω_x is periodic.

Example 13:   A body rotates about a point O under the action of no forces. Suppose $3A$, $5A$ and $6A$ are the principal moments of inertia at O and initially the angular velocity has components $\omega_x = n$, $\omega_y = 0$, $\omega_z = n$ about the corresponding principal axes. Show that at any time t $\omega_y = \dfrac{3n}{\sqrt{5}}\tan\left(\dfrac{nt}{\sqrt{5}}\right)$.

Solution: The rigid body rotates under the action of no forces. Hence the Euler's equations of motion in the absence of forces are given in the set of equations (78). The principal moments of inertia about the fixed point of the body and are given by

$$I_1 = 3A,\ I_2 = 5A,\ I_3 = 6A.$$

Substituting these values in the equation (78) we get

$$3\dot{\omega}_x = -\omega_y\omega_z, \quad 5\dot{\omega}_y = 3\omega_x\omega_z, \quad 3\dot{\omega}_z = -\omega_x\omega_y. \qquad \text{...(104)}$$

Multiplying first and the second equations of the set (104) by ω_x and ω_y respectively and simplifying we get

$$9\omega_x\dot{\omega}_x + 5\omega_y\dot{\omega}_y = 0.$$

Integrating we get

$$9\omega_x^2 + 5\omega_y^2 = c_1, \qquad \text{...(105)}$$

where c_1 is a constant of integration. To find the value of the constant, we apply the initial conditions. viz., initially $\omega_x = n$, $\omega_y = 0 \Rightarrow c_1 = 9n^2$ Hence equation (105) becomes

$$9\omega_x^2 + 5\omega_y^2 = 9n^2. \qquad \text{...(106)}$$

Similarly, from the first and the third equations of the set of equations (104) we have

$$\omega_x\dot{\omega}_x - \omega_z\dot{\omega}_z = 0,$$

$$\Rightarrow \qquad \frac{d}{dt}\left(\omega_x^2 - \omega_z^2\right) = 0.$$

Integrating we get $\omega_x^2 - \omega_z^2 = c_2$. By applying the initial conditions we readily obtain $c_2 = 0$. Hence we have

$$\omega_x^2 = \omega_z^2. \qquad \text{...(107)}$$

Using equations (106) and (107) in second equation in (104) gives

$$15\dot{\omega}_y = \left(9n^2 - 5\omega_y^2\right),$$

i.e.,

$$15\,\frac{d\omega_y}{\left(9n^2 - 5\omega_y^2\right)} = dt.$$

Integrating we get

$$t = \frac{\sqrt{5}}{n}\,\tanh^{-1}\left(\frac{\sqrt{5}}{3n}\,\omega_y\right)$$

$$\frac{\sqrt{5}}{3n}\,\omega_y = \tanh\left(\frac{nt}{\sqrt{5}}\right),$$

$$\omega_y = \frac{3n}{\sqrt{5}}\,\tanh\left(\frac{nt}{\sqrt{5}}\right).$$

Example 14: The principal moments of inertia of a body about the center of mass are A, $3A$, $6A$. The body is so rotated that its angular velocities about the axes are $3n$, $2n$, n respectively. If the body rotates under no force then show that the angular velocities about the principal axes at any time t latter are given by

$$\omega_x = 3\omega_z = \frac{9n}{\sqrt{n}} \operatorname{sech}\left(3nt + \frac{1}{2}\log 5\right),\ \omega_y = 3n \tanh\left(3nt + \frac{1}{2}\log 5\right).$$

Solution: Let I_1, I_2, I_3 be the principal moments of inertia about the center of gravity of the body, and are given by $I_1 = A$, $I_2 = 3A$ $I_3 = 6A$. It is also given that the body rotates with initial angular velocities $\omega_x = 3n$, $\omega_y = 2n$, $\omega_z = n$. Using the values of the principal moments of inertia in the torque free Euler's equations of motion (78), we get

$$\dot{\omega}_x = -3\omega_y\omega_z,\ 3\dot{\omega}_y = 5\omega_x\omega_z,\ 3\dot{\omega}_z = -\omega_x\omega_y. \qquad \ldots(108)$$

Multiplying the first equation of (108) by ω_x ,the second equation of (108) by ω_y and the third equation of (108) by ω_z we get respectively

$$\omega_x\dot{\omega}_x = -3\omega_x\omega_y\omega_z,\ 3\omega_y\dot{\omega}_y = 5\omega_x\omega_y\omega_z\ \ 3\omega_z\dot{\omega}_z = -\omega_x\omega_y\omega_z.$$

From these equations we have

$$\omega_x\dot{\omega}_x = 9\omega_z\dot{\omega}_z,\text{ and } 3\omega_y\dot{\omega}_y = -15\omega_z\dot{\omega}_z.$$

Integrating these equations we get

$$\omega_x^2 - 9\omega_z^2 = c_1,\text{ and } \omega_y^2 + 5\omega_z^2 = c_2,$$

where c_1 and c_2 are constants of integration. To find the values of the constants, we apply the initial values $\omega_x = 3n$, $\omega_y = 2n$, $\omega_z = n$, we find $c_1 = 0$ and $c_2 = 9n^2$. Hence we have

$$\omega_x = 3\omega_z \text{ and } \omega_y^2 + 5\omega_z^2 = 9n^2. \qquad \ldots(109)$$

Now using equation (109) in the second equation of (108) we obtain

$$\dot{\omega}_y = 5\omega_z^2,$$

$$\Rightarrow \qquad \frac{d\omega_y}{dt} = 9n^2 - \omega_y^2,$$

$$\Rightarrow \qquad dt = \frac{d\omega_y}{9n^2 - \omega_y^2}.$$

Integrating we get

$$t = \frac{1}{3n} \tanh^{-1}\left(\frac{\omega_y}{3n}\right) + k,$$

where k is a constant of integration. Initially when $t = 0$, $\omega_y = 2n$. This gives

$$k = \frac{1}{3n} \tanh^{-1}\left(\frac{2}{3}\right),$$

Hence we have

$$t = \frac{1}{3n} \tanh^{-1}\left(\frac{\omega_y}{3n}\right) - \frac{1}{3n} \tanh^{-1}\left(\frac{2}{3}\right).$$

However, on using the formula $\tanh^{-1}\left(\frac{x}{a}\right) = \frac{1}{2} \log\left(\frac{a+x}{a-x}\right)$ we obtain

$$t = \frac{1}{3n} \tanh^{-1}\left(\frac{\omega_y}{3n}\right) - \frac{1}{3n} \log \sqrt{5},$$

$$\Rightarrow \qquad 3nt + \log\sqrt{5} = \tanh^{-1}\left(\frac{\omega_y}{3n}\right)$$

$$\omega_y = 3n \tanh (3nt + \log \sqrt{5}). \qquad\qquad ...(110)$$

Now from equations (109) and (110), we have

$$5\omega_z^2 = 9n^2 - 9n^2 \tanh^2 (3nt + \log\sqrt{5}),$$

$$\omega_z^2 = \frac{9}{5} n^2 \operatorname{sech}^2 (3nt + \log\sqrt{5}),$$

$$\omega_z = \frac{3}{\sqrt{5}} n \operatorname{sech} (3nt + \log \sqrt{5}).$$

Thus we have $\omega_x = 3\omega_z = \dfrac{9n}{\sqrt{5}} \operatorname{sech} (3nt + \log\sqrt{5})$,

$$\omega_y = 3n \tanh (3nt + \log \sqrt{5}).$$

Cayley-Klein Parameters

Introduction: Caley-Klein parameters are another set of variables which are used for the description of rigid body with one point fixed. We have seen that the Eulerian angles are used to describe an orientation of a rigid body with one point fixed. However, it is found that these angles are difficult to use in the numerical computation, because of the large number of trigonometric functions involved. Various other groups of variables have also been used to describe the orientation of a rigid body. Klien's set of four complex parameters is one of them. He introduced the set of four parameters bearing his name to facilitate the integration of complicated gyroscopic problems. These parameters are much better adapted for use on computers. Furthermore, these four parameters are of great theoretical interest in modern branches of physics. Cayley-Klein parameters are the set of four complex numbers used to describe the orientation of a rigid body in space. These four complex terms involved all together eight quantities. We will first show how these eight quantities can be reduced to only three independent quantities to describe the orientation of the rigid body with one point fixed. We make use of these parameters and obtain the matrix of transformation which describes the motion of a rigid body with one point fixed. We also establish the relation between the Eulerian angles and the Cayley-Klein parameters. However, we begin with the definition of some terms which will help the reader in understanding the description of rigid body by Caley-Klein parameters.

Some Definitions

1. **Conjugate matrix:** The matrix obtained from any given matrix A by replacing its elements by the corresponding conjugate complex numbers is called the conjugate of A and it is denoted by A^*.

2. **Trace of a matrix:** Let A be a square matrix of order n. The sum of the elements of A lying along the principal diagonal is called the trace of A.

3. **Transposed conjugate of a matrix:** The transpose of the conjugate of a matrix Q is called transposed conjugate of Q and it is denoted by $Q^\dagger$. It is also called as adjoint of Q. Thus $Q^\dagger = (Q^*)^T = (Q^T)^*$.

4. **Unitary matrix:** A square matrix Q with complex elements is said to be unitary if $Q^\dagger Q = I = QQ^\dagger$. Unitary condition expect that $|Q| = 1 \Rightarrow Q$ is invertible and it is given by $Q^{-1} = Q^\dagger$. This also gives that

$$Q^{-1} = Q^\dagger = adjQ.$$

5. **Self adjoint:** A linear operator which is identical with its adjoint operator is called self-adjoint. If P is self-adjoint then $P = adjP$.

6. **Hermitian matrix:** A square matrix $A = (a_{ij})$ is said to be Hermitian if $a_{ij} = a^*_{ji}$ for every $i, j. \Rightarrow$ If $\tilde{A} = A^* \Rightarrow A$ is Hermitian.

7. **Similar matrices:** Let A and B two square matrices of the same order. Then A and B are said to be similar if there exists a non-singular matrix P such that $AP = PB. \Rightarrow A = PBP^{-1}$.

8. **Property:** Under similar transformation the self-adjoint (Hermitian) property of the matrix and the trace of the matrix are invariant.

WORKED EXAMPLES

Example 15: Explain how the eight quantities of Cayley-Klein parameters can be reduced to only three independent quantities to describe the orientation of a rigid body?

Solution: Consider a 2-dimensional complex space with u and v as complex axis. A general linear transformation in such a space is given by

$$u' = \alpha u + \beta v,$$

$$v' = \gamma u + \delta v, \qquad \qquad \dots(111)$$

where

$$Q = \begin{pmatrix} \alpha & \beta \\ \gamma & \delta \end{pmatrix} \qquad \qquad \dots(112)$$

is the rotation matrix in 2-dimensional complex plane, and α, γ, β, δ are four complex parameters known as Cayley-Klein parameters. There are eight quantities in four complex parameters. However, the minimum number of independent quantities needed to specify the orientation of a rigid body is only three. Thus to reduce eight quantities in equation (112) into three independent quantities, the matrix Q is restricted by imposing an additional condition that it is unitary. The unitary condition implies that

$$Q^{\dagger}Q = I = QQ^{\dagger}, \qquad \qquad \dots(113)$$

where $Q^{\dagger}$ is transposed conjugate of Q known as the adjoint matrix. As Q is unitary, the determinant of Q must be unity. *i.e.,* $|Q| = 1$. This gives

$$\alpha\delta - \beta\gamma = 1. \qquad \qquad \dots(115)$$

Expanding the equation (113) we get

$$\begin{pmatrix} \alpha & \beta \\ \gamma & \delta \end{pmatrix} \begin{pmatrix} \alpha^* & \gamma^* \\ \beta^* & \delta^* \end{pmatrix} = I$$

$$\begin{pmatrix} \alpha\alpha^* + \beta\beta^* & \alpha\gamma^* + \beta\delta^* \\ \gamma\alpha^* + \delta\beta^* & \gamma\gamma^* + \delta\delta^* \end{pmatrix} = \begin{pmatrix} 1 & 0 \\ 0 & 1 \end{pmatrix}$$

$$\Rightarrow \qquad \alpha\alpha^* + \beta\beta^* = 1,$$

$$\gamma\gamma^* + \delta\delta^* = 1,$$

$$\alpha\gamma^* + \beta\delta^* = 0. \qquad \qquad \dots(116)$$

we notice that the first two equations in (116) are real while the third is complex. Therefore equation (116) gives four conditions. These four conditions plus one unitary condition given in (115) are totally five conditions on eight quantities.

Therefore we are left with only three independent quantities, which are used to describe the orientation of the body. To calculate these three independent quantities, divide equation (115) by $\alpha\gamma$ to get

$$\frac{\delta}{\gamma} - \frac{\beta}{\alpha} = \frac{1}{\alpha\gamma}. \qquad \qquad ...(117)$$

Now from the last equation in (116), we have

$$\frac{\delta}{\gamma} = -\frac{a^*}{\beta^*}. \qquad \qquad ...(118)$$

Thus from equations (117) and (118), we have

$$-\frac{\alpha^*}{\beta^*} - \frac{\beta}{\alpha} = \frac{1}{\alpha\gamma}$$

$$\Rightarrow \qquad \frac{-(\alpha\alpha^* + \beta\beta^*)}{\beta^*} = \frac{1}{\gamma}$$

$$\Rightarrow \qquad \gamma = -\beta^*, \ \delta = \alpha^*. \qquad \qquad ...(119)$$

As a result of (119), the matrix Q takes the form

$$Q = \begin{pmatrix} \alpha & \beta \\ -\beta^* & \alpha^* \end{pmatrix} \qquad \qquad ...(120)$$

with the unitary condition

$$\alpha\alpha^* + \beta\beta^* = 1. \qquad \qquad ...(121)$$

Hence Q involves only 3 independent quantities, which are used to describe the orientation of a rigid body.

Matrix of Transformation in Terms of Cayley-Klein Parameters

Theorem 7 Obtain the matrix of transformation in Cayley-Klein parameters, which specify the orientation of a rigid body.

OR

Show that to each unitary matrix Q in the 2-dimensional complex space there is associated some real orthogonal matrix of transformation in ordinary 3-dimensional space.

Proof: Let P be a matrix operator in a specialized u, v complex co-ordinate system in a particular form given by

$$P = \begin{pmatrix} z & x - iy \\ x + iy & -z \end{pmatrix}. \qquad \qquad ...(122)$$

The matrix P is trace free and is self-adjoint. *i.e.*, $P = P^\dagger$. The variables x, y, z are real quantities taken as co-ordinates of a point in space. Suppose the matrix P is transformed to the matrix P' by means of the unitary matrix Q in the following way

$$P' = QPQ^\dagger, \qquad \qquad ...(123)$$

where

$$Q = \begin{pmatrix} \alpha & \beta \\ \gamma & \delta \end{pmatrix} \qquad \qquad ...(124)$$

is the matrix of transformation in 2-dimensional complex space and $Q^\dagger$ is a complex transposed conjugate of Q. *i.e.*, $Q^\dagger = (Q^T)^*$. Since Q is unitary, from the unitary property of Q, we have $QQ^\dagger = I$. This implies that adjoint of Q is same as its inverse. *i.e.*,

$$Q^{-1} = Q^\dagger = adj\ (Q). \qquad \qquad ...(125)$$

Therefore, the equation (123) becomes

$$P' = QPQ^{-1}. \qquad \qquad ...(126)$$

This shows that P' is the similarity transformation of P. It is well known that the self-adjoint and the trace free property of the matrix are invariant under similarity transformation. As P is self-adjoint and trace-free, therefore P' must be like wise self-adjoint and trace free. Thus P' can have the form

$$P' = \begin{pmatrix} z' & x' - iy' \\ x' + iy' & -z' \end{pmatrix},$$

where x', y', z' are to be determined. Let us denote

$$x - iy = x_-,\ x + iy = x_+.$$

Therefore, equation (126) with the help of equations (124) and (125) becomes

$$\begin{pmatrix} z' & x'_- \\ x'_+ & -z' \end{pmatrix} = \begin{pmatrix} \alpha & \beta \\ \gamma & \delta \end{pmatrix} \begin{pmatrix} z & x_- \\ x_+ & -z \end{pmatrix} \begin{pmatrix} \alpha^* & \gamma^* \\ \beta^* & \delta^* \end{pmatrix}.$$

We have proved that α, β, γ, δ are not independent but are related by the equations

$$\alpha^* = \delta,\ \beta^* = -\gamma.$$

Therefore, P' matrix becomes

$$P' = \begin{pmatrix} z' & x'_- \\ x'_+ & -z' \end{pmatrix} = \begin{pmatrix} \alpha & \beta \\ \gamma & \delta \end{pmatrix} \begin{pmatrix} z & x_- \\ x_+ & -z \end{pmatrix} \begin{pmatrix} \delta & -\beta \\ -\gamma & \alpha \end{pmatrix}$$

$$P' = \begin{pmatrix} z' & x'_- \\ x'_+ & -z' \end{pmatrix} = \begin{pmatrix} \alpha & \beta \\ \gamma & \delta \end{pmatrix} \begin{pmatrix} z\delta - \gamma x_-, & -\beta z + \alpha x_- \\ \delta x_+ + \gamma z, & -x_+\beta - z\gamma \end{pmatrix}$$

$$\begin{pmatrix} z' & x'_- \\ x'_+ & -z' \end{pmatrix} = \begin{pmatrix} z(\alpha\delta + \beta\gamma) - \alpha\gamma x_- + \beta\delta x_+, & \alpha^2 x_- - 2\alpha\beta z - \beta^2 x_+ \\ 2\gamma\delta z - \gamma^2 x_- + \delta^2 x_+, & \alpha\gamma x_- - (\alpha\delta + \gamma\beta)z - \beta\delta x_+ \end{pmatrix}.$$

This is the matrix transformation equation in complex 2-plane. Obviously the matrix on the right and side is hermitian, proving that the hermitian property is invariant under any unitary similar transformation. Equating the corresponding elements on both the sides we get

$$x'_+ = \delta^2 x_+ - \gamma^2 x_- + 2\gamma\delta z,$$

$$x'_- = -\beta^2 x_+ + \alpha^2 x_- - 2\alpha\beta z,$$

$$z' = \beta\delta x_+ - \alpha\gamma x_- + (\alpha\delta + \beta\gamma)z. \qquad \text{...(127)}$$

In matrix notation, we write these equations as

$$\begin{pmatrix} x'_+ \\ x'_- \\ z' \end{pmatrix} = \begin{pmatrix} \delta^2 & -\gamma^2 & 2\gamma\delta \\ -\beta^2 & \alpha^2 & -2\alpha\beta \\ \beta\delta & -\alpha\gamma & \alpha\delta + \beta\gamma \end{pmatrix} \begin{pmatrix} x_+ \\ x_- \\ z \end{pmatrix} \qquad \text{...(128)}$$

Explicitly, we write equations (127) as

$$\Rightarrow \qquad x' + iy' = (\delta^2 - \gamma^2)x + i\,(\delta^2 + \gamma^2)y + 2\gamma\delta z,$$

$$x' - iy' = (\alpha^2 - \beta^2)\,x - i\,(\alpha^2 + \beta^2)y - 2\alpha\beta z,$$

$$z' = (\beta\delta - \alpha\gamma)x + i\,(\alpha\gamma + \beta\delta)y + (\alpha\delta + \beta\gamma)z.$$

Solving these equations for x', y', z' we readily obtain

$$\begin{pmatrix} x' \\ y' \\ z' \end{pmatrix} = \begin{pmatrix} \frac{1}{2}(\alpha^2 - \beta^2 + \delta^2 - \gamma^2), & \frac{i}{2}(-\alpha^2 - \beta^2 + \delta^2 + \gamma^2), & \gamma\delta - \alpha\beta \\ \frac{i}{2}(\alpha^2 - \beta^2 - \delta^2 + \gamma^2), & \frac{1}{2}(\alpha^2 + \beta^2 + \delta^2 + \gamma^2), & -i(\alpha\beta + \gamma\delta) \\ \beta\delta - \alpha\gamma, & i(\alpha\gamma + \beta\delta), & \alpha\delta + \beta\gamma \end{pmatrix} \begin{pmatrix} x \\ y \\ z \end{pmatrix}$$

$$\text{...(129)}$$

where the coefficient matrix is A given by

$$A = \left| \begin{array}{ccc} \frac{1}{2}(\alpha^2 - \beta^2 + \delta^2 - \gamma^2), & \frac{i}{2}(-\alpha^2 - \beta^2 + \delta^2 + \gamma^2), & \gamma\delta - \alpha\beta \\[2ex] \frac{i}{2}(\alpha^2 - \beta^2 - \delta^2 + \gamma^2), & \frac{1}{2}(\alpha^2 + \beta^2 + \delta^2 + \gamma^2), & -i(\alpha\beta + \gamma\delta) \\[2ex] \beta\delta - \alpha\gamma, & i(\alpha\gamma + \beta\delta), & \alpha\delta + \beta\gamma \end{array} \right| \quad \ldots(130)$$

This is the matrix of transformation in terms of Cayley-Klein parameters. This matrix specifies the orientation of a rigid body. Hence the Cayley-Klein parameters specify the orientation of a rigid body.

Relation Between the Cayley-Klein Parameters and Eulerian Angles

Theorem 8 Establish the relation between the Eulerian angles ϕ, θ, ψ and the Cayley-Klein parameters α, β, γ, δ.

or

Obtain Q matrices Q_ϕ, Q_ϕ, Q_ψ in complex 2-plane corresponding to the separate successive rotations through an angle ϕ, θ, ψ in 3×3 real space. Hence obtain the orthogonal matrix of complete rotation.

Proof: The Eulerian angles ϕ, θ, ψ are the three successive angles of rotations about the specified axes, such that the space set of co-ordinates (x, y, z) coincide with the body set of co-ordinates (x', y', z'). These angles are used to describe the orientation of a rigid body. Similarly, Cayley-Klein parameters α, β, γ, δ. are also used to describe the orientation of rigid body. Now to find the relation between ϕ, θ, ψ and α, β, γ, δ we first construct Q matrices say Q_ϕ, Q_θ, Q_ψ corresponding to the separate successive rotations ϕ, θ, ψ and then combine them to form the complete matrix $Q = Q_\psi$, Q_θ, Q_ϕ of rotation.

First Rotation: Let $Q_\phi = \begin{pmatrix} \alpha & \beta \\ \gamma & \delta \end{pmatrix}$ be the matrix in 2-complex plane corresponding to the first rotation through an angle ϕ in 3×3 real space. This rotation through an angle ϕ is performed about z-axis. Hence the transformation equations are given by

$$x' = x \cos\phi + y \sin\phi,$$
$$y' = -x\sin\phi + y \cos\phi,$$
$$z' = z.$$

We write these equations as

$$x'_+ = x' + iy' = xe^{-i\phi} + iye^{-i\phi},$$
$$\Rightarrow x'_+ = e^{-i\phi}x_+, \quad x'_- = e^{i\phi} x_-, \quad z' = z. \qquad \ldots(131)$$

Comparing the coefficients of the equations (131) and (127) we have

$$\alpha^2 = e^{i\phi}, \ \delta^2 = e^{-i\phi}, \ \beta = 0 = \gamma.$$

Therefore, the Q_ϕ matrix corresponding to the rotation through an angle ϕ becomes

$$Q_\phi = \begin{pmatrix} e^{\frac{i\phi}{2}} & 0 \\ 0 & e^{-\frac{i\phi}{2}} \end{pmatrix}. \qquad \ldots(132)$$

Second Rotation: Let $Q_\theta = \begin{pmatrix} \alpha & \beta \\ \gamma & \delta \end{pmatrix}$ be the matrix in 2-complex plane corresponding to the second rotation through an angle θ in 3×3 real space. This rotation through an angle θ is performed about new x-axis. Hence the transformation equations are given by

$$x' = x,$$
$$y' = y \cos \theta + z \sin \theta,$$
$$z' = -y \sin \theta + z \cos \theta.$$

These equations can be put in the form

$$x'_+ = x' + iy' = x \left(\cos^2\left(\frac{\theta}{2}\right) + \sin^2\left(\frac{\theta}{2}\right) \right) + iy \cos \theta + iz \sin \theta,$$

$$x'_- = x' - iy' = x \left(\cos^2\left(\frac{\theta}{2}\right) + \sin^2\left(\frac{\theta}{2}\right) \right) - iy \cos \theta - iz \sin \theta,$$

$$z' = -y \sin \theta + z \cos \theta.$$

$$x'_+ = x \left(\cos^2\left(\frac{\theta}{2}\right) + \sin^2\left(\frac{\theta}{2}\right) \right) + iy \left(\cos^2\left(\frac{\theta}{2}\right) - \sin^2 \left(\frac{\theta}{2}\right) \right) + 2 \, iz \sin \left(\frac{\theta}{2}\right) \cos\left(\frac{\theta}{2}\right).$$

This can also be written in the form

$$x'_+ = x_+ \cos^2\left(\frac{\theta}{2}\right) + x_- \sin^2\left(\frac{\theta}{2}\right) + 2iz \sin\left(\frac{\theta}{2}\right) \cos\left(\frac{\theta}{2}\right).$$

Similarly, we find

$$x'_- = x_+ \sin^2\left(\frac{\theta}{2}\right) + x_- \cos^2 \left(\frac{\theta}{2}\right) - 2iz \sin\left(\frac{\theta}{2}\right) \cos \left(\frac{\theta}{2}\right),$$

$$z' = ix_+ \sin \left(\frac{\theta}{2}\right) \cos \left(\frac{\theta}{2}\right) - ix_- \sin \left(\frac{\theta}{2}\right) \cos \left(\frac{\theta}{2}\right) + z \left(\cos^2\left(\frac{\theta}{2}\right) - \sin^2\left(\frac{\theta}{2}\right) \right). \qquad \ldots(133)$$

Comparing the coefficients of the equations (133) with (127), we get

$$\alpha^2 = \delta^2 = \cos^2\left(\frac{\theta}{2}\right), \ \beta^2 = \gamma^2 = -\sin^2\left(\frac{\theta}{2}\right).$$

Hence Q_θ matrix becomes

$$Q_\theta = \begin{pmatrix} \cos\left(\frac{\theta}{2}\right), & i\,\sin\left(\frac{\theta}{2}\right) \\ i\,\sin\left(\frac{\theta}{2}\right), & \cos\left(\frac{\theta}{2}\right) \end{pmatrix}. \qquad \qquad ...(134)$$

Third Rotation: Let $Q_\psi = \begin{pmatrix} \alpha & \beta \\ \gamma & \delta \end{pmatrix}$ be the matrix in 2-complex plane corresponding to the third rotation through an angle ψ in 3×3 real space. This rotation through an angle ψ is performed about new z-axis. This rotation is affected by the transformation equations

$$x' = x \cos\psi + y \sin\psi,$$

$$y' = -x \sin\psi + y \cos\psi,$$

$$z' = z.$$

We write these equations in the form

$$x'_+ = e^{-i\psi}x_+,$$

$$x'_- = e^{i\psi}x_-,$$

$$z' = z. \qquad \qquad ...(135)$$

Hence comparing equation (135) with (127) we obtain the matrix Q_ψ corresponding to the third rotation through an angle ψ about new z-axis as

$$Q_\psi = \begin{pmatrix} e^{\frac{i\psi}{2}}, & 0 \\ 0, & e^{\frac{-i\psi}{2}} \end{pmatrix}. \qquad \qquad ...(136)$$

Hence the orthogonal matrix for the complete transformation from space set of axes to the body set of axes is obtained by taking the product of separate Q matrices for each of the three successive rotations ϕ, θ, ψ. Thus we obtain

$$Q = Q_\psi\, Q_\theta Q_\phi$$

$$Q = \begin{pmatrix} e^{\frac{i\psi}{2}}, & 0 \\ 0, & e^{\frac{-i\psi}{2}} \end{pmatrix} \begin{pmatrix} \cos\left(\frac{\theta}{2}\right), & i\,\sin\left(\frac{\theta}{2}\right) \\ i\sin\left(\frac{\theta}{2}\right), & \cos\left(\frac{\theta}{2}\right) \end{pmatrix} \begin{pmatrix} e^{\frac{i\phi}{2}}, & 0, \\ 0, & e^{\frac{-i\phi}{2}} \end{pmatrix}$$

$$Q = \begin{pmatrix} \alpha & \beta \\ \gamma & \delta \end{pmatrix} = \begin{vmatrix} e^{\frac{i}{2}(\psi + \phi)} \cos\left(\frac{\theta}{2}\right), & ie^{\frac{i}{2}(\psi - \phi)} \sin\left(\frac{\theta}{2}\right) \\ ie^{\frac{i}{2}(\phi - \psi)} \sin\left(\frac{\theta}{2}\right), & e^{\frac{-i}{2}(\psi + \phi)} \cos\left(\frac{\theta}{2}\right) \end{vmatrix} \qquad \text{...(137)}$$

$$\alpha = e^{\frac{i}{2}(\psi + \phi)} \cos\left(\frac{\theta}{2}\right), \quad \beta = ie^{\frac{i}{2}(\psi - \phi)} \sin\left(\frac{\theta}{2}\right),$$

$$\gamma = ie^{\frac{i}{2}(\phi - \psi)} \sin\left(\frac{\theta}{2}\right), \quad \delta = e^{\frac{-i}{2}(\psi + \phi)} \cos\left(\frac{\theta}{2}\right). \qquad \text{...(138)}$$

These are the required relations between the Eulerian angles and the Cayley-Klein parameters.

Note: Note from equations (132), (134) and (136) that the trace of any Q matrix through an angle say χ about some axis is $2\cos\left(\dfrac{\chi}{2}\right)$.

Example 16: With usual notations show that $\cos\left(\dfrac{\chi}{2}\right) = \cos\left(\dfrac{\phi + \psi}{2}\right) \cos\left(\dfrac{\theta}{2}\right)$.

Solution: This example is solved in Example (4). However, by using the relation between Eulerian angles and Cayley-Klein parameters we find the solution. The Q matrix $\begin{pmatrix} \alpha & \beta \\ \gamma & \delta \end{pmatrix}$ in terms of Eulerian angles is given by

$$Q = \begin{pmatrix} \alpha & \beta \\ \gamma & \delta \end{pmatrix} = \begin{vmatrix} e^{\frac{i}{2}(\psi + \phi)} \cos\left(\frac{\theta}{2}\right), & ie^{\frac{i}{2}(\psi - \phi)} \sin\left(\frac{\theta}{2}\right), \\ ie^{\frac{i}{2}(\phi - \psi)} \sin\left(\frac{\theta}{2}\right), & e^{-\frac{i}{2}(\psi + \phi)} \cos\left(\frac{\theta}{2}\right) \end{vmatrix}, \qquad \text{...(139)}$$

where α, β, α, δ, are Cayley-Klein parameters such that $\alpha^* = \delta$, $\beta^* = -\gamma$. This is the matrix of rotation, which describes the orientation of the rigid body with one point fixed. Let this matrix be equivalent to the matrix B obtained by rotating the co-ordinate axes through an angle χ about any axis with the same origin. Then the matrix B is given by

$$B = \begin{pmatrix} \cos\chi & \sin\chi & 0 \\ -\sin\chi & \cos\chi & 0 \\ 0 & 0 & 1 \end{pmatrix} \text{ or } B = \begin{pmatrix} 1 & 0 & 0 \\ 0 & \cos\chi & \sin\chi \\ 0 & -\sin\chi & \cos\chi \end{pmatrix}. \qquad \text{...(140)}$$

Then the Q matrix in complex 2-dimensional plane corresponding to the matrix B is similarly obtain in the form

$$Q_\chi = \begin{bmatrix} e^{\frac{i\chi}{2}} & 0 \\ 0 & e^{-\frac{i\chi}{2}} \end{bmatrix} \text{ or } Q_\chi = \begin{bmatrix} \cos\left(\frac{\chi}{2}\right), & i\sin\left(\frac{\chi}{2}\right) \\ i\sin\left(\frac{\chi}{2}\right), & \cos\left(\frac{\chi}{2}\right) \end{bmatrix}. \qquad \text{...(141)}$$

From equation (139), we have

$$\alpha = e_0 + ie_3 = e^{\frac{i}{2}(\psi+\phi)} \cos\left(\frac{\theta}{2}\right), \quad \beta = e_2 + ie_1 = ie^{\frac{i}{2}(\psi-\phi)} \sin\left(\frac{\theta}{2}\right),$$

$$\delta = e_0 - ie_3 = e^{\frac{-i}{2}(\psi+\phi)} \cos\left(\frac{\theta}{2}\right), \quad \gamma = -e_2 + ie_1 = ie^{-\frac{i}{2}(\psi-\phi)} \sin\left(\frac{\theta}{2}\right). \qquad \text{...(142)}$$

Solving these equations, we find

$$e_0 = \frac{\alpha+\delta}{2} = \cos\left(\frac{\theta}{2}\right)\cos\left(\frac{\phi+\psi}{2}\right), \quad e_1 = \frac{-i}{2}(\beta+\gamma) = \sin\left(\frac{\theta}{2}\right)\cos\left(\frac{\psi-\phi}{2}\right),$$

$$e_2 = \frac{\beta-\gamma}{2} = -\sin\left(\frac{\theta}{2}\right)\sin\left(\frac{\psi-\phi}{2}\right), \quad e_3 = \frac{\alpha-\delta}{2} = \cos\left(\frac{\theta}{2}\right)\sin\left(\frac{\phi+\psi}{2}\right) \qquad \text{...(143)}$$

We see that

$$\text{trace of} \qquad Q = \alpha + \delta = 2\cos\left(\frac{\theta}{2}\right)\cos\left(\frac{\phi+\psi}{2}\right). \qquad \text{...(144)}$$

Similarly, from equation (141) we have

$$\text{trace of} \qquad Q_\chi = 2\cos\left(\frac{\chi}{2}\right). \qquad \text{...(145)}$$

Since trace of the matrix is invariant under similar transformation, therefore, from equations (144) and (145), we have

$$\cos\left(\frac{\chi}{2}\right) = \cos\left(\frac{\phi+\psi}{2}\right)\cos\left(\frac{\phi}{2}\right).$$

Example 17: Find a real matrix of orthogonal transformation in the 3-dimensional space corresponding to the unitary matrix

$$Q = \begin{bmatrix} \cos\left(\frac{\theta}{2}\right), & i\sin\left(\frac{\theta}{2}\right) \\[2mm] i\sin\left(\frac{\theta}{2}\right), & \cos\left(\frac{\theta}{2}\right) \end{bmatrix}$$

in 2-dimensional complex plane.

Solution: Take the matrix P in a particular form which is trace free and self adjoint and following the method illustrated in the Theorem 7, we transform it to another matrix by means of unitary matrix Q. This gives

$$P' = QPQ^{-1}. \qquad \text{...(146)}$$

This shows that P' is obtained from P by means of similarity transformation. Consequently, P' must be like wise self-adjoint and trace free. Thus P' can have the form

$$P' = \begin{pmatrix} z' & x'_- \\ x'_+ & -z' \end{pmatrix} \qquad \qquad \text{...(147)}$$

where x', y', z' are to be determined.

Therefore equation (146) becomes

$$\begin{pmatrix} z' & x'_- \\ x'_+ & -z' \end{pmatrix} = \begin{bmatrix} \cos\left(\dfrac{\theta}{2}\right) & i\sin\left(\dfrac{\theta}{2}\right) \\ i\sin\left(\dfrac{\theta}{2}\right) & \cos\left(\dfrac{\theta}{2}\right) \end{bmatrix} \begin{pmatrix} z & x_- \\ x_+ & -z \end{pmatrix} \begin{bmatrix} \cos\left(\dfrac{\theta}{2}\right) & -i\sin\left(\dfrac{\theta}{2}\right) \\ -i\sin\left(\dfrac{\theta}{2}\right) & \cos\left(\dfrac{\theta}{2}\right) \end{bmatrix}.$$

Solving the right hand side of this equation we find

$$\begin{pmatrix} z' & x'_- \\ x'_+ & -z' \end{pmatrix} =$$

$$= \begin{bmatrix} z\left(\cos^2\left(\dfrac{\theta}{2}\right) - \sin^2\left(\dfrac{\theta}{2}\right)\right) + ix_+ \cos\left(\dfrac{\theta}{2}\right)\sin\left(\dfrac{\theta}{2}\right) - ix_- \cos\left(\dfrac{\theta}{2}\right)\sin\left(\dfrac{\theta}{2}\right), \\ x_+ \cos^2\left(\dfrac{\theta}{2}\right) + x_- \sin^2\left(\dfrac{\theta}{2}\right) + 2iz\cos\left(\dfrac{\theta}{2}\right)\sin\left(\dfrac{\theta}{2}\right), \\ x_+ \sin^2\left(\dfrac{\theta}{2}\right) + x_- \cos^2\left(\dfrac{\theta}{2}\right) - 2iz\cos\left(\dfrac{\theta}{2}\right)\sin\left(\dfrac{\theta}{2}\right), \\ z\left(\sin^2\left(\dfrac{\theta}{2}\right) - \cos^2\left(\dfrac{\theta}{2}\right)\right) + ix_- \cos\left(\dfrac{\theta}{2}\right)\sin\left(\dfrac{\theta}{2}\right) - ix_+ \cos\left(\dfrac{\theta}{2}\right)\sin\left(\dfrac{\theta}{2}\right) \end{bmatrix}$$

On equating the corresponding elements of the matrix we get

$$x'_+ = x' + iy' = 2iz\cos\left(\dfrac{\theta}{2}\right)\sin\left(\dfrac{\theta}{2}\right) + (x+iy)\cos^2\left(\dfrac{\theta}{2}\right) + (x-iy)\sin^2\left(\dfrac{\theta}{2}\right),$$

$$x'_- = x' - iy' = -2iz\cos\left(\dfrac{\theta}{2}\right)\sin\left(\dfrac{\theta}{2}\right) + (x-iy)\cos^2\left(\dfrac{\theta}{2}\right) + (x+iy)\sin^2\left(\dfrac{\theta}{2}\right),$$

$$z' = z\left(\cos^2\left(\dfrac{\theta}{2}\right) - \sin^2\left(\dfrac{\theta}{2}\right)\right) - i(x-iy)\cos\left(\dfrac{\theta}{2}\right)\sin\left(\dfrac{\theta}{2}\right) +$$

$$+ i(x+iy)\cos\left(\dfrac{\theta}{2}\right)\sin\left(\dfrac{\theta}{2}\right)$$

$$\Rightarrow \qquad \begin{aligned} x' + iy' &= x + iy\cos\theta + iz\sin\theta, \\ x' - iy' &= x - iy\cos\theta - iz\sin\theta, \\ z' &= z\cos\theta - y\sin\theta. \end{aligned}$$

Solving for x', y', z', we obtain

$$x' = x,$$
$$y' = y \cos \theta + z \sin \theta,$$
$$z' = -y \sin \theta + z \cos \theta.$$

i.e.,
$$\begin{pmatrix} x' \\ y' \\ z' \end{pmatrix} = \begin{pmatrix} 1 & 0 & 0 \\ 0 & \cos \theta & \sin \theta \\ 0 & -\sin \theta & \cos \theta \end{pmatrix} \begin{pmatrix} x \\ y \\ z \end{pmatrix}$$

This shows that corresponding to the given Q matrix in 2-dimensional complex plane, there exits 3×3 real matrix $A = \begin{pmatrix} 1 & 0 & 0 \\ 0 & \cos \theta & \sin \theta \\ 0 & -\sin \theta & \cos \theta \end{pmatrix}$ in 3-dimensional real space.

Example 18: Find a real matrix of orthogonal transformation in the 3-dimensional space corresponding to the unitary matrix

$$Q = \begin{pmatrix} e^{\frac{i\phi}{2}} & 0 \\ 0 & e^{\frac{-i}{2}\phi} \end{pmatrix}$$

in 2-dimensional complex plane.

Solution: For the given Q matrix in 2-dimensional complex plane, we find in the same way as illustrated in the Example 17 the matrix equation

$$\begin{pmatrix} z' & x'_- \\ x'_+ & -z' \end{pmatrix} = \begin{pmatrix} e^{\frac{i\phi}{2}} & 0 \\ 0 & e^{-\frac{i\phi}{2}} \end{pmatrix} \begin{pmatrix} z & x_- \\ x_+ & -z \end{pmatrix} \begin{pmatrix} e^{-\frac{i\phi}{2}} & 0 \\ 0 & e^{\frac{i\phi}{2}} \end{pmatrix},$$

$$\begin{pmatrix} z' & x'_- \\ x'_+ & -z' \end{pmatrix} = \begin{pmatrix} z & x_- e^{i\phi} \\ x_+ e^{-i\phi} & -z \end{pmatrix}.$$

$\Rightarrow \qquad x' + iy' = (x + iy)(\cos\phi - i \sin \phi),$

$$x' - iy' = (x - iy)(\cos \phi + i \sin \phi),$$

$$z' = z.$$

Solving these equations for x', y', z' we readily get

$\Rightarrow \qquad x' = x \cos \phi + y \sin \phi,$

$$y' = -x \sin \phi + y \cos \phi$$

$$z' = z.$$

$$\Rightarrow \quad \begin{pmatrix} x' \\ y' \\ z' \end{pmatrix} = \begin{pmatrix} \cos\phi, & \sin\phi, & 0 \\ -\sin\phi, & \cos\phi, & 0 \\ 0, & 0, & 1 \end{pmatrix} \begin{pmatrix} x \\ y \\ z \end{pmatrix}.$$

This shows that $A = \begin{pmatrix} \cos\phi & \sin\phi & 0 \\ -\sin\phi & \cos\phi & 0 \\ 0 & 0 & 1 \end{pmatrix}$ is the required matrix in 3-dimensional

space corresponding to the given Q matrix in 2-dimensional complex plane.

Example 19: Find a real matrix of orthogonal transformation in the 3-dimensional space corresponding to the unitary matrix

$$Q = \begin{pmatrix} e^{\frac{i\psi}{2}} & 0 \\ 0 & e^{-\frac{i\psi}{2}} \end{pmatrix}$$

in 2-dimensional complex plane.

Solution: In the same way we transform the trace-free and self adjoint matrix P to the matrix P' by means of the given Q matrix. The transformation equation (146) in this case becomes

$$\begin{pmatrix} z' & x'_- \\ x'_+ & -z' \end{pmatrix} = \begin{pmatrix} e^{\frac{i\psi}{2}} & 0 \\ 0 & e^{-\frac{i\psi}{2}} \end{pmatrix} \begin{pmatrix} z & x_- \\ x_+ & -z \end{pmatrix} \begin{pmatrix} e^{-\frac{i\psi}{2}} & 0 \\ 0 & e^{\frac{i\psi}{2}} \end{pmatrix}$$

$$\begin{pmatrix} z' & x'_- \\ x'_+ & -z' \end{pmatrix} = \begin{pmatrix} z & x_- e^{i\psi} \\ x_+ e^{-i\psi} & -z \end{pmatrix}.$$

$$\Rightarrow \quad x' + iy' = (x + iy)(\cos\psi - i\sin\psi),$$

$$x' - iy' = (x - iy)(\cos\psi + i\sin\psi),$$

$$z' = z.$$

Solving these equations for we x', y', z' readily get

$$\Rightarrow \quad x' = x\cos\psi + y\sin\psi,$$

$$y' = -x\sin\psi + y\cos\psi,$$

$$z' = z$$

$$\Rightarrow \quad \begin{pmatrix} x' \\ y' \\ z' \end{pmatrix} = \begin{pmatrix} \cos\psi & \sin\psi & 0 \\ -\sin\psi & \cos\psi & 0 \\ 0 & 0 & 1 \end{pmatrix} \begin{pmatrix} x \\ y \\ z \end{pmatrix}.$$

This shows there exists 3×3 real matrix $A = \begin{pmatrix} \cos\psi & \sin\psi & 0 \\ -\sin\psi & \cos\psi & 0 \\ 0 & 0 & 1 \end{pmatrix}$ in 3-dimensional

space corresponding to the given Q matrix in 2-complex plane.

Note: The set of complex unitary matrices and the set of real orthogonal matrices are homomorphic.

Exercise

1. Show that the trace of a matrix and the self-adjoint property are invariant under any similarity transformation.
2. Find a real matrix of orthogonal transformation in the 3-dimensional space corresponding to the unitary matrix in $Q = \begin{pmatrix} e^{\frac{i\psi}{2}} & 0 \\ 0 & e^{-\frac{i\psi}{2}} \end{pmatrix}$ in 2-dimensional complex plane.
3. Consider the force free motion of a rigid body about a fixed point O. Suppose $3A$, $5A$ and $6A$ are the principal moments of inertia at O and initially the angular velocity has components $\omega_1 = \sqrt{5}$, $\omega_2 = 0$, $\omega_3 = \sqrt{5}$ about the corresponding principal axes, then show that $5\omega_2^2 + 9\omega_1^2 = 45$, $\omega_1^2 = \omega_3^2$ and the body ultimately rotates about the mean axis.
4. Solve the Euler's equation in force free motion and show that
$\omega_1 = a \cos(\Omega t + b)$, $\omega_2 = - a \sin (\omega t + b)$, $\omega_3 = $ const., where
$$\Omega = \frac{I_1 - I_3}{I_1} \omega_3.$$

6 Canonical Transformation

INTRODUCTION

A set of generalized coordinates is used in the Chapter 1 to describe the motion of a particle or system of particles. However, a given system can be described by more than one set of generalized coordinates. For example, to discuss the motion of a particle in a plane, one can use either Cartesian coordinates x, y or the polar coordinates r and θ as generalized coordinates. We have also seen in the Chapter 3 that the second set of generalized coordinates is more convenient to describe the motion of a planet under central force, because θ is happened to be cyclic coordinate. Consequently, the corresponding generalized momentum is conserved. Hence the solution of the problem becomes easy as there involved only one coordinates r in the description by the Hamiltonian procedure. Since the choice of the generalized coordinates is quite arbitrary, and if we can transform from a given set of generalized coordinates q_j to another new set of generalized coordinates say Q_j such that all the coordinates are cyclic then Hamilton's equations of motion can be solved by just integration. Hence it is often important to transform from one set of generalized coordinates q_j to another set of generalized coordinates Q_j in which some of the new coordinates are cyclic. In this chapter we discuss a procedure for transforming one set of generalized coordinates in to another set which may be more convenient in the discussion of motion of a dynamical system.

Point transformation   The transformation from one set of generalized coordinates q_j to a new set of generalized coordinate Q_j represented by the equation

$$Q_j = Q_j(q_j, t)$$

is called point transformation. This transformation can be referred to as the transformation of the configuration space as only the n generalized coordinates q_j spanned the configuration space.

Gauge function and gauge transformation   We know that the Lagrangian L and the Hamiltonian H are related through the relation

$$H(q_j, p_j, t) = \sum_j p_j \dot{q}_j - L(q_j, \dot{q}_j, t), \qquad \qquad \ldots(1)$$

where $L = T - V$. Since the choice of the generalized coordinates is not unique hence the kinetic energy and the potential energy functions which depend upon the choice of the generalized coordinates and the generalized velocities are not unique. Consequently, the Lagrangian and therefore, the Hamiltonian are also not uniquely defined. If F is an arbitrary function of generalized coordinates q_j and time t then $\dfrac{dF}{dt}$ satisfies the Lagrange's equations of motion. We have proved (refer the Example 13 of Chapter 1) that the new Lagrangian defined by

$$L' = L + \frac{d}{dt} F(q_j, t) \qquad \qquad \ldots(2)$$

always satisfy the Lagrange's equations of motion identically. Equation (2) asserts that the Lagrangian L' is always uncertain by the term $\dfrac{dF}{dt}$. The function $F = F(q_j, t)$ is called the gauge function and the transformation (2) which keeps the dynamical equations of motion unchanged is called the gauge transformation. However, we have also proved in the Example 8 of Chapter 3 that, under the gauge transformation the canonical momenta change and the new Lagrangian L' keeps the Hamilton's principle invariant.

Canonical transformation Let a dynamical system be described by the Hamiltonian $H(q_j, p_j, t)$, where q_j and p_j, $j = 1, 2, 3, \ldots, n$ are position and momentum coordinates respectively. The Hamilton's equations of motion are given by $\dot{q}_j = \dfrac{\partial H}{\partial p_j}$ and $\dot{p}_j = -\dfrac{\partial H}{\partial q_j}$. The transformation from a set of $2n$ variables q_j and p_j, $j = 1, 2, 3, \ldots, n$ to another set of $2n$ variables Q_j and P_j, $j = 1, 2, 3, \ldots, n$ given by the equations

$$Q_j = Q_j(q_i, p_i, t),$$

$$P_j = P_j(q_i, p_i, t) \qquad \qquad \ldots(3)$$

is called canonical transformation if there exists a transformed Hamiltonian

$$K = K(Q_j, P_j, t) \qquad \qquad \ldots(4)$$

in the new coordinates with the Hamilton's canonical equations of motion

$$\dot{Q}_j = \frac{\partial K}{\partial p_j} \text{ and } \dot{P}_j = -\frac{\partial K}{\partial Q_j}. \qquad \qquad \ldots(5)$$

The coordinates Q_j and P_j are called as the canonical coordinates; which are respectively new positions coordinates and new momentum coordinates. The

Hamiltonian functions H and K in terms of the two sets of coordinates are given by

$$H = \sum_j p_j \dot{q}_j - L(q_j, \dot{q}_j, t),$$

$$K = \sum_j p_j \dot{Q}_j - L'(Q_j, \dot{Q}_j, t). \qquad \ldots(6)$$

The transformation equations (3) which leave the form of Hamilton's canonical equations of motion invariant are required to satisfy the invariance of Hamilton's principle. This implies that

$$\delta \int_{t_0}^{t_1} \left[\sum_j p_j \dot{q}_j - H(q_j, p_j, t) \right] dt = 0,. \qquad \ldots(7)$$

$$\Rightarrow \qquad \delta \int_{t_0}^{t_1} \left[\sum_j P_j \dot{Q}_j - K(Q_j, P_j, t) \right] dt = 0. \qquad \ldots(8)$$

The simultaneous validity of equations (7) and (8) does not mean that the integrands of the two integrals are equal. The two integrals may differ by the total time derivative of the function $F = F(q_j, p_j, t)$ because

$$\delta \int_{t_0}^{t_1} \frac{dF}{dt}\, dt = \left[\delta F(q_j, p, t) \right]_{t_0}^{t_1} = \left(\sum_j \frac{\partial F}{\partial q_j} \delta q_j \right)_{t_0}^{t_1} + \left(\sum_j \frac{\partial F}{\partial p_j} \delta p_j \right)_{t_0}^{t_1} = 0.$$

This is because in the Hamilton's principle, the variation in generalized coordinates q_j and generalized momenta p_j vanishes at the end points. Hence the two integrands are generally related by

$$\delta \int_{t_0}^{t_1} \left[\left(\sum_j p_j \dot{q}_j - H(q_j, p_j, t) \right) - \alpha \left(\sum_j P_j \dot{Q}_j - K(Q_j, P_j, t) \right) - \frac{dF}{dt} \right] dt = 0.$$

It follows that

$$\sum_j P_j \dot{q}_j - H(q_j, p_j, t) = \alpha \left(\sum_j P_j \dot{Q}_j - K(Q_j, P_j, t) \right) + \frac{dF}{dt}, \qquad \ldots(9)$$

where α is a constant. If $\alpha = 1$ then the transformation (9) is called ordinary canonical transformation or simply canonical transformation. If $\alpha \neq 1$, then the transformation (9) is called extended canonical transformation. Thus we have

$$\sum_j P_j \dot{q}_j - H(q_j, p_j, t) = \sum_j P_j \dot{Q}_j - K(Q_j, P_j, t) + \frac{dF}{dt}. \qquad \ldots(10)$$

We know that a coordinate which is cyclic in the Lagrangian is also cyclic in the Hamiltonian and the generalized momentum corresponding to the cyclic coordinate is conserved. Thus if q_j is cyclic then $p_j = \text{const.} = \alpha_j$. This is the solution of the Hamilton's equation $\dfrac{\partial H}{\partial q_j} = 0$. Other Hamilton's equation gives

$$\dot{q}_j = \frac{\partial H}{\partial p_j} = \frac{\partial H}{\partial \alpha_j} = \omega_j.$$

Integrating we obtain $q_j = \omega_j t + \beta_j$, where β_j are determined from the initial conditions. Thus if we can transform a given set of generalized coordinates q_j to another new set of generalized coordinates Q_j such that all the coordinates are cyclic then the new equations of motion can be integrated more easily to obtain the solution. (refer Example 32, of Chapter 3).

Generating function We see from the equation (10) that the left hand side of the equation is a function of original generalized coordinates q_j and generalized momenta p_j; while the first term on the right hand side depends on the new generalized coordinates Q_j and momenta P_j. Hence in general the function F must be function of both the original and the new set of coordinates and momenta. Thus in general the function F is a function of $4n$ variables besides the time coordinate t. However, out of these $4n$ variables only $2n$ variables are linearly independent and remaining variables are connected by $2n$ equations of constraints defined in the transformation equations (3). Since F is a function of both old and new set of coordinates and therefore, out of $2n$ variables, n variables can be taken from the set of new variables and the remaining n variables can be taken from the set of old variables. Thus one variable should be out of q_j and p_j and the other should be from Q_j and P_j. Accordingly, we have four following forms of the function F: $F_1(q_j, Q_j, t)$, $F_2(q_j, P_j, t)$, $F_3(p_j, Q_j, t)$, $F_4(p_j, P_j, t)$.

The transformation equation can be derived by the knowledge of the function. Hence these functions are called as the generating functions. Thus for the given generating function we can uniquely construct the transformation equations and the transformed Hamiltonian. The method is illustrated in the following example.

Example 1: Derive the canonical transformations and the transformed Hamiltonian corresponding to the generating functions $F_1(q_j, Q_j, t)$, $F_2(q_j, P_j, t)$, $F_3(p_j, Q_j, t)$ and $F_4(p_j, P_j, t)$.

Solution: First Form: Let the function be given by $F_1 = F_1(q_j, Q_j, t)$. Then its time derivative gives

$$\frac{dF_1}{dt} = \sum_j \frac{\partial F_1}{\partial q_j} \dot{q}_j + \sum_j \frac{\partial F_1}{\partial Q_j} \dot{Q}_j + \frac{\partial F_1}{\partial t}.$$

Hence we obtain from equation (10) that

$$\sum_j \left(p_j - \frac{\partial F_1}{\partial q_j} \right) \dot{q}_j - \sum_j \left(P_j + \frac{\partial F_1}{\partial Q_j} \right) \dot{Q}_j - H + K - \frac{\partial F_1}{\partial t} = 0.$$

Since the old and the new generalized coordinates q_j and Q_j are independent, hence by equating the coefficients of $\dot{q}_j$ and $\dot{Q}_j$ to zero separately we obtain

$$p_j = \frac{\partial F_1}{\partial q_j}, \qquad \qquad \text{...(11)}$$

$$P_j = -\frac{\partial F_1}{\partial Q_j}, \qquad \qquad \text{...(12)}$$

$$K(P_j, Q_j, t) = H(p_j, q_j, t) + \frac{\partial F_1}{\partial t}. \qquad \qquad \text{...(13)}$$

On solving equation (11) we find

$$Q_j = Q_j(q_j, p_j, t). \qquad \qquad \text{...(14)}$$

Substituting this in the equation (12) we obtain the relation

$$P_j = P_j(q_j, p_j, t). \qquad \qquad \text{...(15)}$$

This gives the new momenta as the function of old coordinates and momenta. Equations (14) and (15) are the desired canonical transformation equations of the generating function F_1. The equation (13) gives the relation between the old Hamiltonian H and the transformed Hamiltonian K.

Second Form: Let $F_2 = F_2(q_j, P_j, t)$. Since the function $F_2(q_j, P_j, t)$ is not a function of Q_j, hence we have $\dfrac{\partial F_2}{\partial Q_j} = 0$. Therefore, from the equation (12) we have

$$P_j + \frac{\partial F_1}{\partial Q_j} = 0.$$

$$\Rightarrow \qquad P_j + \frac{\partial F_1}{\partial Q_j} = \frac{\partial F_2}{\partial Q_j}.$$

On integrating this equation with respect to Q_j we obtain

$$\sum_j P_j Q_j + F_1(q_j, Q_j, t) = F_2(q_j, P_j, t) \qquad \qquad \text{...(16)}$$

Since F_2 is arbitrary function hence an addition of the term $-\dfrac{d}{dt}\left(\sum_j P_j Q_j\right)$ to the right hand side of the equation (10) will not affect the value of F_2 because

$$\delta \int_{t_0}^{t_1}\left(\frac{d}{dt}\sum_j P_j Q_j\right) dt = \delta\left(\sum_j P_j Q_j\right)\Bigg|_{t_0}^{t_1},$$

$$\delta \int_{t_0}^{t_1}\left(\frac{d}{dt}\sum_j P_j Q_j\right) dt = \left(\sum_j \delta P_j Q_j\right)\Bigg|_{t_0}^{t_1} + \left(\sum_j P_j \delta Q_j\right)\Bigg|_{t_0}^{t_1} = 0,$$

as δQ_j and δP_j at the end points vanish in the Hamilton's principle. Hence equation (10) yields

$$\sum_j p_j\dot{q}_j - H = \sum_j P_j\dot{Q}_j - K + \sum_j \frac{\partial F_2}{\partial q_j}\dot{q}_j + \sum_j \frac{\partial F_2}{\partial P_j}\dot{P}_j + \frac{\partial F_2}{\partial t} - \sum_j \dot{P}_j Q_j - \sum_j P_j\dot{Q}_j,$$

$$\sum_j \left(p_j - \frac{\partial F_2}{\partial q_j}\right)\dot{q}_j + \sum_j \left(Q_j - \frac{\partial F_2}{\partial P_j}\right)\dot{P}_j - H + K - \frac{\partial F_2}{\partial t} = 0.$$

Since Q_j and P_j are independent, therefore, above equation is satisfied only if

$$p_j = \frac{\partial F_2}{\partial q_j}, \qquad\qquad\qquad ...(17)$$

$$Q_j = \frac{\partial F_2}{\partial P_j}, \qquad\qquad\qquad ...(18)$$

$$K = H + \frac{\partial F_2}{\partial t}. \qquad\qquad\qquad ...(19)$$

Solving equation (17) for P_j we obtain

$$P_j = P_j(q_j, p_j, t). \qquad\qquad\qquad ...(20)$$

Substituting this in the equation (18) and solving we obtain the relation

$$Q_j = Q_j(q_j, p_j, t). \qquad\qquad\qquad ...(21)$$

Equations (20) and (21) are the desired transformation equations and the corresponding transformed Hamiltonian is given in the equation (19).

Third Form: Let $F_3 = F_3(p_j, Q_j, t)$. To find the canonical transformation equations, we proceed in the same way as discussed above. In this case we add the term $\frac{d}{dt}\left(\sum_j P_j q_j\right)$ to the right hand side of the equation (10) as it will not affect the function F_3. Thus we have

$$\sum_j p_j\dot{q}_j - H = \sum_j P_j\dot{Q}_j - K + \frac{d}{dt}\left[F_3(p_j, Q_j, t) + \sum_j P_j q_j\right],$$

$$\sum_j p_j\dot{q}_j - H = \sum_j P_j\dot{Q}_j - K + \sum_j \frac{\partial F_3}{\partial p_j}\dot{p}_j + \sum_j \frac{\partial F_3}{\partial Q_j}\dot{Q}_j + \frac{\partial F_3}{\partial t} + \sum_j \dot{P}_j q_j + \sum_j P_j\dot{q}_j,$$

$$\sum_j \left(P_j + \frac{\partial F_3}{\partial Q_j}\right)\dot{Q}_j + \sum_j \left(q_j + \frac{\partial F_3}{\partial p_j}\right)\dot{q}_j + H - K + \frac{\partial F_3}{\partial t} = 0.$$

This yields the desired transformation equations and the new Hamiltonian function

$$P_j = -\frac{\partial F_3}{\partial Q_j}, \qquad \ldots(22)$$

$$q_j = -\frac{\partial F_3}{\partial p_j}, \qquad \ldots(23)$$

$$K = H + \frac{\partial F_3}{dt}. \qquad \ldots(24)$$

Fourth Form: Let $F_4 = F_4(p_j, P_j, t)$. In this case adding the terms $\dfrac{d}{dt}\left(\sum_j p_j q_j\right)$ and $-\dfrac{d}{dt}\left(\sum_j P_j Q_j\right)$ to the right hand side of the equation (10) and simplifying we readily obtain the equations

$$q_j = -\frac{\partial F_4}{\partial p_j}, \qquad \ldots(25)$$

$$Q_j = \frac{\partial F_4}{\partial P_j}, \qquad \ldots(26)$$

$$K = H + \frac{\partial F_4}{\partial t}. \qquad \ldots(27)$$

We note that the relation between the old Hamiltonian H and the new Hamiltonian K in all the cases has the same form. If the generating function F does not contain time t explicitly, then we have $K = H$ and the equation (10) reduces to

$$\sum_j p_j dq_j = \sum_j P_j dQ_j + dF,$$

$$\sum_j (p_j dq_j - P_j dQ_j) = dF. \qquad \ldots(28)$$

Thus the condition for a transformation to be canonical is that $\sum_j (p_j dq_j - P_j dQ_j)$ must be a total differential. More generally, we use this condition to prove the given transformation is canonical.

Example 2: Let the generating be given by (i) $F = \sum_j q_j P_j$ and (ii) $F = \sum_j q_j Q_j$, show that the canonical transformation is the identity transformation.

Solution: (i) The given transformation is of the form F_2. Then using the corresponding transformation equations (17) and (18) we find $p_j = P_j$, and $Q_j = q_j$. This proves that the function F generates identity transformation.

(ii) In this case the function corresponds to the generating function F_1. Using the corresponding transformation equations (13) and (14) we get $p_j = Q_j$ and $P_j = -q_j$. In both the cases the function does not involve time t explicitly, hence $H = K$.

Example 3:   Show that the transformation $p = \dfrac{1}{Q}$, $q = PQ^2$ is canonical and find the generating function.

Solution: From the given transformation equations we find the differential

$$dq = Q^2 dP + 2PQ\,dQ.$$

Consider
$$pdq - PdQ = \frac{1}{Q}\left(Q^2 dP + 2PQ\,dQ\right) - PdQ,$$

$$pdq - PdQ = QdP + PdQ = d(PQ).$$

The shows that $pdq - PdQ$ is an exact differential of PQ, proving that the given transformation equations are canonical transformations. The corresponding generating function is obtain by solving the equations (11) and (12) viz.,

$$p_j = \frac{\partial F_1}{\partial q_j}, \; P_j = -\frac{\partial F_1}{\partial Q_j}. \text{ This yields}$$

$$\frac{1}{Q} = \frac{\partial F_1}{\partial q} \text{ and } \frac{q}{Q^2} = -\frac{\partial F_1}{\partial Q} \Rightarrow dF_1 = \frac{1}{Q}\,dq = -\frac{q}{Q^2}\,dQ.$$

Integration gives $F_1 = \displaystyle\int \frac{1}{Q}\,dq = -\int \frac{q}{Q^2}\,dQ$. Both these integrals give $F_1 = \dfrac{q}{Q}$. This is required generating function.

Example 4:   Prove that the transformation equations between two sets of coordinates given by

$$p = 2\left(1 + q^{\frac{1}{2}}\cos p\right) q^{\frac{1}{2}} \sin p, \; Q = \log\left(1 + q^{\frac{1}{2}}\cos p\right)$$

are canonical transformation. Also find the function which generates the transformations.

Solution: From the given transformation equation we find

$$dQ = \frac{\cos p\,dq - 2q \sin p\,dp}{2q^{\frac{1}{2}}\left(1 + q^{\frac{1}{2}}\cos p\right)}.$$

Now consider

$$pdq - PdQ = pdq - \frac{2\left(1 + q^{\frac{1}{2}}\cos p\right) q^{\frac{1}{2}} \sin p(\cos p\,dq - 2q \sin p\,dp)}{2q^{\frac{1}{2}}\left(1 + q^{\frac{1}{2}}\cos p\right)}$$

$$pdq - PdQ = pdq - \sin p \cos p dq + 2q \sin^2 p dp,$$

$$= \left(p - \frac{1}{2} \sin 2p\right) dq + q(1 - \cos 2p) dp,$$

$$= pdq + qdp - \left(\frac{1}{2} \sin 2pdq + q \cos 2p\right) dp,$$

$$= d(pq) - d\left(\frac{q}{2} \sin 2p\right)$$

$$pdq - PdQ = d\left(pq - \frac{q}{2} \sin 2p\right).$$

$\Rightarrow pdq - pdQ$ is an exact differential of $q\left(p - \dfrac{1}{2} \sin 2p\right)$. Hence the given transformations are canonical transformations. In order to find the generating function, we solve the second transformation equation and find

$$e^Q = 1 + q^{\frac{1}{2}} \cos p,$$

$$q^{\frac{1}{2}} = \frac{e^Q - 1}{\cos p}. \qquad\qquad ...(29)$$

Substituting this in the first transformation equation, we get

$$P = 2\left(1 + \frac{e^Q - 1}{\cos p} \cos p\right) \frac{e^Q - 1}{\cos p} \sin p,$$

$$P = 2e^Q(e^Q - 1) \tan p. \qquad\qquad ...(30)$$

From the transformation equations (22) and (23), we have

$$P = \frac{\partial F_3}{\partial Q}, \; q = - \frac{\partial F_3}{\partial p}.$$

Using equations (29) and (30) we find

$$\frac{\partial F_3}{\partial p} = -\left(\frac{e^Q - 1}{\cos p}\right)^2, \frac{\partial F_3}{\partial Q} = -2e^Q(e^Q - 1) \tan p,$$

$$dF_3 = -\left(\frac{e^Q - 1}{\cos p}\right)^2 dp, \; dF_3 = -2e^Q(e^Q - 1) \tan p dQ.$$

Integrating we obtain

$$F_3 = -(e^Q - 1)^2 \int \sec^2 pdp, \; F_3 = -2 \tan p \int e^Q(e^Q - 1) \, dQ.$$

The constants of integration are taken to be zero. Both the integrals yield the desired generating function $F_3 = -(e^Q - 1)^2 \tan p$.

Example 5: Find the value of α and β so that the equations $Q = q^{\alpha} \cos (\beta p)$, $P = q^{\alpha} \sin (\beta p)$ represent a canonical transformation. Also find the generating function.

Solution: The given transforations are cononical if $pdq - PdQ$ is an exact differential.

Therefore, we find

$$dQ = -q^{\alpha} \sin (\beta p) \cdot \beta dp + \alpha q^{\alpha-1} \cos (\beta p)dq.$$

$$pdq - PdQ = pdq - q^{\alpha} \sin (\beta p)\left(\alpha q^{\alpha-1} \cos (\beta p)dq - \beta q^{\alpha} \sin (\beta p)dq\right),$$

$$pdq - PdQ = \left(p - \alpha q^{2\alpha-1} \sin (\beta p) \cos (\beta p)\right) dq + \beta q^{2\alpha} \sin^2 (\beta p)dp. \quad ...(31)$$

The right hand side of the equation (31) will be an exact differential equation if we have

$$\frac{\partial}{\partial p}\left(p - \alpha q^{2\alpha-1} \sin (\beta p) \cos (\beta p)\right) = \frac{\partial}{\partial q}\left(\beta q^{2\alpha} \sin^2 (\beta p)\right),$$

$$1 - \alpha q^{2\alpha-1} \beta\left(\cos^2 (\beta p) - \sin^2 (\beta p)\right) = 2\alpha\beta q^{2\alpha-1} \sin^2 (\beta p),$$

$$\Rightarrow \qquad 1 - \alpha\beta q^{2\alpha-1} = 0,$$

$$\Rightarrow \qquad \alpha\beta q^{2\alpha-1} = 1.$$

This is true if $\alpha = \dfrac{1}{2}$, $\beta = 2$. Thus for $\alpha = \dfrac{1}{2}$ and $\beta = 2$ the given transformation is canonical.

Thus we have $Q = q^{\frac{1}{2}} \cos 2p$, $P = q^{\frac{1}{2}} \sin 2p$. $\qquad ...(32)$

Solving this we find

$$q = \left(\frac{Q}{\cos 2p}\right)^2, \; P = Q \tan 2p. \qquad ...(33)$$

These transformation equations are of the form given in equations (22) and (23)

$$P = -\frac{\partial F_3}{\partial Q}, \; q = -\frac{\partial F_3}{\partial p},$$

This implies $\qquad \dfrac{\partial F_3}{\partial Q} = -Q \tan 2p$, and $\dfrac{\partial F_3}{\partial p} = -\left(\dfrac{Q}{\cos 2p}\right)^2$,

$$dF_3 = -Q \tan 2pdQ, \; dF_3 = -\left(\frac{Q}{\cos 2p}\right)^2 dp.$$

Integration these equations we get

$$F_3 = -\int Q \tan 2p \, dQ = -\int \left(\frac{Q}{\cos 2p} \right)^2 dp.$$

This integration yields the desired generating function $F_3 = -\dfrac{Q^2}{2} \tan 2p$.

Example 6: For the harmonic oscillator the Hamiltonian H is given by $H = \dfrac{p^2}{2m} + \dfrac{m\omega^2 q^2}{2}$. By using the canonical transformation whose generating function is $F = -\dfrac{m\omega q^2}{2} \cot Q$. show that

$$q = \sqrt{\frac{2E}{m\omega^2}} \sin(\omega t + \alpha) \text{ and } p = \sqrt{2mE} \cos(\omega t + \alpha)$$

is the solution of the harmonic oscillator.

Solution: The generating function of the transformation is of the form

$$F = F(q, Q) = -\frac{m\omega q^2}{2} \cot Q. \qquad \qquad ...(34)$$

In this case the transformation equations are given by the equations (11) and (12)

$$p = \frac{\partial F}{\partial q}, P = -\frac{\partial F}{\partial Q}. \qquad \qquad ...(35)$$

From equations (34) and (35), we have

$$p = \frac{\partial F}{\partial q} = m\omega q \cot Q, \text{ and} \qquad \qquad ...(36)$$

$$P = -\frac{\partial F}{\partial Q} = \frac{m}{2} \omega q^2 \text{cosec}^2 Q, $$

$$q = \sqrt{\frac{2P}{m\omega}} \sin Q. \qquad \qquad ...(37)$$

Hence equation (36) becomes

$$p = \sqrt{2Pm\omega} \cos Q. \qquad \qquad ...(38)$$

Using equations (37) and (38) the Hamiltonian of the harmonic oscillator becomes

$$H = \frac{p^2}{2m} + \frac{m\omega^2 q^2}{2} = P\omega. \qquad \qquad ...(39)$$

We see that Q is cyclic in H and time t is absent in H hence $H = E$-a constant of motion. Hence equation (39) gives

$$P = \frac{E}{\omega}. \qquad \qquad ...(40)$$

The Hamilton's canonical equations of motion in new variables are given by

$$\dot{Q} = \frac{\partial H}{\partial P} = \omega \text{ and } \dot{P} = -\frac{\partial H}{\partial Q} = 0.$$

Integration gives $Q = \omega t + \alpha$. Substituting this in the equations (37) and (38) we get the solutions of the harmonic oscillator in the form

$$q = \sqrt{\frac{2E}{m\omega^2}} \, \sin(\omega t + \alpha) \text{ and } p = \sqrt{2mE} \, \cos(\omega t + \alpha). \qquad ...(41)$$

Example 7: Show that the transformation defined by $e^Q = \frac{1}{q}\sin p$ and $P = q\cot p$ is a canonical transformation. Further, if the Hamiltonian $H = \frac{p^2}{2m} + \frac{kq^2}{2}$, find the generating function $F_1(q, Q, t)$.

Solution: The transformation equations are given by

$$Q = \log\left(\frac{1}{q}\sin p\right) \text{ and } P = q\cot p. \qquad ...(42)$$

From which we find $dQ = \dfrac{1}{q\sin p}(q\cos p\, dp - \sin p\, dq).$

Consider

$$p\,dq - P\,dQ = p\,dq - \frac{1}{q\sin p}\, q\cot p\,(q\cos p\, dp - \sin p\, dq),$$

$$= p\,dq - q\cot^2 p\, dp + \cot p\, dq,$$

$$= p\,dq + q\left(\operatorname{cosec}^2 p - 1\right)dp + \cot p\, dq,$$

$$= p\,dq + q\,dp - q\operatorname{cosec}^2 p\, dp + \cot p\, dq,$$

$$p\,dq - P\,dQ = d(pq) + d(q\cot p),$$

$$p\,dq - P\,dQ = d[q(p + \cot p)].$$

$\Rightarrow pdq - PdQ$ is an exact differential of $q(p + \cot p)$.

This implies that the transformations defined in the equation (42) are canonical. From equation (42) we find

$$\sin p = qe^Q \Rightarrow \cos p = \left(1 - \sin^2 p\right)^{\frac{1}{2}} = \left(1 - q^2 e^{2Q}\right)^{\frac{1}{2}}.$$

Hence we have

$$\cot p = \left(q^{-2}e^{-2Q} - 1\right)^{\frac{1}{2}}. \qquad ...(43)$$

Substituting this in the second equation of (42) we get

$$P = q(q^{-2}e^{-2Q} - 1)^{\frac{1}{2}} \Rightarrow P^2 = e^{-2Q} - q^2.$$

Solving for q we obtain $q = (e^{-2Q} - P^2)^{\frac{1}{2}}.$ \hfill ...(44)

Further, from the equations (42) and (43), we get

$$P = (e^{-2Q} - q^2)^{\frac{1}{2}}.$$ \hfill ...(45)

Also from the equation (43), we have

$$p = \cot^{-1}\left[\frac{(e^{-2Q} - q^2)}{q}\right]^{\frac{1}{2}}.$$ \hfill ...(46)

For the transformation equations (45) and (46) the corresponding generating function is obtained from equations (11) and (12) given by

$$p = \frac{\partial F_1}{\partial q}, \; P = -\frac{\partial F_1}{\partial Q}.$$

These equations from the equations (45) and (46) become

$$dF_1 = \sin^{-1}(qe^{Q})\, dq = -\, q \cot p \; dQ.$$

Integrating we get

$$F_1 = \int \sin^{-1}(qe^{Q})\, dq + E(Q),$$

where E is a constant of integration and is a function of Q only. Let us put

$$x = \sin^{-1}(qe^{Q}) \Rightarrow \sin x = qe^{Q},$$

$$\cos x\, dx = e^{Q} dq \Rightarrow dq = e^{-Q} \cos x \; dx.$$

Hence the above integral becomes

$$F_1 = e^{-Q} \int x \cos x\, dx + E(Q),$$

$$F_1 = e^{-Q}\left[qe^{Q} \sin^{-1}(qe^{Q}) + \sqrt{1 - q^2 e^{2Q}}\,\right] + E(Q),$$

$$F_1 = q \sin^{-1}(qe^{Q}) + e^{-Q} \sqrt{1 - q^2 e^{2Q}} + E(Q).$$

The function E seems to be independent of Q as can be seen evaluating $\frac{\partial F_1}{\partial Q}$. Thus we finally get $F_1 = q \sin^{-1}(qe^{Q}) + e^{-Q} \sqrt{1 - q^2 e^{2Q}}.$

Example 8: Show that under the transformation $q_j = q_j(Q_1, Q_2, ..., Q_n)$ the Lagrange's equations of motion retain the same form in the new generalized coordinates Q_j.

Solution: Let the generalized coordinate q_j be transformed to another generalized coordinate Q_j by the transformation equation given by

$$q_j = q_j(Q_1, Q_2, ..., Q_n). \qquad ...(47)$$

Under this transformation the Lagrangian function L transforms to a new Lagrangian function L' such that

$$L(q_j, \dot{q}_j, t) = L'(Q_j, \dot{Q}_j, t). \qquad ...(48)$$

From equation (48), we find

$$\frac{\partial L'}{\partial \dot{Q}_j} = \sum_k \frac{\partial L}{\partial q_k} \frac{\partial q_k}{\partial \dot{Q}_j} + \sum_k \frac{\partial L}{\partial \dot{q}_k} \frac{\partial \dot{q}_k}{\partial \dot{Q}_j},$$

Therefore,

$$\frac{d}{dt}\left(\frac{\partial L'}{\partial \dot{Q}_j}\right) = \frac{d}{dt}\left[\sum_k \frac{\partial L}{\partial q_k} \frac{\partial q_k}{\partial \dot{Q}_j} + \sum_k \frac{\partial L}{\partial \dot{q}_k} \frac{\partial \dot{q}_k}{\partial \dot{Q}_j}\right], \qquad ...(49)$$

where $\qquad \dot{q}_j = \sum_k \frac{\partial q_j}{\partial Q_k} \dot{Q}_k,$ and $\dfrac{\partial \dot{q}_j}{\partial \dot{Q}_k} = \dfrac{\partial q_j}{\partial Q_k}. \qquad ...(50)$

Consequently, the equation (49) becomes

$$\frac{d}{dt}\left(\frac{\partial L'}{\partial \dot{Q}_j}\right) = \frac{d}{dt}\left(\sum_k \frac{\partial L}{\partial \dot{q}_k} \frac{\partial q_k}{\partial Q_j}\right),$$

$$\frac{d}{dt}\left(\frac{\partial L'}{\partial \dot{Q}_j}\right) = \sum_k \frac{d}{dt}\left(\frac{\partial L}{\partial \dot{q}_k}\right) \frac{\partial q_k}{\partial Q_j} + \sum_k \frac{\partial L}{\partial \dot{q}_k} \frac{d}{dt}\left(\frac{\partial q_k}{\partial Q_j}\right).$$

However, from the Lagrange's equations of motion we have

$$\frac{d}{dt}\left(\frac{\partial L}{\partial \dot{q}_k}\right) = \frac{\partial L}{\partial q_k}.$$

Also the time derivative and the partial derivative with respect to the generalized coordinate are commutative. It means that

$$\frac{d}{dt}\left(\frac{\partial f}{\partial q_j}\right) = \frac{\partial}{\partial q_j} \frac{df}{dt}.$$

Hence above equation becomes

$$\frac{d}{dt}\left(\frac{\partial L'}{\partial \dot{Q}_j}\right) = \sum_k \frac{\partial L}{\partial q_k}\frac{\partial q_k}{\partial Q_j} + \sum_k \frac{\partial L}{\partial \dot{q}_k}\frac{\partial \dot{q}_k}{\partial Q_j},$$

$$\frac{d}{dt}\left(\frac{\partial L'}{\partial \dot{Q}_j}\right) = \frac{\partial L}{\partial Q_j}, \qquad\qquad ...(51)$$

where by the chain rule of differential calculus

$$\frac{\partial L}{\partial Q_j} = \sum_k \frac{\partial L}{\partial q_k}\frac{\partial q_k}{\partial Q_j} + \sum_k \frac{\partial L}{\partial \dot{q}_k}\frac{\partial \dot{q}_k}{\partial Q_j}. \qquad\qquad ...(52)$$

Also from the equation (48), we find

$$\frac{\partial L'}{\partial Q_j} = \sum_k \frac{\partial L}{\partial q_k}\frac{\partial q_k}{\partial Q_j} + \sum_k \frac{\partial L}{\partial \dot{q}_k}\frac{\partial \dot{q}_k}{\partial Q_j}. \qquad\qquad ...(53)$$

From equations (51), (52) and (53), we have the desired result

$$\frac{d}{dt}\left(\frac{\partial L'}{\partial \dot{Q}_j}\right) - \frac{\partial L'}{\partial Q_j} = 0.$$

2. The Poisson bracket   A Poisson bracket is a special type of relation between a pair of dynamical variables of a system. The Poisson brackets are used to construct new integrals of motion from the known ones. Poisson brackets are found to be very useful tool in the field theory and quantum mechanics. In this chapter we will define the Poisson bracket and study its properties. We will express Hamilton's canonical equations of motion in terms of Poisson brackets and study a dynamical system. We will also show that Poisson brackets are invariant under canonical transformations.

Definition: Let $u = u(q_j, p_j, t)$, $v = v(q_j, p_j, t)$ be two given dynamical variables which are functions of $2n$ variables q_j, p_j and time t. Then the Poisson bracket of u and v is defined as

$$[u, v]_{(q, p)} = \sum_{j=1}^{n}\left(\frac{\partial u}{\partial q_j}\frac{\partial v}{\partial p_j} - \frac{\partial u}{\partial p_j}\frac{\partial v}{\partial q_j}\right), \qquad\qquad ...(54)$$

where the suffix (q, p) refers to the set (q_j, p_j, t) of independent variables. Thus suffix (q, p) can be dropped.

Properties of Poisson Bracket

(i) The Poisson bracket of any two dynamical variables is anti-commutative.
   i.e., $[u, v] = -[v, u]$. $\qquad\qquad ...(55)$

(ii) The Poisson bracket of a function with itself is identically zero.
   $[u, u] = 0$.

(iii) If c is a constant *i.e.*, not a function of (q_j, p_j, t) then

$[cu, v] = [u, cv] = c[u, v]$.

(iv) The Poisson brackets satisfy the distributive property, *i.e.*,

(a) $[u + v, w] = [u, w] + [v, w]$,   (b) $[u, vw] = [u, v] w + v [u, w]$.

(v) Partial derivative of any Poisson bracket satisfies

$$\frac{\partial}{\partial t}[u, v] = \left[\frac{\partial u}{\partial t}, v\right] + \left[u, \frac{\partial v}{\partial t}\right].$$

(vi) (a) $[p_j, p_k] = [q_j, q_k] = 0$,   (b) $[u, q_j] = -\dfrac{\partial u}{\partial p_j}$,

     (c) $[u, p_j] = -\dfrac{\partial u}{\partial q_j}$,   (d) $[q_j, p_k] = \delta_{jk}$.

(vii) Jacobi Identity: $[u, [v, w]] + [v, [w, u]] + [w, [u, v]] = 0$.

(viii) Let $w_1, w_2, w_3, ..., w_n$ be a set of dynamical quantities which are functions of p, q, t and let $F(w_1, w_2, w_3, ..., w_n)$ be a differentiable function of $w_1, w_2, w_3, ..., w_n$ then show that

$$\left[u, F(w_1, w_2, w_3, ..., w_n)\right] = \frac{\partial F}{\partial w_1} [u, w_1] + \frac{\partial F}{\partial w_2} [u, w_2] + ... + \frac{\partial F}{\partial w_n}[u, w_n]$$

Proof: (i) The proof of first three properties is obvious and hence left to the readers, (iv) (a) By virtue of the definition (54) we have,

$$[u + v, w] = \sum_{j=1}^{n} \left(\frac{\partial}{\partial q_j} (u + v) \frac{\partial w}{\partial p_j} - \frac{\partial}{\partial p_j} (u + v) \frac{\partial w}{\partial q_j}\right),$$

$$= \sum_{j=1}^{n} \left[\left(\frac{\partial u}{\partial q_j} + \frac{\partial v}{\partial q_j}\right) \frac{\partial w}{\partial p_j} - \left(\frac{\partial u}{\partial p_j} + \frac{\partial v}{\partial p_j}\right) \frac{\partial w}{\partial q_j}\right],$$

$$= \sum_{j=1}^{n} \left(\frac{\partial u}{\partial q_j} \frac{\partial w}{\partial p_j} - \frac{\partial u}{\partial p_j} \frac{\partial w}{\partial q_j}\right) + \sum_{j=1}^{n} \left(\frac{\partial v}{\partial q_j} \frac{\partial w}{\partial p_j} + \frac{\partial v}{\partial p_j} \frac{\partial w}{\partial q_j}\right),$$

$$[u + v, w] = [u, w] + [v, w]. \qquad ...(56)$$

(b) Now the definition (54) yields

$$[u, vw] = \sum_{j=1}^{n} \left(\frac{\partial u}{\partial q_j} \frac{\partial (vw)}{\partial p_j} - \frac{\partial u}{\partial p_j} \frac{\partial (vw)}{\partial q_j}\right),$$

$$= \sum_{j=1}^{n} \left(\frac{\partial u}{\partial q_j} \left(\frac{\partial v}{\partial p_j} w + v \frac{\partial w}{\partial p_j}\right) - \frac{\partial u}{\partial p_j} \left(\frac{\partial v}{\partial q_j} w + v \frac{\partial w}{\partial q_j}\right)\right),$$

$$= \sum_{j=1}^{n} \left(\frac{\partial u}{\partial q_j} \frac{\partial v}{\partial p_j} - \frac{\partial u}{\partial p_j} \frac{\partial v}{\partial q_j} \right) w + v \sum_{j=1}^{n} \left(\frac{\partial u}{\partial q_j} \frac{\partial w}{\partial p_j} - \frac{\partial u}{\partial p_j} \frac{\partial w}{\partial q_j} \right),$$

$$[u, vw] = [u, v] w + v[u, w]. \qquad \qquad \dots(57)$$

(v) By virtue of the definition of Poisson bracket (54) we have

$$\frac{\partial}{\partial t}[u, v] = \sum_{j=1}^{n} \frac{\partial}{\partial t} \left(\frac{\partial u}{\partial q_j} \frac{\partial v}{\partial p_j} - \frac{\partial u}{\partial p_j} \frac{\partial v}{\partial q_j} \right),$$

$$\frac{\partial}{\partial t}[u, v] = \sum_{j=1}^{n} \left[\frac{\partial}{\partial t} \left(\frac{\partial u}{\partial q_j} \right) \frac{\partial v}{\partial p_j} + \frac{\partial u}{\partial q_j} \frac{\partial}{\partial t} \left(\frac{\partial v}{\partial p_j} \right) - \frac{\partial}{\partial t} \left(\frac{\partial u}{\partial p_j} \right) \frac{\partial v}{\partial q_j} - \frac{\partial u}{\partial p_j} \frac{\partial}{\partial t} \left(\frac{\partial v}{\partial q_j} \right) \right],$$

$$\frac{\partial}{\partial t}[u, v] = \sum_{j=1}^{n} \left(\frac{\partial}{\partial q_j} \left(\frac{\partial u}{\partial t} \right) \frac{\partial v}{\partial p_j} - \frac{\partial}{\partial p_j} \left(\frac{\partial u}{\partial t} \right) \frac{\partial v}{\partial q_j} \right) +$$

$$+ \sum_{j=1}^{n} \left(\frac{\partial u}{\partial q_j} \frac{\partial}{\partial p_j} \left(\frac{\partial v}{\partial t} \right) - \frac{\partial u}{\partial p_j} \frac{\partial}{\partial q_j} \left(\frac{\partial v}{\partial t} \right) \right),$$

$$\frac{\partial}{\partial t}[u, v] = \left[\frac{\partial u}{\partial t}, v \right] + \left[u, \frac{\partial v}{\partial t} \right]. \qquad \qquad \dots(58a)$$

Similarly, one can prove that $\dfrac{d}{dt}[u, v] = \left[\dfrac{du}{dt}, v \right] + \left[u, \dfrac{dv}{dt} \right].$ $\qquad \dots(58b)$

(vi) (a) Applying the definition (54) and using the fact that q_j and p_j are independent we readily show that

$$[p_j, p_k] = [q_j, q_k] = 0. \qquad \qquad \dots(59)$$

(b) Now replacing $u = q_i$ and $v = p_k$ in the definition (54) we get

$$[q_i, p_k] = \sum_{j=1}^{n} \left(\frac{\partial q_i}{\partial q_j} \frac{\partial p_k}{\partial p_j} - \frac{\partial q_i}{\partial p_j} \frac{\partial p_k}{\partial q_j} \right),$$

$$[q_i, p_k] = \sum_{j=1}^{n} \delta_{ij} \delta_{jk},$$

$$[q_i, p_k] = \delta_{ik}. \qquad \qquad \dots(60)$$

(c) Now we put $v = q_i$ in the definition (54) we obtain $[u, q_i] = -\dfrac{\partial u}{\partial p_i}.$

(d) Similarly, replacing $v = p_i$ in the equation (54) yields $[u, p_i] = \dfrac{\partial u}{\partial q_i}.$

(vii) We use the definition (54) and write

$$[u, [v, w]] = \sum_{j=1}^{n} \left(\frac{\partial u}{\partial q_j} \frac{\partial}{\partial p_j} [v, w] - \frac{\partial u}{\partial p_j} \frac{\partial}{\partial q_j} [v, w] \right),$$

$$[u, [v, w]] = \sum_{j=1}^{n} \left(\frac{\partial u}{\partial q_j} \frac{\partial}{\partial p_j} \sum_{k=1}^{n} \left(\frac{\partial v}{\partial q_k} \frac{\partial w}{\partial p_k} - \frac{\partial v}{\partial p_k} \frac{\partial w}{\partial q_k} \right) - \frac{\partial u}{\partial p_j} \frac{\partial}{\partial q_j} \sum_{k=1}^{n} \left(\frac{\partial v}{\partial q_k} \frac{\partial w}{\partial p_k} - \frac{\partial v}{\partial p_k} \frac{\partial w}{\partial q_k} \right) \right),$$

$$[u, [v, w]] = \sum_{j=1}^{n} \left(\frac{\partial u}{\partial q_j} \frac{\partial^2 v}{\partial p_j \partial q_k} \frac{\partial w}{\partial p_k} + \frac{\partial u}{\partial q_j} \frac{\partial v}{\partial q_k} \frac{\partial^2 w}{\partial p_j \partial p_k} - \right.$$

$$- \frac{\partial u}{\partial q_j} \frac{\partial^2 v}{\partial p_j \partial p_k} \frac{\partial w}{\partial q_k} - \frac{\partial u}{\partial q_j} \frac{\partial v}{\partial p_k} \frac{\partial^2 w}{\partial p_j \partial q_k} - \frac{\partial u}{\partial p_j} \frac{\partial^2 v}{\partial q_j \partial q_k} \frac{\partial w}{\partial p_k} -$$

$$\left. - \frac{\partial u}{\partial p_j} \frac{\partial v}{\partial q_k} \frac{\partial^2 w}{\partial q_j \partial p_k} + \frac{\partial u}{\partial p_j} \frac{\partial^2 v}{\partial q_j \partial p_k} \frac{\partial w}{\partial q_k} + \frac{\partial u}{\partial p_j} \frac{\partial v}{\partial p_k} \frac{\partial^2 w}{\partial q_j \partial q_k} \right) \quad ...(61)$$

Similarly, we find

$$[v,[w, u]] = \sum_{j=1}^{n} \left(\frac{\partial v}{\partial q_j} \frac{\partial^2 w}{\partial p_j \partial q_k} \frac{\partial u}{\partial p_k} + \frac{\partial v}{\partial q_j} \frac{\partial w}{\partial q_k} \frac{\partial^2 u}{\partial p_j \partial p_k} - \right.$$

$$- \frac{\partial v}{\partial q_j} \frac{\partial^2 w}{\partial p_j \partial p_k} \frac{\partial u}{\partial q_k} - \frac{\partial u}{\partial q_j} \frac{\partial v}{\partial p_k} \frac{\partial^2 w}{\partial p_j \partial q_k} -$$

$$- \frac{\partial v}{\partial p_j} \frac{\partial^2 w}{\partial q_j \partial q_k} \frac{\partial u}{\partial p_k} - \frac{\partial v}{\partial p_j} \frac{\partial w}{\partial q_k} \frac{\partial^2 u}{\partial q_j \partial p_k} +$$

$$\left. + \frac{\partial v}{\partial p_j} \frac{\partial^2 w}{\partial q_j \partial p_k} \frac{\partial u}{\partial q_k} + \frac{\partial v}{\partial p_j} \frac{\partial w}{\partial p_k} \frac{\partial^2 u}{\partial q_j \partial q_k} \right), \quad ...(62)$$

$$[v, [w, u]] = \sum_{j=1}^{n} \left(\frac{\partial w}{\partial q_j} \frac{\partial^2 u}{\partial p_j \partial q_k} \frac{\partial v}{\partial p_k} + \frac{\partial w}{\partial q_j} \frac{\partial u}{\partial q_k} \frac{\partial^2 v}{\partial p_j \partial p_k} - \right.$$

$$- \frac{\partial w}{\partial q_j} \frac{\partial^2 u}{\partial p_j \partial p_k} \frac{\partial v}{\partial q_k} - \frac{\partial w}{\partial q_j} \frac{\partial u}{\partial p_k} \frac{\partial^2 v}{\partial p_j \partial q_k} -$$

$$- \frac{\partial w}{\partial p_j} \frac{\partial^2 u}{\partial q_j \partial q_k} \frac{\partial v}{\partial p_k} - \frac{\partial w}{\partial p_j} \frac{\partial u}{\partial q_k} \frac{\partial^2 v}{\partial q_j \partial p_k} +$$

$$\left. + \frac{\partial w}{\partial p_j} \frac{\partial^2 u}{\partial q_j \partial p_k} \frac{\partial v}{\partial q_k} + \frac{\partial w}{\partial p_j} \frac{\partial u}{\partial p_k} \frac{\partial^2 v}{\partial q_j \partial q_k} \right). \quad ...(63)$$

Adding the above equations (61), (62) and (63) we readily get the Jacobi identity

$$[u, [v, w]] + [v, [w, u]] + [w, [u, v]] = 0, \qquad \qquad ...(64)$$

(viii) Using the definition (54) we have

$$[u, F] = \sum_{j=1}^{n} \left(\frac{\partial u}{\partial q_j} \frac{\partial F}{\partial p_j} - \frac{\partial u}{\partial p_j} \frac{\partial F}{\partial q_j} \right),$$

where $F = F(w_1, w_2, w_3, ..., w_n)$ and $w_i = w_i(q, p, t)$. Therefore, we write the above definition as

$$[u, F] = \sum_{j=1}^{n} \left(\frac{\partial u}{\partial q_j} \left[\sum_{i=1}^{n} \frac{\partial F}{\partial w_i} \frac{\partial w_i}{\partial p_j} \right] - \frac{\partial u}{\partial p_j} \left[\sum_{i=1}^{n} \frac{\partial F}{\partial w_i} \frac{\partial w_i}{\partial q_j} \right] \right),$$

$$[u, F] = \sum_{i=1}^{n} \left[\frac{\partial F}{\partial w_i} \sum_{j=1}^{n} \left(\frac{\partial u}{\partial q_j} \frac{\partial w_i}{\partial p_j} - \frac{\partial u}{\partial p_j} \frac{\partial w_i}{\partial q_j} \right) \right],$$

$$[u, F] = \sum_{i=1}^{n} \frac{\partial F}{\partial w_i} [u, w_i].$$

This is equivalent to

$$[u, F(w_1, w_2, w_3, ..., w_n)] = \frac{\partial F}{\partial w_1} [u, w_1] + \frac{\partial F}{\partial w_2} [u, w_2] + ... + \frac{\partial F}{\partial w_n} [u, w_n]. \quad ...(65)$$

Invariance of Poisson Bracket under Canonical Transformation:

Theorem 1   Prove that Poisson brackets are invariant under canonical transformations.

Proof: Let q_j, p_j be the set of old generalized coordinates and momenta. Let these be transformed to new set of generalized coordinates and momenta Q_j, P_j respectively, by the transformation equations

$$u = u(Q_i, P_i), \, v = v(Q_i, P_i). \qquad \qquad ...(66)$$

From definition of the Poisson bracket we have

$$[u, v]_{(q, p)} = \sum_{j=1}^{n} \left(\frac{\partial u}{\partial q_j} \frac{\partial v}{\partial p_j} - \frac{\partial u}{\partial p_j} \frac{\partial v}{\partial q_j} \right), \qquad \qquad ...(67)$$

By using the chain rule of differential calculus we write the equation (67) as

$$[u, v]_{(q, p)} = \sum_{j=1}^{n} \left(\frac{\partial u}{\partial q_j} \frac{\partial v}{\partial p_j} - \frac{\partial u}{\partial p_j} \frac{\partial v}{q_j} \right),$$

$$[u, v]_{(q, p)} = \sum_{j=1}^{n} \left(\sum_{i,k} \left(\frac{\partial u}{\partial Q_j} \frac{\partial Q_i}{\partial q_j} + \frac{\partial u}{\partial P_i} \frac{\partial P_i}{\partial q_j} \right) \left(\frac{\partial v}{\partial Q_k} \frac{\partial Q_k}{\partial p_j} + \frac{\partial v}{\partial P_k} \frac{\partial P_k}{\partial p_j} \right) \right.$$

$$\left. - \sum_{i,k} \left(\frac{\partial u}{\partial Q_i} \frac{\partial Q_i}{\partial p_j} + \frac{\partial u}{\partial P_i} \frac{\partial P_i}{\partial p_j} \right) \left(\frac{\partial v}{\partial Q_k} \frac{\partial Q_k}{\partial q_j} + \frac{\partial v}{\partial P_k} \frac{\partial P_k}{\partial q_j} \right) \right),$$

$$[u, v]_{(q, p)} = \sum_{j=1}^{n} \left(\sum_{i,k} \left[\frac{\partial u}{\partial Q_i} \frac{\partial v}{\partial Q_k} \left(\frac{\partial Q_i}{\partial q_j} \frac{\partial Q_k}{\partial p_j} - \frac{\partial Q_i}{\partial p_j} \frac{\partial Q_k}{\partial q_j} \right) + \right. \right.$$

$$\left. + \frac{\partial u}{\partial P_i} \frac{\partial v}{\partial P_k} \left(\frac{\partial P_i}{\partial q_j} \frac{\partial P_k}{\partial p_j} - \frac{\partial P_i}{\partial p_j} \frac{\partial P_k}{\partial q_j} \right) \right] +$$

$$+ \sum_{i,k} \left[\frac{\partial u}{\partial Q_i} \frac{\partial v}{\partial P_k} \left(\frac{\partial Q_i}{\partial q_j} \frac{\partial P_k}{\partial p_j} - \frac{\partial Q_i}{\partial p_j} \frac{\partial P_k}{\partial q_j} \right) + \right.$$

$$\left. \left. + \frac{\partial u}{\partial P_i} \frac{\partial v}{\partial Q_k} \left(\frac{\partial P_i}{\partial q_j} \frac{\partial Q_k}{\partial p_j} - \frac{\partial P_i}{\partial p_j} \frac{\partial Q_k}{\partial q_j} \right) \right] \right).$$

By interchanging $i \leftrightarrow k$ in the fourth term of the above expression we write this as

$$[u, v]_{(q, p)} = \sum_{i,k} \left[\frac{\partial u}{\partial Q_i} \frac{\partial v}{\partial Q_k} [Q_i, Q_k]_{(q, p)} + \frac{\partial u}{\partial P_i} \frac{\partial v}{\partial P_k} [P_i, P_k]_{(q, p)} \right] +$$

$$+ \sum_{i,k} \left[\sum_j \left(\frac{\partial Q_i}{\partial q_j} \frac{\partial P_k}{\partial p_j} - \frac{\partial Q_i}{\partial p_j} \frac{\partial P_k}{\partial q_j} \right) \left(\frac{\partial u}{\partial Q_i} \frac{\partial v}{\partial P_k} - \frac{\partial u}{\partial P_k} \frac{\partial v}{\partial Q_i} \right) \right].$$

Since we know from the formula (59) that $[Q_i, Q_k]_{(q, p)} = 0$, $[P_i, P_k]_{(q, p)} = 0$. Hence we have

$$[u, v]_{(q, p)} = \sum_{i,k} \left[\sum_j \left(\frac{\partial Q_i}{\partial q_j} \frac{\partial P_k}{\partial p_j} - \frac{\partial Q_i}{\partial p_j} \frac{\partial P_k}{\partial q_j} \right) \left(\frac{\partial u}{\partial Q_i} \frac{\partial v}{\partial P_k} - \frac{\partial u}{\partial P_k} \frac{\partial v}{\partial Q_i} \right) \right],$$

$$[u, v]_{(q, p)} = \sum_{i,k} \left[\left(\frac{\partial u}{\partial Q_i} \frac{\partial v}{\partial P_k} - \frac{\partial u}{\partial P_k} \frac{\partial v}{\partial Q_i} \right) [Q_i, P_k]_{(q, p)} \right].$$

But we know $[Q_i, P_k]_{(q, p)} = \delta_{ik}$, hence the above equation becomes

$$[u, v]_{(q, p)} = \sum_{i,k} \left[\left(\frac{\partial u}{\partial Q_i} \frac{\partial v}{\partial P_k} - \frac{\partial u}{\partial P_k} \frac{\partial v}{\partial Q_i} \right) \delta_{ik} \right],$$

$$[u, v]_{(q, p)} = \sum_i \left[\left(\frac{\partial u}{\partial Q_i} \frac{\partial v}{\partial P_i} - \frac{\partial u}{\partial P_i} \frac{\partial v}{\partial Q_i} \right) \right]$$

$$[u, v]_{(q, p)} = [u, v]_{(Q, p)}. \qquad \qquad ....(68)$$

This proves that Poisson brackets are invariant under canonical transformations.

Hamilton's Canonical Equations of Motion in Poisson Brackets

A function depending on coordinates, momenta and time is taken as the Hamiltonian of a system. *i.e.*,

$$H = u = u(q_j, p_{,j}, t).$$

We have by definition of the Poisson bracket

$$[q_k, H] = \sum_{j=1}^{n} \left(-\frac{\partial H}{\partial q_j} \frac{\partial q_k}{\partial p_j} + \frac{\partial H}{\partial p_j} \frac{\partial q_k}{\partial q_j} \right).$$

This implies
$$\dot{q}_j = \frac{\partial H}{\partial p_j} = [q_j, H]. \qquad \qquad ...(69)$$

Similarly, we obtain
$$\dot{p}_j = -\frac{\partial H}{\partial q_j} = [p_j, H]. \qquad \qquad ...(70)$$

Equations (69) and (70) are the Hamilton's canonical equations of motion in terms of Poisson bracket.

1st Poisson Theorem The total time rate of evolution of any dynamical variable $u(q_j, p_j, t)$ is given by $\dfrac{du}{dt} = \dfrac{\partial u}{\partial t} + [u, H].$

Proof: Let u be a dynamical variable function of generalized coordinates, momenta and time. *i.e.*, $u = u(q_j, p_{,j}, t)$. Taking the time derivative of this dynamical variable, we get

$$\frac{du}{dt} = \sum_j \frac{\partial u}{\partial q_j} \dot{q}_j + \sum_j \frac{\partial u}{\partial p_j} \dot{p}_j + \frac{\partial u}{\partial t}.$$

Using the Hamilton's canonical equations of motion we have

$$\frac{du}{dt} = \sum_j \left(\frac{\partial u}{\partial q_j} \frac{\partial H}{\partial p_j} - \frac{\partial u}{\partial p_j} \frac{\partial H}{\partial q_j} \right) + \frac{\partial u}{\partial t},$$

$$\frac{du}{dt} = [u, H] + \frac{\partial u}{\partial t}. \qquad \qquad ...(71)$$

If u is a constant of motion then we have $\dfrac{du}{dt} = 0$. Then from the Poisson's Theorem (71) we have

$$[u, H] + \frac{\partial u}{\partial t} = 0. \qquad\qquad\qquad ...(72)$$

Further, if u does not contain time t then equation (72) yields $[u, H] = 0$. We summarise the result in the following way: "if u does not involve time t explicitly and is a constant of motion then its Poisson's bracket with Hamiltonian vanishes".

Also "if u is independent of time t explicitly then the time derivative of u is simply the Poisson bracket with H".

In the case when the dynamical variable u be taken as the Hamiltonian $H(q_j, p_{,j}, t)$, then we have from the equation (71)

$$\frac{dH}{dt} = \frac{\partial H}{\partial t}. \qquad\qquad\qquad ...(73)$$

The usual meaning of the Hamiltonian function that "if it does not contain time t explicitly then it represents a constant of motion" can also be inferred from the Poisson's theorem.

2nd Poisson Theorem   If u and v are any two constants of motion of any given holonomic dynamical system, then their Poisson bracket $[u, v]$ is also constant of motion.

Proof: Replacing the dynamical variable u by the Poisson bracket $[u, v]$ in the 1st Poisson's theorem (71) we get

$$\frac{d}{dt}[u, v] = [[u, v], H] + \frac{\partial}{\partial t}[u, v].$$

By using the property (58) we write

$$\frac{d}{dt}[u, v] = [[u, v], H] + \left[\frac{\partial u}{\partial t}, v\right] + \left[u, \frac{\partial v}{\partial t}\right]. \qquad ...(74)$$

Using the Jacobi identity (63) for $w = H$ we have

$$[u, [v, H]] + [v, [H, u]] + [H, [u, v]] = 0.$$

$$[H, [u, v]] = -[u, [v, H]] - [v, [H, u]],$$

Using the skew symmetric property of the Poisson bracket we write this equation as

$$[[u, v], H] = -[[v, H], u] - [[H, u], v].$$

Substituting this value in the equation (74), we get

$$\frac{d}{dt}[u, v] = -[[v, H], u] - [[H, u], v] + \left[\frac{\partial u}{\partial t}, v\right] + \left[u, \frac{\partial v}{\partial t}\right],$$

$$\frac{d}{dt}[u, v] = \left[\frac{\partial u}{\partial t}, v\right] + [[u, H], v] + \left[u, \frac{\partial v}{\partial t}\right] + [u, [v, H]].$$

Using the distributive property of the Poisson bracket we write this equation

$$\frac{d}{dt}[u, v] = \left[\frac{\partial u}{\partial t} + [u, H], v\right] + \left[u, \frac{\partial v}{\partial t} + [v, H]\right].$$

On using 1st Poisson's theorem (71), we get

$$\frac{d}{dt}[u, v] = \left[\frac{du}{dt}, v\right] + \left[u, \frac{dv}{dt}\right]. \qquad \qquad ...(75)$$

It is given that u and v are constants of motion. Hence we have $\dfrac{du}{dt} = 0$ and $\dfrac{dv}{dt} = 0$. Consequently, equation (75) reduces to

$$\frac{d}{dt}[u, v] = 0 \Rightarrow [u, v] = \text{const.}$$

This proves that the Poisson bracket of two constants of motion is also constant of motion.

Example 9: For the Hamiltonian $H = \dfrac{1}{2}\left(p^2 + q^2\right)$, find $[\dot{p}, H]$ and $[\dot{q}, H]$, solve the equation of motion and show that energy is conserved.

Solution: The Hamilton's canonical equations of motion in terms of Poisson's bracket are defined in equations (69) and (70). These are given by

$$\dot{q}_j = \frac{\partial H}{\partial p_j} = [q_j, H], \ \dot{p}_j = -\frac{\partial H}{\partial q_j} = [p_j, H]. \qquad ...(76)$$

Differentiating these equations with respect to time t we obtain

$$\ddot{q}_j = [\dot{q}_j, H] + \left[q_j, \frac{dH}{dt}\right], \ \ddot{p}_j = [\dot{p}_j, H] + \left[p_j, \frac{dH}{dt}\right],$$

where

$$\frac{dH}{dt} = \sum_j \frac{\partial H}{\partial q_j} \dot{q}_j + \sum_j \frac{\partial H}{\partial p_j} \dot{p}_j$$

$$\frac{dH}{dt} = \sum_j \left(\frac{\partial H}{\partial q_j} \frac{\partial H}{\partial p_j} - \frac{\partial H}{\partial p_j} \frac{\partial H}{\partial q_j}\right),$$

$$\Rightarrow \qquad \frac{dH}{dt} = 0.$$

$$\Rightarrow \qquad \ddot{q}_j = [\dot{q}_j, H] \text{ and } \ddot{p}_j = [\dot{p}_j, H]. \qquad ...(77)$$

The Hamilton canonical equations of motion give

$$\dot{p}_j = -\frac{\partial H}{\partial q_j} \Rightarrow \dot{p} = -\frac{\partial}{\partial q}\frac{1}{2}(p^2 + q^2) \Rightarrow \dot{p} = -q.$$

$$\dot{q}_j = \frac{\partial H}{\partial p_j} \Rightarrow \dot{q} = \frac{\partial}{\partial p}\frac{1}{2}(p^2 + q^2) = p \Rightarrow \dot{q} = p.$$

Hence the equations (77) on using equations (76) become

$$\Rightarrow \qquad \ddot{q} = [\dot{q}, H] = [p, H] = \dot{p} = -q \text{ and}$$

$$\ddot{p} = [\dot{p}, H] = [-q, H] = -\dot{q} = -p. \qquad \qquad \text{...(78)}$$

$$\ddot{q} + q = 0, \ddot{p} + p = 0. \qquad \qquad \text{...(79)}$$

We see that for the choice $p = \cos t$ and $q = \sin t$ the equation (78) is satisfied. Hence the total energy of the particle is given by $E = \frac{1}{2}(p^2 + q^2) = \frac{1}{2}$. Consequently, the energy is conserved.

Example 10: For the Lagrangian $L = \frac{1}{2}\dot{q}^2 - q\dot{q} + q^2$ find p in terms of p and $\dot{q}$ and obtain the value of $[p, \dot{q}^2]$.

Solution: From the given Lagrangian we see that q is the only generalized coordinate. Hence the corresponding generalized momentum gives

$$p = \frac{\partial L}{\partial \dot{q}} = \dot{q} - q \Rightarrow \dot{q} = p + q. \qquad \qquad \text{...(80)}$$

The Poisson bracket of two dynamical variables is defined as

$$[u, v] = \frac{\partial u}{\partial q}\frac{\partial v}{\partial p} - \frac{\partial u}{\partial p}\frac{\partial v}{\partial q},$$

$$[p, \dot{q}^2] = \frac{\partial p}{\partial q}\frac{\partial \dot{q}^2}{\partial p} - \frac{\partial p}{\partial p}\frac{\partial \dot{q}^2}{\partial q},$$

$$[p, \dot{q}^2] = -2\dot{q}.$$

Example 11: Let the Hamiltonian of the system be given by $H = \frac{p^2}{2m} + V(q)$, find the Poisson's bracket (i) $[p, [p, H]]$ and show that

(ii) $[[\dot{q}, \dot{p}], H] + [[\dot{p}, H], \dot{q}] + [[H, \dot{q}], \dot{p}] = 0$.

Solution: (i) For the given Hamiltonian, we find

$$[p, H] = \frac{\partial p}{\partial q}\frac{\partial H}{\partial p} - \frac{\partial p}{\partial p}\frac{\partial H}{\partial q},$$

$$[p, H] = -\frac{\partial V}{\partial q}.$$

This gives
$$[p, [p, H]] = \left[p, -\frac{\partial V}{\partial q}\right] = \frac{\partial^2 V}{\partial q^2}.$$

(ii) From the Hamilton's canonical equations of motion, we have

$$\dot{p} = -\frac{\partial H}{\partial q} = -\frac{\partial V}{\partial q}, \quad \dot{q} = \frac{\partial H}{\partial p} = \frac{p}{m}.$$

Thus the values of the Poisson bracket $[\dot{p}, H] = -\left[\frac{\partial V}{\partial q}, H\right] = -\frac{p}{m}\frac{\partial^2 V}{\partial q^2}$. Hence we find

$$[\dot{q}, \dot{p}] = \left[\frac{p}{m}, -\frac{\partial V}{\partial q}\right] = -\frac{1}{m}\left[p, \frac{\partial V}{\partial q}\right] = \frac{1}{m}\frac{\partial^2 V}{\partial q^2}.$$

$$[[\dot{q}, \dot{p}], H] = \frac{1}{m}\left[\frac{\partial^2 V}{\partial q^2}, H\right] = \frac{1}{m}\frac{\partial^3 V}{\partial q^3}\frac{\partial H}{\partial p} = \frac{p}{m^2}\frac{\partial^3 V}{\partial q^3}.$$

Next we find

$$[[\dot{p}, H], \dot{q}] = -\frac{1}{m^2}\left[p\frac{\partial^2 V}{\partial q^2}, p\right],$$

Using the property (57) we write this as

$$[[\dot{p}, H], \dot{q}] = -\frac{1}{m^2}\left\{\frac{\partial^2 V}{\partial q^2}[p, p] + p\left[\frac{\partial^2 V}{\partial q^2}, p\right]\right\}$$

$$[[\dot{p}, H], \dot{q}] = -\frac{p}{m^2}\left[\frac{\partial^2 V}{\partial q^2}, p\right],$$

$$[[\dot{p}, H], \dot{q}] = -\frac{p}{m^2}\frac{\partial^3 V}{\partial q^3}.$$

Finally we obtain

$$[H, \dot{q}] = \left[H, \frac{p}{m}\right] = \frac{1}{m}[H, p] = \frac{1}{m}\frac{\partial H}{\partial q} = \frac{1}{m}\frac{\partial V}{\partial q},$$

$$[[H, \dot{q}], \dot{p}] = \left[\frac{1}{m}\frac{\partial V}{\partial q}, -\frac{\partial V}{\partial q}\right] = -\frac{1}{m}\left[\frac{\partial V}{\partial q}, \frac{\partial V}{\partial q}\right] = 0.$$

Adding these relations we readily get identity

$$[[\dot{q}, \dot{p}], H] + [[\dot{p}, H], \dot{q}] + [[H, \dot{q}], \dot{p}] = 0.$$

Exercise

1. Show that the transformations $Q_j = p_j \tan t$, $P_j = -q_j \cot t$ are canonical.

2. Show that the following transformation equations

   (i) $Q = \sqrt{q} \cos 2p$, $P = \sqrt{q} \sin 2p$,

   (ii) $P = \dfrac{(p^2 + q^2)}{2}$, $Q = \tan\left(\dfrac{q}{p}\right)$,

   (iii) $Q = \sqrt{2q}\,e^{\alpha} \cos p$, $P = \sqrt{2q}\,e^{-\alpha} \sin p$,

   (iv) $q = \sqrt{2P} \sin Q$, $p = \sqrt{2P} \cos Q$,

   (v) $Q = q \tan p$, $P = \log(\sin p)$,

   (vi) $q = P^2 + Q^2$, $p = \dfrac{1}{2} \tan^{-1}\left(\dfrac{P}{Q}\right)$,

   are canonical transformations hence find the generating functions.

3. For the Lagrangian $L = \dfrac{1}{2}\dot{q}^2 - q\dot{q} + q^2$ find p. Show that the Hamiltonian H reduces to $\dfrac{1}{2} P^2 - Q^2$ under the canonical transformations $p = P - Q$; $q = Q$.

4. Let the Hamiltonian H be given by $H = p\dot{q} - L$, calculate the Poisson bracket $[p, H]$ and show that it is $\dot{p}$.

5. Show that the transformations $p = m\omega q \cot Q$, $P = \dfrac{m\omega q^2}{2} \csc^2\theta$ are canonical, and find the corresponding generating function.

6. Given the generating function $F_1(q, Q, t) = \dfrac{1}{2} m\omega\left(q - \dfrac{f(t)}{m\omega^2}\right) \cot Q$, find the transformation equations.

Some Useful Formulae

1. Mean Value Theorem of Integral Calculus

If a function f is continuous on $[a, b]$, then there exists a number x_0 in $[a, b]$ such that

$$\int_a^b f(t)\, dt = f(x_0)(b - a), \text{ where } x_0 = a + \theta(b - a),\ 0 < \theta < 1.$$

2. Differentiation of Integral

$$\frac{d}{dx} \int_{p(x)}^{Q(x)} f(x, y)\, dy = \int_{p(x)}^{Q(x)} \frac{\partial}{\partial x} f(x, y)\, dy + f(x, Q)\frac{dQ}{dx} - f(x, P)\frac{dP}{dx},$$

provided $\dfrac{\partial f}{\partial x}$ is a continuous function of x and y in $P(x) \leq x \leq Q(x)$.

3. Taylor's Theorem for two and three variables

$$f(x, y + h, z + k) = f(x, y, z) + \left(h\frac{\partial f}{\partial y} + k\frac{\partial f}{\partial z} \right) +$$

$$+ \frac{1}{2}\left(h^2\frac{\partial^2 f}{\partial y^2} + 2hk\frac{\partial^2 f}{\partial z\partial y} + k^2\frac{\partial^2 f}{\partial z^2} \right) + \dots$$

$$f(x + h, y + k, z + l) = f(x, y, z) + \left(h\frac{\partial f}{\partial x} + k\frac{\partial f}{\partial y} + l\frac{\partial f}{\partial z} \right) +$$

$$+ \frac{1}{2}\left(h^2\frac{\partial^2 f}{\partial x^2} + 2hk\frac{\partial^2 f}{\partial x\partial y} + k^2\frac{\partial^2 f}{\partial y^2} + 2kl\frac{\partial^2 f}{\partial y\partial z} + l^2\frac{\partial^2 f}{\partial z^2} + 2hl\frac{\partial^2 f}{\partial x\partial z} \right) + \dots$$

4. Formula for surface Integral

If z is a continuously differentiable function of x and y, then the area of a portion of the surface is given by

$$I = \iint_D \left[1 + \left(\frac{\partial z}{\partial x} \right)^2 + \left(\frac{\partial z}{\partial y} \right)^2 \right] dx dy,$$

where the integration is carried out over the domain D of the xy-plane on to which the given portion of the surface projects.

5. Formula for the integral

$$\int \frac{dx}{\sqrt{ax^2 + bx + c}} = \frac{1}{\sqrt{-a}} \cos^{-1} \left[-\frac{2ax + b}{\sqrt{b^2 - 4ac}} \right].$$

6. Formula for partial Derivative

Let $f = f(x_1, x_2, \ldots x_n, t)$ and $x_i = x_i(t)$, then

(a) $$\frac{df}{dt} = \frac{\partial f}{\partial x_1} \frac{dx_1}{dt} + \frac{\partial f}{\partial x_2} \frac{dx_2}{dt} + \cdots + \frac{\partial f}{\partial x_n} \frac{dx_n}{dt} + \frac{\partial f}{\partial t}$$

i.e., $$\frac{df}{dt} = \sum_{i=1}^{n} \frac{\partial f}{\partial x_i} \frac{dx_i}{dt} + \frac{\partial f}{\partial t}.$$

(b) If $f = f(x_1, x_2, \ldots x_n)$ and $x_i = x_i(r, s, \ldots, t)$, then

$$\frac{\partial f}{\partial r} = \frac{\partial f}{x_1} \frac{\partial x_1}{\partial r} + \frac{\partial f}{\partial x_2} \frac{\partial x_2}{\partial r} + \cdots + \frac{\partial f}{\partial x_n} \frac{\partial x_n}{\partial r}$$

i.e., $$\frac{\partial f}{\partial r} = \sum_{i=1}^{n} \frac{\partial f}{\partial x_i} \frac{\partial x_i}{\partial r}.$$

Similarly, we write the other derivatives as

$$\frac{\partial f}{\partial s} = \sum_{i=1}^{n} \frac{\partial f}{\partial x_i} \frac{\partial x_i}{\partial s} \quad \text{and} \quad \frac{\partial f}{\partial t} = \sum_{i=1}^{n} \frac{\partial f}{\partial x_i} \frac{\partial x_i}{\partial t}.$$

7. Euler Theorem

If $f = f(x_1, x_2, \ldots x_n)$ is a homogeneous differentiable function of n independent variables $x_1, x_2, \ldots, x_n$ of degree n then

$$\sum_i x_i \frac{\partial f}{\partial x_i} = nf.$$

8. Spherical Co-ordinate System

We are familiar with the relations between the Cartesian co-ordinate system and spherical co-ordinate system which are cited on next page:

$$x = r \sin\theta \cos\phi, \; y = r \sin\theta \sin\phi, \; z = r \cos\theta. \qquad ...(A1.1)$$

The metric in terms of r, θ, ϕ is given by

$$ds^2 = dr^2 + r^2 d\theta^2 + r^2 \sin^2\theta d\phi^2. \qquad ...(A1.2)$$

(I) If r is constant, the set of relations (A1.1) represents a sphere, and the metric on the surface of the sphere reduces to

$$ds^2 = r^2 d\theta^2 + r^2 \sin^2\theta d\phi^2. \qquad ...(A1.3)$$

(II) If θ is constant such that $\sin\theta = a$, $\cos\theta = b \Rightarrow a^2 + b^2 = 1$. In this case the set (A1.1) reduces to

$$x = ra \cos\phi, \; y = ra \sin\phi, \; z = rb. \qquad ...(A1.4)$$

Eliminating ϕ from equations (A1.4) we obtain

$$x^2 + y^2 = z^2 \tan^2\theta. \qquad ...(A1.5)$$

This is the standard equation of a cone, and the metric on the surface of the cone becomes

$$ds^2 = dr^2 + r^2 a^2 d\phi^2. \qquad ...(A1.6)$$

(III) Now if $r \sin\theta = a$ (constant), and $z = z$ then the set of equations (A1.1) becomes

$$x = a \cos\phi, \; y = a \sin\phi, \; z = z. \qquad ...(A1.7)$$

This set of equations represents a cylinder, and the metric on the surface of the cylinder becomes

$$ds^2 = a^2 d\phi^2 + dz^2. \qquad ...(A1.8)$$

9. Vector triple product

(i) $\bar{a} \times (\bar{b} \times \bar{c}) = (\bar{a} \cdot \bar{c})\,\bar{b} - (\bar{a} \cdot \bar{b})\,\bar{c}$

(ii) $\bar{a} \cdot (\bar{b} \times \bar{c}) = \bar{b} \cdot (\bar{c} \times \bar{a}) = \bar{c} \cdot (\bar{a} \times \bar{b})$.

10. Torque (Moment of a force)

Torque is denoted by $\bar{N}$ and is defined as

$$\bar{N} = \bar{r} \times \bar{F} = \frac{d\bar{L}}{dt},$$

where $\bar{L}$ is the angular momentum defined by $\bar{L} = \bar{r} \times \bar{p}$ and $\bar{p}$ the linear momentum.

11. Curve in 3-dimentional space

A locus of points in 3-dimentional space whose co-ordinates are expressible as a function of single parameter is called *a* curve.

Mathematically,

(i) *Gaussian representation:* $y = f(x)$ represent a plane curve.

(ii) *Parametric representation:* $x = x(t)$, $y = y(t)$, $z = z(t)$ represent space curve.

12. Polar equation of the conic: $\dfrac{1}{r} = c[1 + e \cos(\theta - \theta_0)]$.

13. Area of an ellipse with semi-major axis a and semi-minor axis b is equal to πab.

14. Relation between the semi-major and semi-minor axes of an ellipse: $b^2 = a^2(1 - e^2)$.

Appendix

2

Lagrange's Method to Find Maxima and Minima of a Function Under Some Constraints

Let

$$u = f(x, y, z) \qquad \qquad ...(A2.1)$$

be a function which is to be extremized under the conditions

$$\phi_1(x, y, z) = 0, \qquad \qquad ...(A2.2)$$

$$\phi_2(x, y, z) = 0. \qquad \qquad ...(A2.3)$$

For the extremum of (A2.1) we have $du = 0$

$$\Rightarrow \qquad \qquad f_x dx + f_y dy + f_z dz = 0. \qquad \qquad ...(A2.4)$$

Now differentiating equations (A2.2) and (A2.3) we obtain

$$\phi_{1x} dx + \phi_{1y} dy + \phi_{1z} dz = 0, \qquad \qquad ...(A2.5)$$

$$\phi_{2x} dx + \phi_{2y} dy + \phi_{2z} dz = 0. \qquad \qquad ...(A2.6)$$

Lagrange's method suggests that, multiply equations (A2.5) and (A2.6) by λ_1, λ_2 and adding to equations (A2.4) we get

$$(f_x + \lambda_1 \phi_{1x} + \lambda_2 \phi_{2x})\, dx + (f_y + \lambda_1 \phi_{1y} + \lambda_2 \phi_{2y})\, dy +$$

$$+ (f_z + \lambda_1 \phi_{1z} + \lambda_2 \phi_{2z})\, dz = 0. \qquad \qquad ...(A2.7)$$

The function $f(x, y, z)$ to be extremized is restricted by two constraints (A2.2) and (A2.3). Solving these relations for x and y we will get a function of single independent variable in z. Thus out of three variables x, y and z, only one variable say z is independent and other two variables x and y are dependent variables. Now we choose the Lagrange's multipliers λ_1 and λ_2 such that coefficients of dx and dy in equation (A2.7) are zero. Thus we have

$$f_x + \lambda_1 \phi_{1x} + \lambda_2 \phi_{2x} = 0, \qquad \qquad ...(A2.8)$$

$$f_y + \lambda_1 \phi_{1y} + \lambda_2 \phi_{2y} = 0. \qquad \qquad \text{...(A2.9)}$$

Consequently we have

$$f_z + \lambda_1 \phi_{1z} + \lambda_2 \phi_{2z} = 0. \qquad \qquad \text{...(A2.10)}$$

Thus we have five equations (A2.2), (A2.3), (A2.8), (A2.9) and (A2.10) and five unknown x, y, z, λ_1 and λ_2. These equations can be solved for x, y and z and extremum value of the function can be obtained. Thus extremization of (A2.1) under the conditions (A2.2) and (A2.3) is equivalent to extremize the function f^*; where f^* is defined by

$$f^* = f + \lambda_1 \phi_1 + \lambda_2 \phi_2.$$

Thus

$$df^* = 0 \Rightarrow f_x^* = f_y^* = f_z^* = 0 \qquad \qquad \text{...(A2.11)}$$

These equations are equivalent to equations (A2.8), (A2.9) and (A2.10).

Moments of Inertia of Some Standard Bodies

The calculation of a moment of inertia is generally a simple matter of integration. For some standard bodies the method of finding the moments of inertia is illustrated in the following examples and some of the results are cited for convenience.

WORKED EXAMPLES

Example 1: Show that the M.I. of a circular disc of radius 'a' and mass m about the axis through its centre perpendicular to its plane is $\dfrac{1}{2} ma^2$.

Solution: Consider a uniform circular disc of mass m and radius 'a' rotating about its centre and perpendicular to its plane. If ρ is the density, then we have ρ = mass per unit area of the disc, then the total mass of the disc is given by

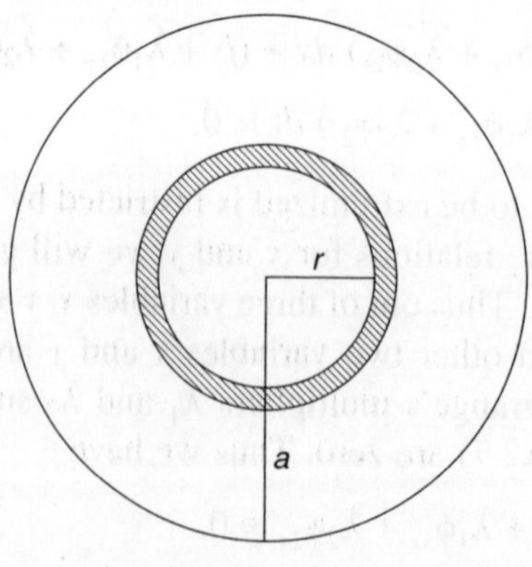

$$m = \rho\pi a^2. \qquad\qquad\qquad ...(A2.12)$$

Let the disc consists of number of thin concentric rings. Imagine one such ring be of mass *dm* and of thickness *dr* and radius *r*, then

$$dm = \rho.\text{ area of the ring} = \rho \cdot 2\pi r\, dr.$$

Hence the M.I. of the ring about the axis through the centre and perpendicular to its plane is defined by

$$\text{M.I.} = r^2\, dr \Rightarrow \text{M.I.} = 2\pi\rho r^3\, dr.$$

Thus the total M.I. of the whole disc about the axis through the centre and perpendicular to its plane is obtain as

$$I = 2\pi\rho \int_0^a r^3 dr \Rightarrow I = \frac{ma^2}{2}.$$

Example 2: Show that the M.I. of a right circular cylinder of radius '*a*' and mass *m* about its axis is $\frac{1}{2} ma^2$.

Solution: Consider a uniform circular cylinder of radius '*a*' and mass *m*. Divide the cylinder in to a large number of thin circular discs parallel to the base. Consider a disc of breadth *dx* at a distance *x* from the base. If ρ is the density then we have ρ = mass per unit area of the disc. Thus the mass of the circular disc is

$$dm = \rho\pi a^2\, dx. \qquad\qquad\qquad ...(A2.13)$$

The total mass of the cylinder becomes

$$m = \rho\pi a^2 h. \qquad\qquad\qquad ...(A2.14)$$

From the above example, the M.I. of the circular disc about the axis of the cylinder is given by

$$\text{M.I.} = \frac{1}{2} ma^2 \Rightarrow \text{M.I.} = \frac{1}{2}(\rho\pi a^2\, dx)\, a^2.$$

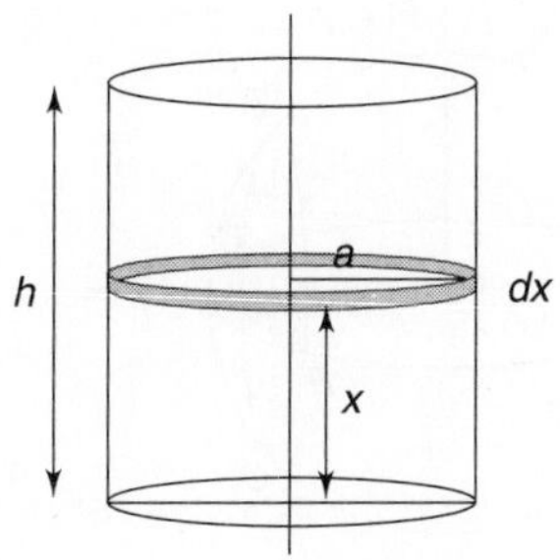

Hence the total M.I. of the cylinder is obtain as

$$I = \int_0^h \frac{1}{2}\,\rho\pi a^4\,dx$$

$$I = \frac{1}{2}\,\rho\pi a^4 \cdot h$$

$$I = (\rho\pi a^2 \cdot h)\left(\frac{a^2}{2}\right)$$

$$I = \frac{ma^2}{2}. \qquad\qquad\qquad ...(A2.15)$$

Note: If the cylinder is hollow, then divide the cylinder in to circular rings. In this case the M.I. of the hollow cylinder is obtain as M.I. $= ma^2$.

Example 3: Show that the M.I. of a solid sphere of radius '*a*' and mass *m* about its diameter is $\frac{2}{5}\,ma^2$.

Solution: Consider a solid sphere $x^2 + y^2 + z^2 = a^2$. Let the sphere be divided by planes parallel to the *YOZ* and let *ABCD* be the circular section at a distance $ON = x$ from the centre of the sphere *O*. Then the radius of the section is $\sqrt{(a^2 - x^2)}$.

Therefore, the area of the section $= \pi(a^2 - x^2)$.

Hence the volume of the slice of thickness $dx = \pi(a^2 - x^2)\,dx$.

However, the whole volume of the sphere $= \frac{4}{3}\,\pi a^3$. If ρ is the density of the sphere, then the whole mass of the solid sphere is given by

$$m = \frac{4\pi a^3}{3}\,\rho \qquad\qquad\qquad ...(A2.16)$$

The mass of the slice at a distance x from $O = \rho\pi(a^2 - x^2)\,dx$.

Now the M.I. of the slice about x-axis $= \frac{1}{2}(\text{mass}) \times (\text{radius})^2$

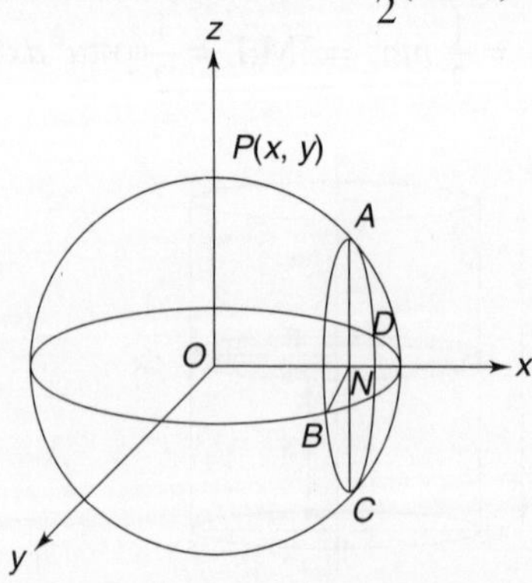

$$M.I. = \frac{1}{2}\left[\rho\pi(a^2 - x^2)\,dx\right](a^2 - x^2),$$

$$M.I. = \frac{1}{2}\,\rho\pi(a^2 - x^2)^2\,dx.$$

Thus the M.I. of the whole sphere about x-axis $= \int_{-a}^{a} \frac{1}{2}\,\rho\pi(a^2 - x^2)^2\,dx$

$$M.I. = \frac{1}{2}\,\rho\pi\int_{-a}^{a}(a^4 - 2a^2x^2 + x^4)\,dx,$$

$$M.I. = \frac{8}{15}\,\rho\pi a^5.$$

On using equation (A2.16) we obtain

$$M.I. = \frac{2}{5}\,ma^2. \qquad\qquad ...(A2.17)$$

Example 4: Show that the M.I. of a hollow sphere of radius 'a' and mass m about its diameter is $\frac{2}{3}\,am^2$.

Solution: Consider a hollow sphere $x^2 + y^2 + z^2 = a^2$. Take O be the origin. Let $P(x, y, z)$ be any point on the sphere. The moment of inertia of a hollow sphere of radius 'a' about the origin O is evidently equal to ma^2, since every point on the sphere is at distant 'a' from O. This can be written as

$$M.I. = \sum m(x^2 + y^2 + z^2) = ma^2. \qquad\qquad ...(A2.18)$$

Now the M.I. of a hollow sphere of radius 'a' about x-axis is given by

$$M.I. = \sum m(y^2 + z^2). \qquad\qquad ...(A2.19)$$

Similarly, M.I. of the hollow sphere about y and z-axes are respectively given by

$$M.I. = \sum m(z^2 + x^2). \qquad\qquad ...(A2.20)$$

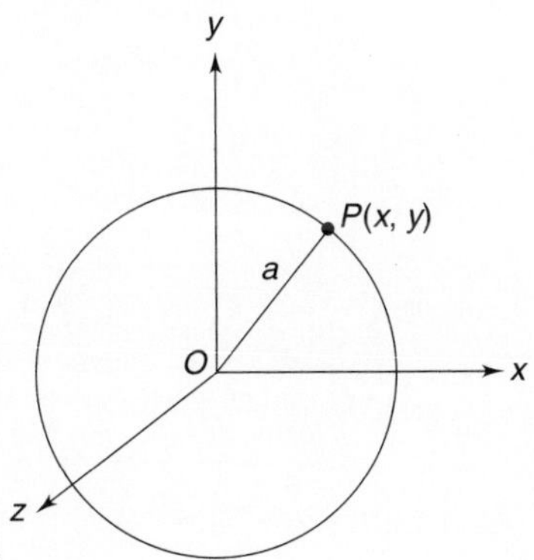

and
$$\text{M.I.} = \Sigma\, m\!\left(x^2 + y^2\right). \qquad\qquad \text{...(A2.21)}$$

Since the sphere is symmetric about any axis, hence the moments of inertia about all diameters are the same. This implies that

$$\Sigma\, m\!\left(y^2 + z^2\right) = \Sigma\, m\!\left(z^2 + x^2\right) = \Sigma\, m\!\left(x^2 + y^2\right). \qquad \text{...(A2.22)}$$

Adding these equations we have each term of the equation is equal to

$$\Sigma\, m\!\left(y^2 + z^2\right) = \frac{1}{3}\,\Sigma\, m\!\left(y^2 + z^2 + z^2 + x^2 + x^2 + y^2\right),$$

$$= \frac{2}{3}\,\Sigma\, m\!\left(x^2 + y^2 + z^2\right).$$

On using equation (A2.18) we have

$$\Sigma\, m\!\left(y^2 + z^2\right) = \frac{2}{3}\, ma^2. \qquad\qquad \text{...(A2.23)}$$

Thus M.I. of the hollow sphere of radius '*a*' about a diameter is

$$\text{M.I} = \frac{2}{3}\, ma^2. \qquad\qquad \text{...(A2.24)}$$

Note: Moments of Inertia of a rectangular parallelepiped of length *l*, breadth *b* and thickness *t*, rotates about its center of gravity and axis parallel to (i) length (ii) breadth and (iii) thickness are respectively

$$\text{(i) M.I} = \frac{m\!\left(b^2 + t^2\right)}{12}, \quad \text{(ii) M.I.} = \frac{m\!\left(l^2 + t^2\right)}{12}, \quad \text{(iii) M.I.} = \frac{m\!\left(l^2 + b^2\right)}{12}.$$

In particular if $l = 2a$, $b = 2a$, $t = 2b$ then moments of inertia are respectively:

$$\text{(i) M.I} = \frac{m\!\left(a^2 + b^2\right)}{3}, \quad \text{(ii) M.I.} = \frac{m\!\left(a^2 + b^2\right)}{3}, \quad \text{(iii) M.I.} = \frac{2ma^2}{3}.$$

References

1. H. Goldstein: Classical Mechanics, Narosa Publishing House, (1997).

2. N. C. Rana and P. S. Jog: Classical Mechanics, Tata McGraw-Hill (1992).

3. V. B. Bhatia: Classical Mechanics with Introduction to Non-linear Oscillations and Chaos, Narosa Publishing House (1997).

4. J. C. Upadhyaya: Classical Mechanics, Himalaya Publishing House (1999).

5. A. S. Gupta: Calculus of Variations with Applications, Prentice-Hall of India (1997).

6. Robert Weinstock: Calculus of Variations with Applications to Physics and Engineering, McGraw-Hill Book Comp. (1952).

7. I. M. Gelfand and S. V. Fomin: Calculus of Variations, Prentice-Hall Inc., (1963).

8. L. N. Katkar: Classical Mechanics, (SIM): Shivaji University, Kolhapur, 2013.